Wireless Networks: From Theory to Applications

Wireless Networks: From Theory to Applications

Edited by Glynn Clermont

Larsen & Keller
www.larsen-keller.com

Wireless Networks: From Theory to Applications
Edited by Glynn Clermont
ISBN: 979-8-88836-070-5 (Hardback)

Larsen & Keller
Published by Larsen and Keller Education,
5 Penn Plaza,
19th Floor,
New York, NY 10001, USA

Cataloging-in-Publication Data

Wireless networks : from theory to applications / edited by Glynn Clermont.
p. cm.
Includes bibliographical references and index.
ISBN 979-8-88836-070-5
1. Wireless communication systems. 2. Wireless sensor networks. 3. Wireless LANs.
4. Multimedia communications. I. Clermont, Glynn.
TK5103.2 .W57 2023
621.384--dc23

For more information regarding Larsen and Keller Education and its products, please visit the publisher's website www.larsen-keller.com

Table of Contents

Preface

A wireless network refers to a computer network that utilizes wireless data connections among network nodes. They are unique as they can connect devices, services, people and applications to a network, without being bound to any wired hardware. Wireless networking enables businesses to create local area networks (LANs), wide area networks (WANs), and metropolitan area networks (MANs) without using cables. A wireless LAN reduces the expense of installing LAN cabling and makes changes to network structure and its relocation easier. Wireless local area networks (WLANs), terrestrial microwave networks, wireless sensor networks, cell phone networks, and satellite communication networks are a few examples of wireless networks. They are a widespread choice for telecommunications, and residential and commercial networks. Various applications of wireless networking include television remote controls, cell phones, computer interface devices, security systems, Wi-Fi, wireless power transmission, and a variety of wireless communication based projects. This book aims to shed light on the theory and applications of wireless networks. It will serve as a reference to a broad spectrum of readers.

This book has been the outcome of endless efforts put in by authors and researchers on various issues and topics within the field. The book is a comprehensive collection of significant researches that are addressed in a variety of chapters. It will surely enhance the knowledge of the field among readers across the globe.

It gives us an immense pleasure to thank our researchers and authors for their efforts to submit their piece of writing before the deadlines. Finally in the end, I would like to thank my family and colleagues who have been a great source of inspiration and support.

Editor

1

On Ad Hoc Communication in Industrial Environments

Christian Sauer [1,*], Marco Schmidt [2] and Eike Lyczkowski [1]

1 SEW-EURODRIVE GmbH & Co. KG, 76646 Bruchsal, Germany; eike.lyczkowski.p@sew-eurodrive.de
2 Bochum University of Applied Sciences, Campus Velbert/Heiligenhaus, 42579 Heiligenhaus, Germany; marco.schmidt@hs-bochum.de
* Correspondence: christian.sauer.w@sew-eurodrive.de

Abstract: Wireless communication is becoming vital in the industrial environment. New communication technologies, including ad hoc communication, are researched for this application. A thorough understanding regarding the connection characteristics of industrial networks could benefit this trend. In this work it was possible to record the time-variant network topology of such a network utilizing a novel method. Using this method and the generated recordings, novel insights into the behavior of industrial ad hoc networks are presented. The recorded time-variant topology, the tools and method of acquisition, and tools for processing and examination are published. This enables researchers and engineers to check their communication technologies in terms of applicability to the industrial use case and record more network topologies in a wide variety of wireless networking scenarios.

Keywords: mobile ad hoc networks; industrial application; nodal encounter pattern; communication characteristics

---

1. Introduction

Trends like Industry 4.0 are shaping the factory of the future. Flexibility and mobility are central requirements for future production facilities. Wireless communication plays a central part in enabling mobility and facilitating flexibility [1]. Automated guided vehicles (AGVs) for example, depend on wireless communication systems. Researchers are investigating the possibility of utilizing ad hoc communication (e.g., Mesh [2] or Delay Tolerant Networks (DTNs) [3]) for communication in AGVs. The research and implementation of wireless and ad hoc networks, in particular, would benefit from a thorough understanding of the wireless channel in industrial environments or the ability to simulate this use case. Additionally, further practical research is needed [4].

Research regarding wireless communication in industrial environments is challenging due to the limited accessibility of industrial facilities; thus, data from such facilities are sparse. This research aims to ease this challenge by publishing a data-set from a running production facility, and this enables researchers to test custom routing algorithms and networking strategies for ad hoc networks in an industrial scenario without accessing a production facility. Additionally, the tools for the creation, examination, and usage of this data-set are provided. The utilized method simplifies the scenario by abstracting the physical layer of the network and focusing on the connection-oriented aspects of communication. This work proposes a new method to record time-variant topologies of wireless networks and emphasizes its advantages over existing methods. The topologies of a mobile wireless network from a running production facility are recorded and published. These data enable researchers and engineers to test and verify networking strategies and routing algorithms. The connection-oriented topology characteristics are examined and compared to non-industrial and static wireless networks. Based on these comparisons, recommendations for the design of industrial ad hoc networks are given.

An encounter is a continuous time frame in which a transmitter can send data to a receiver and is the basic building block of a wireless network topology. A nodal encounter pattern (NEP) is a recording of the encounters in a specific scenario using a specific communication technology. The NEP denotes the time variant topology of the network and abstracts the physical layer of the communication interface. Properties of the NEP can therefore be used to determine connection characteristics of the used communication technologies in an examined scenario. This work contains a new method for recording and analyzing NEPs that were recorded in a running production facility. The NEPs are additionally published so the research community can use them for further analysis and as a basis for the simulation of wireless industrial networks. Although the NEPs recorded in this work are limited to the examined combination of the factory environment and communication technology, the proposed methods are applicable to a wide variety of applications and use cases. In terms of communication technology, the proposed method is limited to technologies that enable wireless broadcasting.

Utilizing NEPs to denote the time-variant topology of an industrial ad hoc network is novel and enables many new use cases. A researcher can, for example, use the published NEP to test a newly developed routing algorithm in regards to its applicability to an industrial use case when utilizing an IEEE802.11 interface. In another example, a research engineer can utilize the provided tools and record an NEP in a different environment, analyze the NEP, and select a communication technology on the basis of his observations.

The following research questions are answered by this work: How can the time-variant topology of a complex network be observed? What are the topology characteristics of an industrial ad hoc network? Additional topics regard the method of observing the topology, the applicability of different routing strategies to the industrial use case, and the ability to emulate a complex network without simulating it in detail.

The contribution of this work is four-fold:

1. The applicability of an existing acquisition method for time-variant network topologies to the industrial use case is analyzed, and the method is found to not be applicable.
2. A new method in the form of a custom protocol for recording time-variant network topologies is described and published. This includes the tools and methods for processing the acquired the data.
3. NEPs describing the time-variant network topology of mobile industrial ad hoc networks are recorded and published for use by other researchers in further examinations and industrial ad hoc network simulations.
4. The acquired NEPs are analyzed. Connection characteristics are extracted from these NEPs, and recommendations for suitable ad hoc networking solutions are given.

Wireless communication in the industrial context and related nodal encounter patterns are surveyed in Section 2. In Section 3 two different methods for NEP acquisition are compared in terms of their ability to represent the true network characteristics. Additionally, metrics are introduced, which characterize a wireless network and can be extracted from a NEP. Section 4 introduces the proposed and utilized tracing protocol and the process of encounter pattern generation from the recordings. The performed measurements and examined network characteristics are described in Section 5. Afterwards, Section 6 concludes this work with a discussion on methods and presented observations.

2. Related Work

Wireless communication is not new in the industrial context, but it has continuously gained importance [5,6]. The industrial environment is particularly challenging for the application of wireless communication technologies [7]. On the one hand, high requirements in terms of latency, throughput, and especially reliability are applied [8,9]. On the other hand, dynamic environments including fading, mobility in the transmitter and receiver, and the permanent presence of interference degrade the performance of wireless communication technologies. Fellan et al. and Zhang et al. [10,11] analyzed these challenging requirements for the communication systems of mobile robots in the Industry 4.0

context. The requirements for different industrial use cases were compared and the applicability of several communication technologies examined.

Relevant communication technologies for the industrial AGV use case include LTE [12], 5G [1,13], WiFi [11,14], ZigBee [11], and specialized solutions like visible light communication (VLC) [15] and radar [16]. These technologies have different advantages and strengths and must be chosen based on the application specific requirements. References [5,10,11,16] compare these strengths to the requirements of emerging industrial trends. This high variability in underlying communication technology makes it necessary to develop methods that do not depend on specific technologies.

References [17–24] performed measurements in different industrial environments and provide valuable insight on the expected channel characteristics at different frequencies. The observed variability of the link over time is a strong indicator for the dynamic nature of the propagation properties within an industrial environment. The measurements were executed with static transmitter and receivers at different positions. Spatial variability is an important factor in scenarios, where the transmitter and/or receiver are mobile [14]. Reference [25] illustrates the lack of measurements in industrial applications. No previous work has specifically characterized the peer-to-peer communication channels in a mobile industrial network. Additionally, the previously published measurements of wireless channel propagation characteristics lack abstraction towards available connections that the utilized method provides. Thus, this works provides a complementary approach to existing work.

Ad hoc communication technologies are an emerging trend in the use case of factory automation and industry in general. Wireless sensor networks (WSN) are possibly the best established use case for this technology [26,27]. These sensor networks are utilized to monitor machines, environmental conditions, and more. Software-Defined WSN (SDWSN) are a particular variant of these networks, surveyed by [4].

Mobile Ad hoc NETworks (MANET) are not yet present in many industrial facilities. However, they are being investigated by researchers in the context of mobile robots (e.g., AGVs) [2]. In this use case, MANETs and other other ad hoc technologies, like delay tolerant networks (DTNs) [3], offer enhanced flexibility, reliability, and independence from network infrastructure. The applicability of existing communication and routing solutions to an industrial MANET have not been researched at this point in time. This work offers, in contrast to previous work, a non-simulation-based, routing-independent view on the mobile industrial peer-to-peer channel. It is therefore useful as a foundation for future research on ad hoc technologies in the industrial environment. In contrast to the presented methods, the observed connection characteristics depend on the chosen combination of examined environment and communication technologies.

NEPs have previously been used to examine mobile ad hoc networks [28]. Hsu et al. used NEPs to examine the properties and applicability of ad hoc data dissemination schemes to campus networks. They acquired NEPs using network traces [29,30] from the campus networks of several universities. These traces contain the complete communication within the examined infrastructure network, including the registration of mobile devices at certain access points. With the assumption that two nodes registered at the same access point can also directly communicate with each other, NEPs are extracted from the network traces [31]. The presented work, in contrast, utilizes a novel protocol to directly record NEPs. This enables a more precise observation of the network without assumptions about the connectivity of clients.

No previous work has analyzed the ad hoc communication channel of mobile clients in an industrial environment. With ongoing trends like Industry 4.0, the importance of mobile clients and wireless communication will increase. Therefore, this investigation is highly relevant. Additionally, this work uses the established tool NEP to examine the ad hoc channel and proposes and publishes a new method to record these NEPs, which can be applied with other communication technologies in industrial scenarios, but also in non-industrial applications.

3. Comparing Methods for the Extraction of NEPs

It is hypothesized that the necessary base assumption for the NEP extraction from network traces is not applicable to the industrial scenario. In this section differences between the indirect generation of NEPs by means of network traces and the direct recording of NEPs by means of a custom protocol are shown. Two primary metrics are examined.

The average number of simultaneous encountered per node $N_{\varnothing}$ and the average duration of these encounters $D_{\varnothing}$ are examined. Both metrics can be directly calculated from a NEP. Both metrics are important when analyzing the characteristics of the ad hoc communication channels. The first is an indicator for the number of reachable destinations from any node, while the second indicates the duration for which these destinations are reachable.

The first NEP acquisition method is the acquisition by means of a custom protocol, further denoted as trace protocol. The protocol is described in Section 4. The second is the acquisition by means of network traces [28]. This second method is based on the assumption that any two nodes encounter each other if they are registered at the same access point. Both approaches have different strengths and weaknesses.

In this section the fundamental behavioral differences of encounters acquired by both methods are observed. The goal is to analyze the behavior of the average number of simultaneous encounters in regards to the number of nodes N and number of access points N_{AP}. Additionally, the average encounter duration in regards to the communication range r of the nodes and their speed v is analyzed. The communication range r results in a node coverage area $A_r = \pi r^2$.

In the following subsections the behavior of the average number of simultaneous encountered per node $N_{\varnothing}^{XX}(N, N_{AP})$ and the average duration of these encounters $D_{\varnothing}^{XX}(r, v)$ are explored. XX describes the type of examined acquisition method. TP is the recording of encounters by means of the tracing protocol, and NT is the acquisition by means of network traces. Models are proposed to emulate the behavior of the different metrics, when observed by the different methods. The goal of these models is to predict the network performance, as best as possible, given the impact of certain network parameters on the network behavior.

3.1. *Average Number of Simultaneous Encounters per Node*

With the assumption of equally distributed, randomly placed access points and randomly moving nodes on area A, $N_{\varnothing}^{XX}(N, N_{AP})$ can be determined for both acquisition methods. For the observation by trace protocol, the number of nodes that a specific node might encounter is calculated. It is expected that the $N-1$ other nodes are randomly distributed on A. The number of simultaneously encountered nodes is therefore expected to be the fraction of $N-1$ that are present in A_r. It follows that

$$N_{\varnothing}^{TP}(N) = \frac{A_r}{A}(N-1) \tag{1}$$

For the acquisition by means of network traces NT, two different cases must be considered. The first case is that the area A is not completely covered by access points. This case is defined by $N_{AP}A_r \leqslant A$. This means that the combined covered area by all access point is smaller than A. Overlap of access point communication ranges is rare, due to the assumed equal distribution of access points. If the area is not completely covered, the average number of simultaneous encounters is equal to the average number of nodes, within range of an access point minus the source node ($\frac{A_r}{A}(N-1)$) times the probability, to be within the range of an access point ($\frac{N_{AP}A_r}{A}$). When assuming complete coverage of the area this probability is 1. Once the area is completely covered, the covered area per access point decreases because the nodes will tend to register at the closest access point. In reality, the chosen access point depends on the applied roaming scheme; most are based on received signal strength. This decline can be formalized with $A_{AP} = \frac{A}{N_{AP}}$, where $A_r \geqslant A_{AP}$. It simplifies to

$$N_{\varnothing}^{NT}(N, N_{AP}) = \begin{cases} \left(\frac{A_r}{A}(N-1)\right) \cdot \frac{N_{AP}A_r}{A}, & \text{if } N_{AP}A_r \leqslant A \\ \frac{N}{N_{AP}} - 1, & \text{otherwise} \end{cases} \tag{2}$$

The model for the acquisition via trace protocol is therefore equal to the model for the acquisition via network trace only for the case $N_{AP}A_r = A$ and a perfect 1:1 coverage of the application area by the access points. Later results show differences even in this case.

3.2. Average Duration of Encounters

The average duration of encounters $D_{\varnothing}^{XX}(r, v)$ is mostly dependent on the mobility of the nodes. Hsu et al. [28] assumed nodes stayed within the range of an access point for a prolonged duration. This assumption minimizes the influence of mobility on the encounter pattern, which is a valid assumption for the examined network traces. In the examined campus networks, students listen to lectures or visit libraries and similar locations for a duration of $\geqslant 1$ h. This assumption, however, is not transferable to the examined use case of AGVs in an industrial environment; therefore, the effect of mobility on the two observation methods must be considered.

The proposed models reduce the dependencies of the encounter duration of two nodes to the speed v of the nodes and the communication range r. On average, a moving node passes the static communication range of an access point along a path of length $\frac{\pi}{2}r$ (see Figure A2) and therefore for a time of $D_{AP} = \frac{\pi}{2}\frac{r}{v}$. This average only accounts for a node passing an AP range. If the destination of the node is within the AP range, the average traveled distance within communication range (average destination is at the AP position; distance to reach the AP and subsequently leave range is $2r$) changes to $2r$. It is assumed that $\sqrt{A} \gg r$; therefore, this special case is subsequently not considered. An encounter between any two nodes persists as long as both are in range of the access point. At the point in time at which any node enters the range, every other node that already is connected to the AP will leave the range after $\delta t \in]0, D_{AP}[$. Therefore, the average encounter duration of two nodes can be reduced to $\frac{D_{AP}}{2}$:

$$D_{\varnothing}^{NT}(r, v) = \begin{cases} \frac{\pi}{4}\frac{r}{v}, & \text{if } N_{AP}A_r \leqslant A \\ \frac{\pi}{4}\frac{\sqrt{\frac{A}{\pi N_{AP}}}}{v}, & \text{otherwise} \end{cases} \tag{3}$$

Analogous to the model of $N_{\varnothing}^{NT}$, the case of complete coverage has to be considered when calculating the duration. This is done with the scaled access point communication radius r_{AP}.

Determining the encounter duration of two mobile nodes is challenging. Therefore, an approximate for the average encounter duration was determined via a fit to data from extensive simulation:

$$D_{\varnothing}^{TP}(r, v) = \frac{r^{\lambda}}{v} \tag{4}$$

For typical values of A the best fit was generated with $\lambda = 1.11$. Both models are highly simplified. For the model of NEP acquisition via network trace, a dependence on N_{AP} can be observed. The duration of encounters in a real ad hoc network, however, does not depend on this parameter.

3.3. Numerical Comparison

The proposed models for $N_{\varnothing}^{NT}$, $N_{\varnothing}^{TP}$, $D_{\varnothing}^{NT}$, and $D_{\varnothing}^{TP}$ were compared to a numerical simulation of an ad hoc network. In this simulation, the nodes use the Random WayPoint Model [32] to move within the confined area. The results of the proposed models and the numerical simulation are presented in Appendix A.

In Appendix A it can be seen that the numerical simulation behaved similarly to the proposed models. This validates that the proposed models are able to estimate the network performance and behavior without simulating the complete network.The plots of the performance predicted by observations from tracing protocol and from network traces, however, are highly dissimilar; this suggests that the assumptions of the NEP generation from network traces are not applicable.

The numeric simulation confirms the behavior expected from the presented models. Models and simulations show the same behavior in regards to parameters and similar network performance. Figure A1 also shows the differences of the two methodologies TP and NT. The two acquisition types showed clear behavioral differences, which are examined in detail in the following subsection. Only for specific configurations of N, N_{AP}, v, and r can similar results in terms of $N_{\varnothing}$ and $D_{\varnothing}$ be obtained from both acquisition methods. All these observations were done under assumptions of random node mobility and random access point placement. More complex distributions of nodes and APs will lead to even more complex relations between all of these parameters and metrics, and even more pronounced differences between both acquisition methods.

3.4. Behavioral Differences

The average number of simultaneous encounters $N_{\varnothing}^{XX}$ and the average encounter duration $D_{\varnothing}^{XX}$ are two metrics that change their behavior in regard to the applied acquisition method. Direct acquisition by means of a tracing protocol enables the more precise examination of a network. Using this method, both metrics behave as expected. The number of encounters increases linearly with the number of mobile nodes. The duration of the encounters is constant when the communication range and node speed are not changed. Both extensive simulation and the proposed model confirm this intuitively expected behavior.

The indirect acquisition by means of network traces shows a different behavior. The number of access points is important when observing the behavior with this method. The duration of encounters is lower than expected, and further declines when the density of access points is increased. A higher density of access points is beneficial when observing the number of encounters. The best fit to the direct observation is at the point of full coverage ($N_{AP}A_r = A$). At higher access point density the number of encounters decreases. In both metrics the decrease is caused by the higher probability for overlapping of the communication ranges of the access points.

The direct acquisition by means of a tracing protocol is therefore recommended. Even if the number of access points is known, a correction of these metrics is hardly possible due to the complex spatial distribution of nodes and access points in real networks [33].

Using a tracing protocol for the acquisition of NEPs also has the advantage that encounters can be directional. This means that node A can send data to node B, while node B cannot send data to node A. This directionality in encounters can not be extracted when a network trace is the basis of the NEP. However, this is of major importance when evaluating the applicability of certain routing protocols.

4. Tracing Protocol

NEPs describe encounters between nodes of a wireless network [28]. The patterns can be used two-fold. Firstly, they can be analyzed independently to determine specific network or channel characteristics, like bidirectionality, encounter duration, and more. Secondly, they can replace the mobility model, signal propagation model, and the lowest layers of the network model in a network simulation. A NEP has the advantage of being extracted from the examined environment; therefore, no complex validation of models is necessary.

The goal of this section is to introduce a protocol that can be executed on mobile nodes (e.g., AGVs in a production facility) and generate a NEP. The protocol and the required processing are described, and a simple implementation based on the Click-Router [34] is published [35]. The tracing protocol has the advantage that the real NEP can be directly recorded, but the protocol must be implemented and running on all observed nodes.

Protocol Description

The basic idea of the protocol is to use beacons to indicate the possibility of data exchange between a transmitter T and a receiver R. A number of nodes is placed in the examined environment. All nodes send beacons with a certain frequency $\frac{1}{dt}$. If any beacon is received by any receiver, it is logged to a log file. After the recording is completed, the log files of all nodes are processed, and a NEP is created.

The nodes are all identical in function. The protocol defines a time step dt. This is the time resolution of the resulting NEP. A smaller dt leads to a higher resolution in the NEP and to a higher bandwidth usage by the protocol. For the channel to be non-changing within dt, dt must be chosen to be smaller than the coherence time of the communication channel. In the case of wireless communication at 2.4 GHz and a node speed of 1.2 m/s, the time resolution should be chosen to be smaller than 25 ms. An address A_n is assigned to every node, and a counter $I_n(t)$ is incremented every time the node sends a beacon. Beacons are send every dt. The beacon contains the address of the transmitter A_T and the current index of the transmitter I_T.

This beacon is send by the wireless interface (e.g., WiFi IEEE802.11 b/g/n) of the node. Any receiver (A_r) logs this beacon as an encounter tuple, with r being the receiver and s the transmitter. All recorded encounter tuples can be concatenated to form an encounter recording L_R, which is a set of encounter tuples recorded by node R.

$$C = \begin{pmatrix} A_R & T_R = I_R \cdot dt & A_T & T_T = I_T \cdot dt \end{pmatrix} \in L_r \tag{5}$$

Any entry in the recording describes the start or persistence of an encounter of the nodes T and R. It is important to note that these encounters are directional. The entry only indicates a connection from T to R, not vice versa. A second entry must indicate the reverse encounter.

With $t = I \cdot dt$ the indices can be converted to time values. All connections have two time values. This is necessary to compensate for time offset and drift between the internal clocks of the nodes. The encounter recordings of all nodes can be concatenated to form L_Δ. The clock offset is compensated by choosing a reference node N_R, and for every other node an offset $O_n(t)$ has to be determined. For every non-reference node n a time pair is extracted from L_Δ, where t_R is the record time of the receiver, while t_T is the send time recorded at the transmitter. The discrete offset function $O_n(t_n) = t_R - t_T$ can be made continuous by assuming, for example, no or linear drift between the clocks of node N_R and n. The offset is subsequently compensated by the following conversion:

$$C_n = \begin{pmatrix} R & t_R & T & t_T \end{pmatrix} \in L_\Delta \to \begin{pmatrix} t_R + O_R(t_R) & T & R \end{pmatrix} = \begin{pmatrix} t & T & R \end{pmatrix} \in L \tag{6}$$

An offset compensated encounter list L is the result, when applying this offset to all entries.

The NEP is subsequently a function of the transmitter T, the receiver R, and the time t. It is defined as

$$C(T,R,t) = \begin{cases} 1, & \text{if } T = R \\ 1, & \text{if } \begin{pmatrix} t & T & R \end{pmatrix} \in L \\ 0, & \text{otherwise} \end{cases} \tag{7}$$

For computational purposes this function is represented as a 3d matrix of the dimension $N \times N \times \frac{t}{dt}$, with N being the number of nodes that were used for the recording, and the last dimension offers one entry per dt time step for the complete measurement time t.

5. Examination of the Industrial Ad Hoc Channels

Measurements with the proposed tracing protocol were performed in different environments and under varying conditions. The goal of the measurements and the analysis is to characterize the industrial environment in terms of effects on ad hoc communication channels. Knowing the characteristics of a communication channel allows for a more effective selection and configuration of applied routing solutions. The networks examined by Hsu et al. [28] are fundamentally different from the network examined in this work. The number of clients, the kind of mobility, and the environment are the most obvious differences. Different metrics for these network characterizations are therefore applied in this work.

Bai et al. [36] showed that after sufficient time all nodes of a network encounter each other if they move randomly on the same area. AGVs do not move randomly, but for the examined small networks the same behavior was observed. It is expected that, in bigger AGV systems, it may not be true all AGVs encounter all other AGVs. Certain AGVs could, for example, exclusively transport goods within specified disjoint areas.

The industrial environment where the measurements took place involves electric drives and gear production that adhere to Industry 4.0 paradigms, although it is a brown-field factory. Thus, a typical industrial environment by means of the amount of mobility and conductive material is present. The AGVs that were equipped with the measurement equipment facilitate intra-logistic processes of half-finished and finished products and drive up to 1.2 m/s. They cover an area of $\approx$25,000 m^2.

5.1. Performed Tests

In order to evaluate the channel characteristics of industrial ad hoc channels, tests in different environments and with different setups were conducted. The goal is to differentiate between the influence of environment and mobility on ad hoc communication and how to extract this information from the NEPs.

A reference test was performed to check the general functionality of the protocol and to deliver a reference for the examined network characteristics. It was performed in an office environment with static nodes.

A static industry test describes the measurement with nodes in an industrial environment. In this test all nodes collectively moved in an industrial environment; hence, they did not experience any relative movement and therefore moved as one group. The goal of this test is to characterize the effect of interference on the industry, while mitigating the effects of mobility and variable signal propagation. The absolute movement of the node group enabled the observation of the spatial variation in the interference.

Lastly a mobile industry test was conducted by utilizing AGVs in an accessible production facility to implement mobility. The nodes were mounted on the AGVs in an unobstructed way. Therefore, two signal attenuation effects influenced the existence of encounters in the resulting NEP. Firstly, large scale fading causes path loss between transmitter and receiver due to the distance between them. Secondly, small-scale fading caused by reflection, refraction, and scattering can be caused by obstacles on the primary propagation path.

The tests are characterized by a number of varying parameters. When comparing the presented results of the measurements, variations in these parameters have to be taken into account. Table 1 compiles and describes the different parameters and their values for the performed experiments.

Table 1. Measurement parameter description and values for measurements.

Parameter Name	Unit	Reference Test	Static Industry Test	Mobile Industry Test	Description
dt	s	0.2 s	0.2 s	0.2 s	Time resolution of the NEP
T	s	>11,800 s	>8800 s	>10,400 s	Run time of measurement
N		6	7	8	Number of nodes
Mobility	Type	None	Group	AGV	Type of mobility
Environment	Type	Office	Industry	Industry	Environment description

Some parameters are restricted by external requirements. The send period *dt* of the trace protocol, for example, had to be adjusted, as a minimal bandwidth impact of the measurement was required. Table 1 compares the measurement parameters of the three measurements.

All tests were performed with IEEE802.11 b/g/n communication interfaces at 20 dbm send power. The protocol was implemented with the Click modular router [34] on a battery-powered single-board computer. The implementation and the resulting NEPs are published in [35].

5.2. Network Connectedness

The network connectedness is the average percentile of neighboring nodes (encountered nodes with which direct data exchange is possible) [33]. A connection can be established if an encounter is registered. It is assumed that two nodes could communicate for the time of 1 dt after an encounter was registered. Within a network of n nodes, $n-1$ connections are simultaneously possible for any node. Hence, nodes cannot connect to themselves. The network connectedness of the network at time t is then defined as

$$N(t) = \frac{1}{N}\sum_{i=1}^{N}\frac{1}{n-1}\left(\sum_{j=1}^{N} C(i,j,t)\right) - 1 \tag{8}$$

where C is the NEP, as described in Section 4, and i and j iterate over N nodes (transmitter and receiver) in the examined network. When de-normalized and averaged over the time of the recording, the network connectedness is equal to the previously used metric $N_\varnothing = \frac{1}{T}\int_0^T nN(t)dt = N_\varnothing^{TP}$. Figure 1 illustrates the meaning of different network connectedness values. A network connectedness $C_N = 0$ describes a network where no nodes are connected. The network connectedness increases once connections are available. If all nodes can reach all other nodes the network connectedness reaches 1.

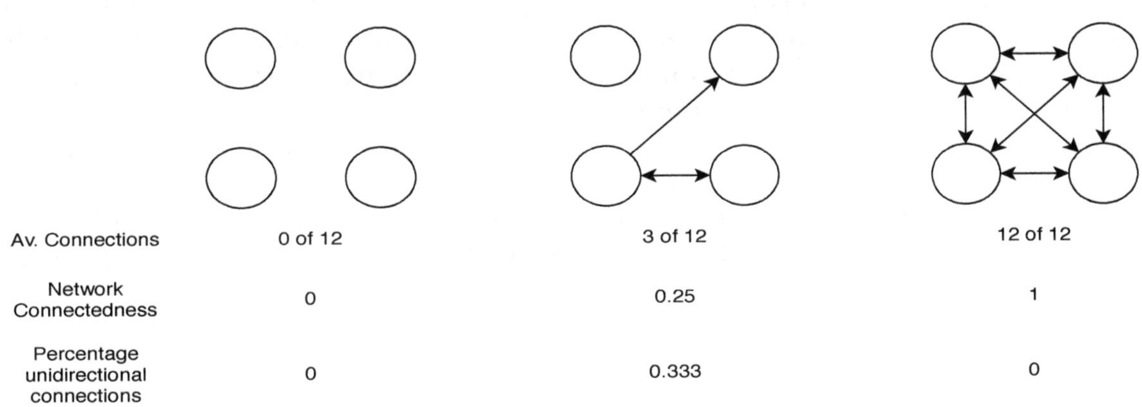

Figure 1. Illustration of network connectedness and probability for unidirectional channels with three example networks.

The network connectedness of the three examined measurement configurations are displayed in Figure 2. On average, the network in the reference measurement is fully connected. This means all nodes can communicate with all other nodes. Short-lived variations occur due to interference between the nodes and interference with other wireless communication systems within the same spectrum. The second measurement set shows slightly lower connectedness within the network. This indicates that the industrial environment might contain more sources for interference than the office environment. The variations in network connectedness indicate spatial correlation. The NEP of mobile nodes in the industrial environment exhibits the lowest network connectedness and even higher variations in connectedness as the static measurement in the industrial environment. In Figure 2 it can be seen that the effect of mobility is far more pronounced than the one of interference.

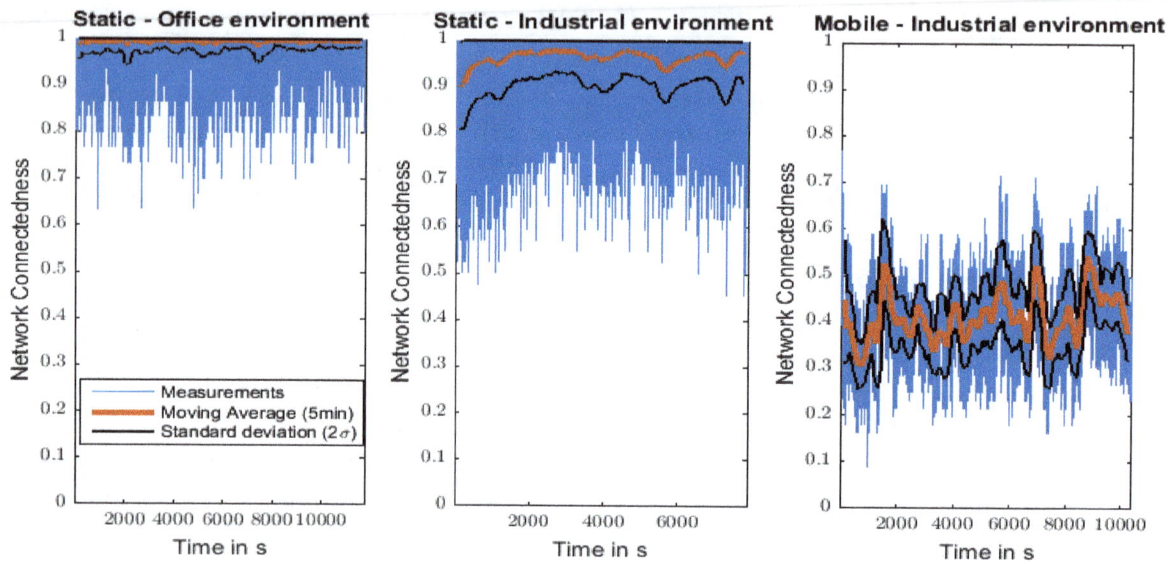

Figure 2. Network connectedness in different scenarios as extracted from the nodal encounter pattern (NEP).

5.3. Directional Channel Probability

Consider a transmitter A sent a message to receiver B at time T_{AB}. In this work a channel is classified as unidirectional if a transmission at time $T_{BA} \in [T_{AB} - dt, T_{AB} + dt]$ from B was not received at A. If the transmission is received, the channel is classified as bidirectional. Possible reasons for unidirectional channels are changes in the propagation path within dt or interference with other communication networks. The office reference test shows that interference within the tracing protocol is unlikely.

Many common routing protocols (e.g., DSR [37], AODV [38]) expect bidirectional connections. Routing protocols can be enhanced to work in the presence of unidirectional channels at the cost of higher overhead [39]. The percentage of unidirectional connection is therefore highly relevant in the evaluation of the applicability of ad hoc routing protocols to the industrial environment. We assume that such protocols need at least about 200 ms for route search and establishment; therefore, the chosen 0.2 s NEP time resolution is sufficient for the examined application.

NEPs that were extracted from the proposed trace protocol can be used to determine this probability of a channel being unidirectional. It is defined by Equation (9) using the same parameters as Equation (8).

$$P_u(t) = \frac{\sum_{i=1}^{N} \sum_{j=1}^{N} |C(i,j,t) - C(j,i,t)|}{2 \cdot \left(\sum_{i=1}^{N} \sum_{j=1}^{N} C(i,j,t) - N \right)} \tag{9}$$

Previously shown, Figure 1 illustrated examples for the percentage of channels that are unidirectional. In the central graph three connections exist. One of these connections has no reverse connection. Therefore, one-third of all connections are unidirectional.

As seen in Figure 3, the percentage of unidirectional connections is much higher in mobile industrial scenarios than in the reference use case. In the static reference measurement, unidirectional connections are very rare and only of short duration. On average only 0.52% of all connections are unidirectional. In contrast, about 33.6% of all connections are unidirectional in the mobile industrial scenario. In the static industrial scenario, on average about 3.8% of all connections are unidirectional. The results therefore support the previous observations that node mobility has a higher impact on the wireless channel than the industrial environment. Overall, such channel characteristics have to be taken into account when selecting or designing a routing protocol for the industrial use case. Another important aspect for this task is the route lifetime.

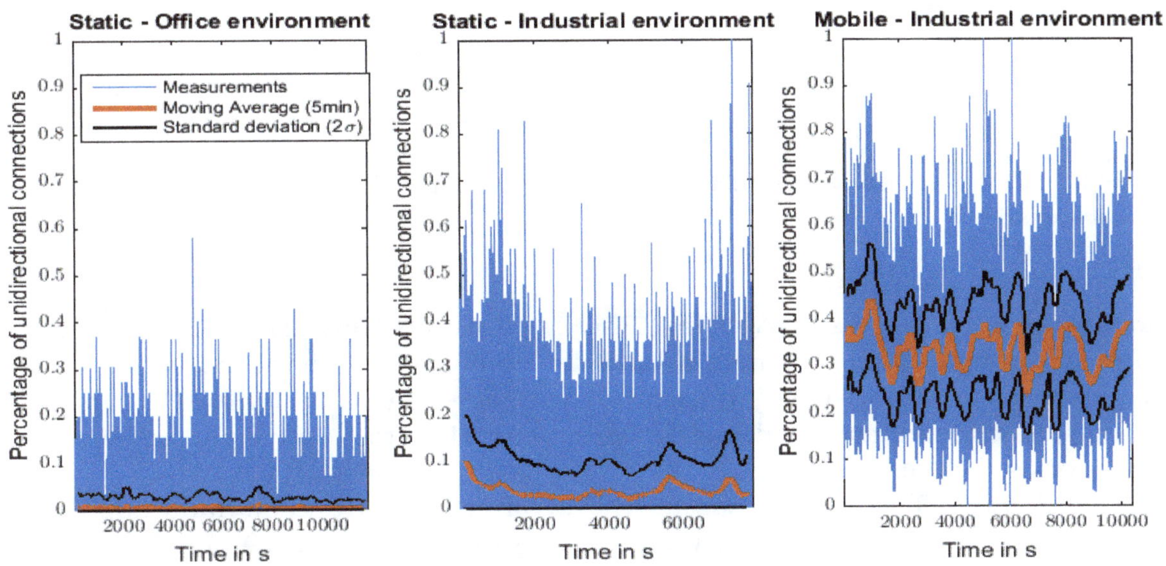

Figure 3. Percentage of unidirectional connections in scenarios.

5.4. Route Lifetime

The route lifetime describes how long a connection between two nodes persists before the ability to transmit data is lost. This is an important parameter in the analysis of applicability for certain network technologies. A low route lifetime would, for example, lead to more route failures and therefore more overhead in a MANET routing protocol. The average route lifetime is equivalent to the previously used parameter $D_{\varnothing}$.

Figure 4 shows the probabilities for route lifetimes in the three scenarios. This metric also iterates the same trend as the previous: From static office scenario over static industrial scenario to mobile industrial scenario the route life time decreases. In the static office scenario an average route life time of about 39 s was observed. Interference between nodes or with other signal sources is rare. This route lifetime significantly decreases when the same setup is observed under industrial conditions. The route lifetime further decreases when observing mobile nodes. Two AGVs are within communication range for far longer than 1.21 s (assuming a communication range $r = 20\,\text{m}$ and an AGV speed $v = 1\,\text{m/s}$; an encounter duration of 20 s follows); therefore, the further decrease in route lifetime cannot be explained by the distance between the AGVs and their communication range. Rather, effects on the primary line-of-sight path or on secondary propagation paths might be the cause of the increased number of disconnections.

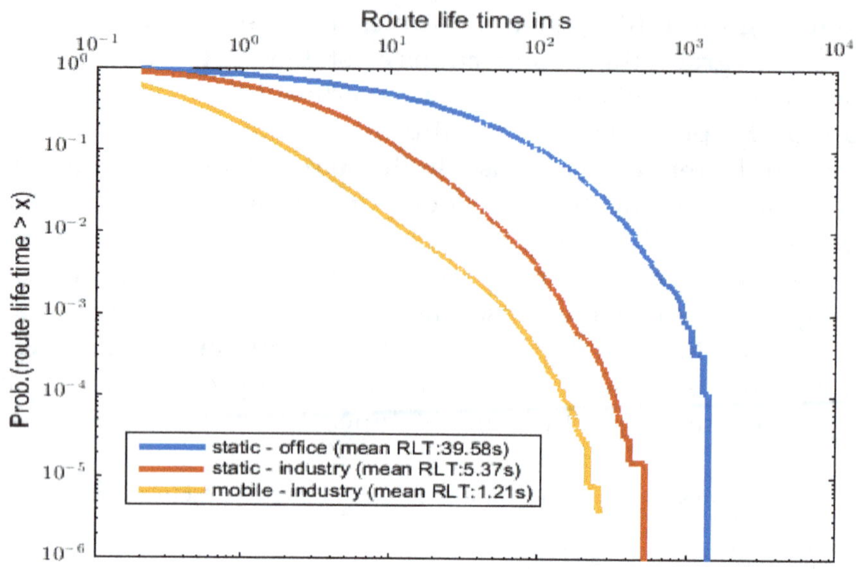

Figure 4. Complementary Cumulative Distribution Function (CCDF) of route lifetime in different scenarios.

5.5. Effects of Multi-Hop Relaying

Mobile Ad hoc NETworks (MANETs) [2], Delay Tolerant Networks (DTNs) [3] and Wireless Sensor Networks (WSNs) [27] are emerging and developing trends in the industrial context. An ad hoc network's major advantage over infrastructure networks (e.g., WiFi) is flexibility and redundancy. They are envisioned to mitigate the dependence on network infrastructure and enhance a combined wireless network structure.

In this work the NEP is used to examine the advantage of redundant multi-hop links between mobile nodes in industrial environments. The previously examined route lifetime is the primary metric for evaluating the improvement. It is envisioned that the utilization of redundant links increases the route lifetime. Another expected improvement will be that multi-hop relaying enhances the network connectedness. In this examination only the mobile industrial measurements are used. Different network sizes, in terms of number of hops, are examined.

The relevant metrics regarding both expectations were examined and are presented in Figure 5. It can be confirmed that the utilization of multi-hop connections is beneficial in the mobile industrial context. Firstly, the network connectedness increases; therefore, more nodes can be reached by any other node. Secondly, the average route lifetime is positively affected. The number of available hops is highly relevant when examining these benefits. For the examinations it must be assumed that finding and establishing a route of length h (in hops) is possible within dt. As illustrated in Figure 5, the first hop is the most effective in increasing network connectedness and route lifetime. It is suspected that in AGV scenarios with more mobile nodes and/or on a bigger area, more hops would be effective in increasing the network connectedness. In the observed scenarios the first and second hops were most effective in enhancing the route lifetime, while only the first hop enhanced the number of reachable nodes.

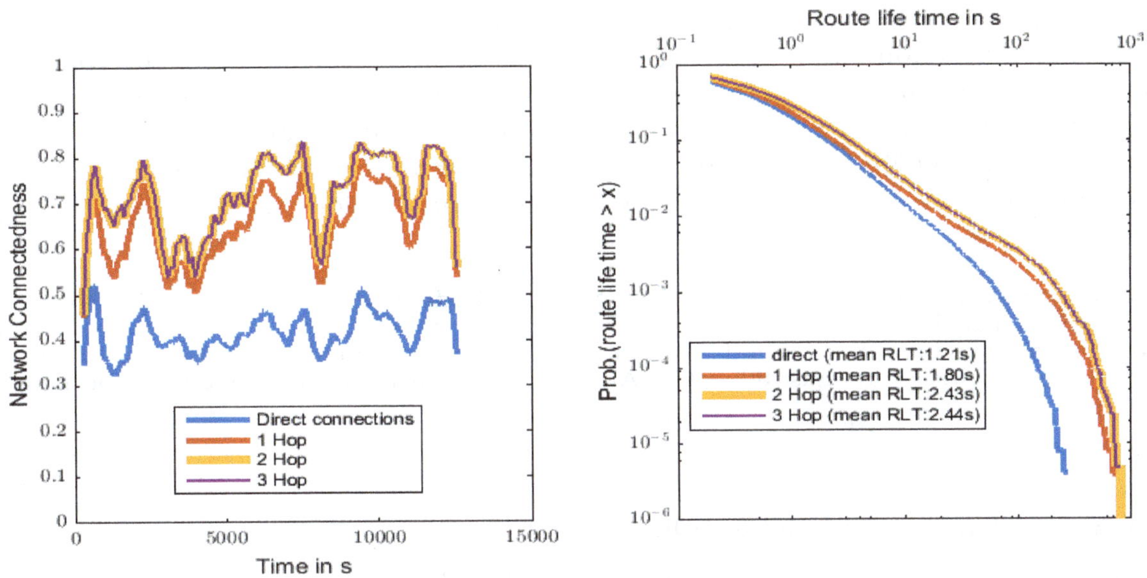

Figure 5. Network connectedness over time and CCDF of route lifetime in a mobile industrial scenario utilizing multi-hop relaying.

6. Conclusions

This work examined the time-variant network topology of ad hoc communication under industrial conditions. Existing methods for the extraction of these topologies from network traces were examined. It is concluded that the direct observation of encounters with a novel custom protocol can more accurately represent the behavior of an ad hoc network compared to the work in [28]. The custom protocol was implemented, tested, and used in an industrial environment. The protocol is made available for researchers and engineers to analyze the behavior of other network applications. The time-variant topologies that were recorded are also made available. The examined production facility followed the principles of Industry 4.0. To the best of our knowledge, we are the first to provide comprehensive measurements that characterize ad hoc network behavior in this context. With these recordings, researchers and engineers (can for the first time) analyze, simulate, and test industry-specific communication solutions for the factory of the future. Additionally, the acquired topologies were analyzed in terms of general network behavior with the goal to give general recommendations for communication system design.

The ad hoc channels in an industrial environment present some challenging characteristics. The observations of network connectedness (sparse vs. fully meshed) suggest that interference impacts the channel availability in the industry. Mobility of the clients, however, has by far higher effects on the availability of channels between nodes. The analysis of the bidirectionality of the available channels suggests that many existing MANET protocols are not applicable to the shop floor. The high percentage of unidirectional connection (30% to 35%) highly impacts the search for routes and increases the resulting routing overhead. The network performance is further impacted by the low route lifetime. For the ad hoc channels between mobile clients in the industrial environment, an average route life time of 1.21 s was observed. In regards to route lifetime, interference in the industrial environment has the higher impact factor, compared to the mobility. Lastly, the effects of multi-hop networks on the network connectedness and route lifetime were observed. Both benefit especially from the inclusion of the first and second relay/hop. This is an interesting observation when considering the availability of technologies like Side-Link for 5G. Even more hops have an even bigger effect, but the benefit decreases. The presented observations are currently limited to the wireless peer-to-peer channel of IEEE802.11 interfaces. However, the proposed methods are applicable to any other broadcast-enabled communication technology.

The presented results illustrate the benefits of industrial MANETs, as well as the challenges. In the future it is planned to acquire more NEPs from a rich set of industrial and other environments and a variety of wireless communication technologies. This data-set will benefit us in the design and testing of industrial MANETs and a unified communication framework for mobile robots in the industry. Additionally, the acquired NEPs shall be used to test different routing protocols, where the results will be validated by experimental MANET implementations in production facilities.

Author Contributions: Conceptualization, C.S. and M.S.; methodology, C.S.; software, C.S.; validation, C.S., M.S. and E.L.; formal analysis, C.S. and E.L.; investigation, E.L. and C.S.; resources, C.S.; data curation, C.S.; writing—original draft preparation, C.S. and E.L.; writing—review and editing, C.S., E.L. and M.S.; visualization, C.S.; supervision, M.S.; project administration, C.S. All authors have read and agreed to the published version of the manuscript.

Appendix A. Simulation and Model Results

Results of numerical simulations are described and analyzed in Section 3.3.

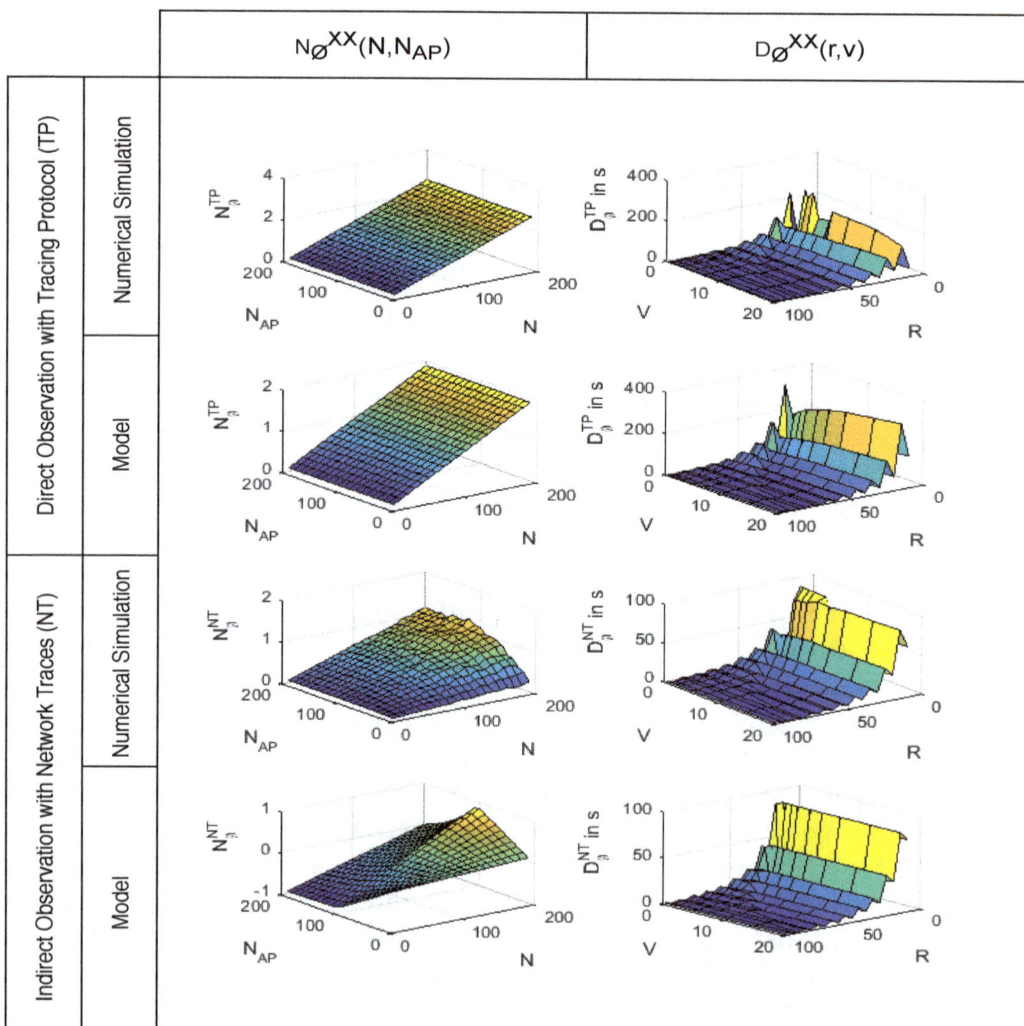

Figure A1. Comparison of the proposed model and numerical simulation.

Appendix B. Theorem of Circle Intersection Length

Theorem A1. *The chord length of a circle intersection depends on the radius r of the circle and the distance s of the chord to the center of the circle. With these two values* α *can be determined as* $\alpha = \arcsin \frac{s}{r}$*. The length of the intersection is subsequently defined as* $I(s,r) = 2r \cdot \cos \arcsin \frac{s}{r}$*. In the scenario of randomly driving nodes through the circle, it is assumed that s is equally distributed over* $s \in [0, r[$*. With this assumption the intersection length can be averaged over the distribution as follows:*

$$I(r) = \frac{1}{s_{max} - s_{min}} \int_{s_{min}}^{s_{max}} I(s,r)ds = \frac{1}{r-0} \int_{0}^{r} 2r \cos \arcsin \frac{s}{r} ds = \frac{1}{r} \frac{\pi r^2}{2} = \frac{\pi}{2} r \quad \text{(A1)}$$

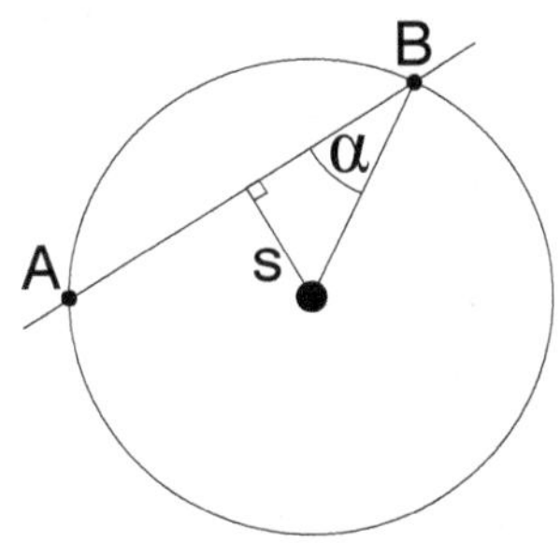

Figure A2. Line intersecting a circle.

References

1. 5G ACIA. *5G for Connected Industries and Automation*; Technical Report; ZVEI—German Electrical and Electronic Manufacturers: Frankfurt, Germany, 2018.
2. Jarchlo, E.A.; Haxhibeqiri, J.; Moerman, I.; Hoebeke, J. To Mesh or not to Mesh: Flexible Wireless Indoor Communication among Mobile Robots in Industrial Environments. In Proceedings of the AdHoc-Now, Lille, France, 4–6 July 2016.
3. Sauer, C.; Schmidt, M.; Sliskovic, M. Delay Tolerant Networks in Industrial Applications. In Proceedings of the IEEE ETFA, Zaragoza, Spain, 10–13 September 2019.
4. Kobo, H.I.; Abu-Mahfouz, A.M.; Hancke, G.P. A Survey on Software-Defined Wireless Sensor Networks: Challenges and Design Requirements. *IEEE Access* **2017**. [CrossRef]
5. Wollschlaeger, M.; Sauter, T.; Jasperneite, J. The Future of Industrial Communication: Automation Networks in the Era of the Internet of Things and Industry 4.0. *IEEE Ind. Electron. Mag.* **2017**. [CrossRef]
6. Ericsson. The Industry Impact of 5G. Self-Published. 2018. Available online: https://files.vogel.de/vogelonline/files/9763.pdf (accessed on 18 December 2020).
7. Lucas-Estañ, M.C.; Maestre, J.L.; Coll-Perales, B.; Gozalvez, J.; Lluvia, I. An Experimental Evaluation of Redundancy in Industrial Wireless Communications. In Proceedings of the 2018 IEEE 23rd International Conference on Emerging Technologies and Factory Automation (ETFA), Turin, Italy, 4–7 September 2018; Volume 1, pp. 1075–1078. [CrossRef]
8. Chen, H.; Abbas, R.; Cheng, P.; Shirvanimoghaddam, M.; Hardjawana, W.; Bao, W.; Li, Y.; Vucetic, B. Ultra-Reliable Low Latency Cellular Networks: Use Cases, Challenges and Approaches. *IEEE Commun. Mag.* **2018**, *56*, 119–125. [CrossRef]
9. 3GPP. *TR22.804 Study on Communication for Automation in Vertical Domains (Release 16)*; Technical Report; 3GPP, Sophia Antipolis Cedex: Valbonne, France, 2018.
10. Fellan, A.; Schellenberger, C.; Zimmermann, M.; Schotten, H.D. Enabling Communication Technologies for Automated Unmanned Vehicles in Industry 4.0. In Proceedings of the ICTC, Jeju, Korea, 17–19 October 2018.
11. Zhang, M.; Yu, K. Wireless Communication Technologies in Automated Guided Vehicles: Survey and Analysis. In Proceedings of the IEEE IECON, Washington, DC, USA, 21–23 October 2018.
12. Lyczkowski, E.; Munz, H.; Kiess, W.; Joshi, P. Performance of Private LTE on the Factory Floor during Production. In Proceedings of the IEEE ICC Workshop, Dublin, Ireland, 7–11 June 2020.
13. Gidlund, M.; Lennvall, T.; Åkerberg, J. Will 5G become yet another Wireless Technology for Industrial Automation? In Proceedings of the IEEE ICIT, Toronto, ON, Canada, 22–25 March 2017. [CrossRef]

14. Lucas-Estañ, M.; Maestre, J.; Coll-Perales, B.; Gozalvez, J.; Lluvia, I. An Experimental Evaluation of Redundancy in Industrial Wireless Communications. In Proceedings of the IEEE ETFA, Turin, Italy, 4–7 September 2018.
15. Pathak, P.H.; Feng, X.; Hu, P.; Mohapatra, P. Visible Light Communication, Networking, and Sensing: A Survey, Potential and Challenges. *IEEE Commun. Surv. Tutor.* **2015**, *17*, 2047–2077. [CrossRef]
16. Lyczkowski, E.; Wanjek, A.; Sauer, C. Wireless Communication in Industrial Applications. In Proceedings of the IEEE ETFA, Zaragoza, Spain, 10–13 September 2019.
17. Willig, A.; Kubisch, M.; Hoene, C.; Wolisz, A. Measurements of a wireless link in an industrial environment using an IEEE 802.11-compliant physical layer. *IEEE Trans. Ind. Electron.* **2002**, *49*, 1265–1282. [CrossRef]
18. Chrysikos, T.; Georgakopoulos, P.; Oikonomou, I.; Kotsopoulos, S.; Zevgolis, D. Channel Measurement and Characterization for a Complex Industrial and Office Topology at 2.4 GHz. In Proceedings of the SKIMA, Malabe, Sri Lanka, 6–8 December 2017. [CrossRef]
19. Chrysikos, T.; Georgakopoulos, P.; Oikonomou, I.; Kotsopoulos, S. Measurement-based Characterization of the 3.5 GHz Channel for 5G-enabled IoT at Complex Industrial and Office Topologies. In Proceedings of the IEEE WTS, Phoenix, AZ, USA, 17–20 April 2018. [CrossRef]
20. Adegoke, E.I.; Edwards, R.M.; Whittow, W.G.; Bindel, A. Characterizing the Indoor Industrial Channel at 3.5GHz for 5G. In Proceedings of the 2019 Wireless Days (WD), Manchester, UK, 24–26 April 2019; pp. 1–4. [CrossRef]
21. Tanghe, E.; Joseph, W.; Verloock, L.; Martens, L.; Capoen, H.; Herwegen, K.V.; Vantomme, W. The industrial indoor channel: Large-scale and temporal fading at 900, 2400, and 5200 MHz. *IEEE Trans. Wirel. Commun.* **2008**, *7*, 2740–2751. [CrossRef]
22. Ai, Y.; Cheffena, M.; Li, Q. Power delay profile analysis and modeling of industrial indoor channels. In Proceedings of the 2015 9th European Conference on Antennas and Propagation (EuCAP), Lisbon, Portugal, 13–17 April 2015; pp. 1–5.
23. Schmieder, M.; Eichler, T.; Wittig, S.; Peter, M.; Keusgen, W. Measurement and Characterization of an Indoor Industrial Environment at 3.7 and 28 GHz. In Proceedings of the 2020 14th European Conference on Antennas and Propagation (EuCAP), Copenhagen, Denmark, 15–20 March 2020; pp. 1–5. [CrossRef]
24. Liu, L.; Zhang, K.; Tao, C.; Zhang, K.; Yuan, Z.; Zhang, J. Channel Measurements and Characterizations for Automobile Factory Environments. In Proceedings of the IEEE ICACT, Chuncheon-si Gangwon-do, Korea, 11–14 February 2018. [CrossRef]
25. Wang, W. On Performance Modeling of 3D Mobile Ad Hoc Networks. In *Graduate School of Systems Information Science*; Future University Hakodate: Hakodate, Japan, 2018.
26. Gungor, V.C.; Hancke, G.P. Industrial Wireless Sensor Networks: Challenges, Design Principles, and Technical Approaches. *IEEE Trans. Ind. Electron.* **2009**, *56*, 4258–4265. [CrossRef]
27. Arvind, R.V.; Raj, R.R.; Raj, R.R.; Prakash, N.K. Industrial Automation using Wireless Sensor Networks. *Indian J. Sci. Technol.* **2016**, *9*, 1–8. [CrossRef]
28. Hsu, W.j.; Helmy, A. On Nodal Encounter Patterns in Wireless LAN Traces. *IEEE Trans. Mob. Comput.* **2010**, *9*, 1563–1577. [CrossRef]
29. CRAWDAD: A Community Resource for Archiving Wireless Data at Dartmouth. Available online: http://crawdad.cs.dartmouth.edu/index.php (accessed on 18 December 2020).
30. MobiLib: Community-Wide Library of Mobility and Wireless Networks Measurements. Available online: https://www.cise.ufl.edu/~helmy/MobiLib.htm (accessed on 18 December 2020).
31. Moon, S.; Helmy, A. Understanding Periodicity and Regularity of Nodal Encounters in Mobile Networks: A Spectral Analysis. In Proceedings of the IEEE GLOBECOM, Miama, FL, USA, 6–10 December 2010.
32. Bettstetter, C.; Hartenstein, H.; Pérez-Costa, X. Stochastic properties of the random waypoint mobility model. *Wirel. Netw.* **2004**, *10*, 555–567. [CrossRef]
33. Maruyama, S.; Nakano, K.; Meguro, K.; Sengoku, M.; Shinoda, S. On Location of Relay Facilities to Improve Connectivity of Multihop Wireless Networks. In Proceedings of the IEEE APCC/MDMC, Beijing, China, 29 August–1 September 2004.
34. Kohler, E.; Morris, R.; Chen, B.; Jannotti, J.; Kaashoek, M.F. The Click Modular Router. *ACM Trans. Comput. Syst. (TOCS)* **2000**, *18*, 263–297. [CrossRef]
35. Sauer, C. Git Repository of Data, Processing Tools and Protocol Implementation. Available online: https://github.com/c17r?tab=repositories (accessed on 18 December 2020)

36. Bai, F.; Helmy, A. Impact of Mobility on Last Encounter Routing Protocols. In Proceedings of the IEEE SECON, San Diego, CA, USA, 18–21 June 2007.
37. Johnson, D.B.; Maltz, D.A.; Broch, J. DSR: The Dynamic Source Routing Protocol for Multi-Hop Wireless Ad Hoc Networks. In *Ad Hoc Networking*; Addison-Wesley Longman Publishing Co., Inc.: Boston, MA, USA, 2001.
38. Perkins, C.; Belding-Royer, E.; Das, S. *Ad Hoc On-demand Distance Vector (AODV) Routing*; Technical Report; Network Working Group: Wilmington, NC, USA, 2003.
39. Asano, T.; Unoki, H.; Higaki, H. LBSR: Routing Protocol for MANETs with Unidirectional Links. In Proceedings of the IEEE AINA, Fukuoka, Japan, 29–31 March 2004.

2

Energy-Balanced Cluster-Routing Protocol based on Particle Swarm Optimization with Five Mutation Operators for Wireless Sensor Networks

Yamin Han [1], Heejung Byun [2,*] and Liangliang Zhang [1]

[1] Department of Computer Science, The University of Suwon, Hwaseong 18323, Korea; amyhan@suwon.ac.kr (Y.H.); leonzhang@suwon.ac.kr (L.Z.)

[2] Department of Information and Technology, The University of Suwon, Hwaseong 18323, Korea

* Correspondence: heejungbyun@suwon.ac.kr

Abstract: Prolonging the network lifetime is one of the fundamental requirements in wireless sensor networks (WSNs). Sensor node clustering is a very popular energy conservation strategy in WSNs, allowing to achieve energy efficiency, low latency, and scalability. According to this strategy, sensor nodes are grouped into several clusters, and one sensor node in each cluster is assigned to be a cluster head (CH). The responsibility of each CH is to aggregate data from the other sensor nodes within its cluster and send these data to the sink. However, the distribution of sensor nodes in the sensing region is often non-uniform, which may lead to an unbalanced number of sensor nodes between clusters and thus unbalanced energy consumption between CHs. This, in turn, may result in a reduced network lifetime. Furthermore, a different number of clusters lead to a different quality of service of a WSN system. To address the problems of unbalanced number of sensor nodes between clusters and selecting an optimal number of clusters, this study proposes an energy-balanced cluster-routing protocol (EBCRP) based on particle swarm optimization (PSO) with five mutation operators for WSNs. The five mutation operators are specially proposed to improve the performance of PSO in optimizing sensor node clustering. A rotation CH selection scheme based on the highest residual energy is used to dynamically select a CH for each cluster in each round. Simulation results show that the proposed EBCRP method performs well in balancing energy consumption and prolonging the network lifetime.

Keywords: wireless sensor networks; cluster; energy balance; particle swarm optimization; mutation operators

---

1. Introduction

Wireless sensor networks (WSNs) compose a large number of sensor nodes, which are scattered in the sensing region, and one or more sink nodes [1]. Generally, sensor nodes are low-cost, energy-limited, static, and randomly deployed in the sensing region. They are used to obtain such data as temperature, humidity, and air pressure in a physical environment. Communication between sensor nodes or between sensor nodes and sink is realized through wireless signals. Sensor nodes can send data obtained from an environment to a sink in one hop or multiple hops. WSNs have been widely used for environment detection, surveillance and defense reconnaissance, and medical monitoring [2–4].

Prolonging the network lifetime is one of the fundamental requirements in WSNs. The lifetime of a network is mainly affected by energy consumption by sensor nodes. A significant amount of the energy consumed by sensor nodes is used for data transmission [5–7]. However, replacing or charging batteries of sensor nodes is not possible in some environments such as mines and battlefields [8].

Therefore, it is crucial to make an effective strategy for transmitting data sensed by sensor nodes to the sink node to prolong the network lifetime.

Sensor node clustering is one of the effective strategies for prolonging the network lifetime [9–12]. According to this strategy, sensor nodes are divided into several groups, referred to as clusters. For each cluster, one sensor node is selected as a cluster head (CH), while the remaining sensor nodes are considered as cluster members. These cluster members send their sensed data to their CH. Then, the CH transmits the data aggregated from all the cluster members to the sink in one hop or multiple hops. The sensor node clustering strategy has many benefits such as energy efficiency, low latency, and scalability [13,14]. Many clustering methods have been proposed to prolong the network lifetime. One of the methods divides the sensing region equally into many grids (subregions) [10–12], with sensor nodes in each grid being regarded as a cluster (Figure 1a). Another popular clustering method called low-energy adaptive clustering hierarchy (LEACH) can prolong the network lifetime [9]. In LEACH, CHs are selected from sensor nodes based on a random probability, with the remaining sensor nodes being assigned to their corresponding nearest CH (Figure 1b). Furthermore, in order to improve the performance of clustering sensor nodes, scholars have improved the LEACH-based methods, for example, modifying CH selection algorithms [15–17], energy aware algorithms [18–20], and optimization in CH selection [21–23].

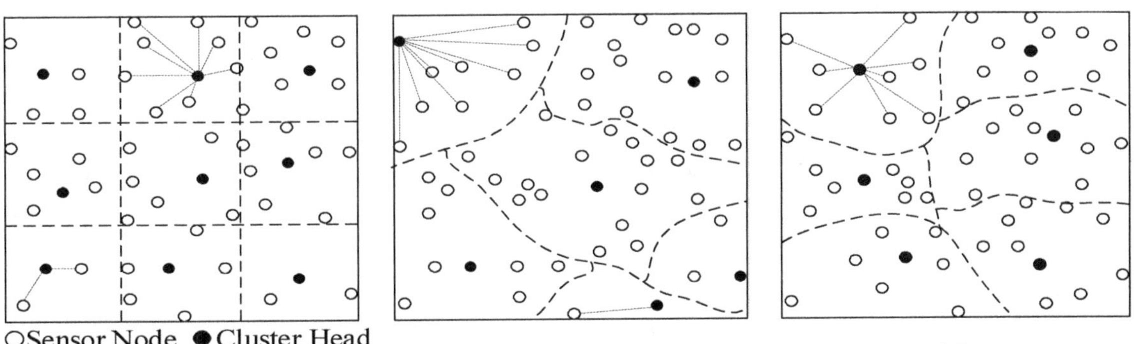

(a) (b) (c)

Figure 1. Simplified diagram of sensor node clustering: (**a**) grid-based clustering method, (**b**) low-energy adaptive clustering hierarchy (LEACH)-based clustering method, and (**c**) adaptive sensor node clustering method for determining the number of clusters and grouping the number of sensor nodes into clusters evenly.

However, the clustering of sensor nodes still faces many challenges. Generally, sensor nodes are randomly scattered in the sensing region. Grid-based clustering or assigning sensor nodes to their corresponding nearest randomly selected CH may cause an imbalance in the number of sensor nodes between clusters, resulting in different energy consumption by CHs. A CH within a cluster containing a large number of sensor nodes would consume a lot of energy and be exhausted quickly. Furthermore, the number of clusters affects the performance of WSNs. On the one hand, a small number of clusters would increase the energy consumption of each CH when gathering data from sensor nodes within its cluster and sensor nodes far away from the CH when sending data to the CH. On the other hand, a large number of clusters would increase the traffic burden in the network. Therefore, a crucial task for increasing the network lifetime is to adaptively determine the number of clusters and group sensor nodes into clusters evenly according to their distribution. This is important to improve the performance of WSN such as the network lifetime.

Recently, computational intelligence has been widely used in WSNs to provide solutions for some problems such as nodes deployment, task allocation, path planning, and movement control [24–26]. Computational intelligence has the ability to deal with imprecise information and find approximate yet good-enough solutions to these problems [27]. Computational intelligence methods are often inspired by biological processes. As one of the technologies in computational intelligence, evolutionary computation [28] imitates the process of natural evolution to provide a near-optimal solution for an optimization problem. The advantages of evolutionary computation are self-organization, self-learning, and parallel search for the global optimal solution. Among many evolutionary algorithms, the particle swarm optimization algorithm (PSO) can find the optimal solution to a problem at a higher velocity by considering previous global and local best experiences of the entire population.

This study proposes an energy-balanced cluster-routing protocol (EBCRP) using PSO with five mutation operators for WSN that groups sensor nodes into clusters evenly. In the step of EBCRP, the number of clusters and their centroids is optimized (Figure 1c), in which some sensor nodes are selected as the centroids of clusters and other sensor nodes are assigned to the nearest centroid. The average distance between sensor nodes belonging to one cluster and the centroids of this cluster, as well as balancing the number of sensor nodes between clusters are used as optimization objectives. Five mutation operators are specially designed to improve the performance of PSO in optimizing the clustering of sensor nodes. According to the clustering results of sensor nodes, in each round, the sensor node with the highest residual energy in each cluster has the priority to be selected as a CH to aggregate data from other sensor nodes in the same cluster. Then, these CHs are responsible for sending aggregated data to the sink. The proposed schemes for balanced clustering scheme and rotation CH selection based on the highest residual energy are helpful in balancing the energy consumption of sensor nodes and thus prolonging the network lifetime. The main contributions of this work are the following:

- An energy-balanced cluster-routing protocol for WSNs is proposed to balance the energy consumption of sensor nodes and prolong the network lifetime.
- An adaptive sensor node clustering scheme based on PSO is proposed to determine the number of clusters and group sensor nodes into clusters evenly.
- Five mutation operators are specially proposed to improve the performance of PSO in optimizing the clustering of sensor nodes.

The rest of the paper is organized as follows. Section 2 introduces related works on clustering sensor nodes and prolonging the network lifetime. The considered network model is described in Section 3. Section 4 outlines the proposed EBCRP method. Simulations and their results are represented in Section 5. Section 6 concludes the study and suggests directions for future work.

2. Related Works

Many clustering methods have been proposed for WSNs, and the effectiveness of sensor nodes clustering for prolonging the network lifetime has been verified. In this section, existing methods for clustering sensor nodes are reviewed from two aspects, namely, non-computational intelligence and computational intelligence.

2.1. Clustering Sensor Nodes Based on Non-Computational Intelligence

Many methods for sensor node clustering based on non-computational intelligence have been proposed. These methods group sensor nodes into several clusters and select a CH for each cluster. The CH can gather data from other sensor nodes within the same cluster and send the aggregated data to the sink. This effectively saves energy and prolongs the network lifetime.

For clustering of sensor nodes based on non-computational intelligence, LEACH is a very popular clustering routing protocol for WSNs proposed by Heinzelman et al. [9]. In LEACH, some sensor nodes are randomly selected as CHs. Each CH aggregates data from the sensor nodes within its own cluster, and then sends the data to the sink. LEACH has the advantage of increasing energy efficiency, but, its problem is to randomly select cluster heads, sensor nodes with more residual energy and less residual energy have the same probability of being selected as CH. If sensor nodes with less residual energy are selected as CHs, then they die quickly. To solve this problem, a stable election protocol (SEP) for clustered heterogeneous WSNs is proposed by Smaragdakis et al. [29]. In SEP, a certain percentage of sensor nodes are equipped with more energy than other sensor nodes. The selection of CHs is based on the election weight according to the remaining energy of each node. However, neither LEACH nor SEP takes into account the location and distance information of sensor nodes. Therefore, a quadrant-based routing protocol (Q-LEACH) is proposed by Manzoor et al. [30]. In Q-LEACH, the sensing region is divided into four quadrants, and the area within each quadrant is regarded as a cluster. The CH of each cluster communicate with each other to determine the shortest route between the source and the destination. The Q-LEACH protocol prolonged the network lifetime while increased congestion in WSNs. Marappan and Rodrigues [31] proposed a cross layer-low energy adaptive clustering hierarchy model (CL-LEACH) to save the energy of sensor nodes. CL-LEACH considered the residual energy of sensor nodes for CH selection, which preserves the overall energy. In addition, some scholars proposed the grid-based sensor node clustering strategy [10–12]. For the grid-based clustering method, the sensing region is regularly divided into several subregions, and each subregion is regarded as a cluster, which can avoid the redundant data and long-distance communications. But the grid-based clustering strategies lead to an imbalance in the number of sensor nodes between clusters, thereby resulting in an imbalance in the energy consumption between CHs.

2.2. Clustering Sensor Nodes Based on Computational Intelligence

Clustering sensor nodes based on computational intelligence is another active research trend [32–39]. Compared with the non-computational intelligence methods, technologies of computational intelligence has the characteristics of self-learning, self-organization, and self-adaptive, which enables them to provide more effective solutions for sensor node clustering. The researches have tried to optimize the sensor node clustering with computational intelligence technologies from many different aspects. For example, Zhou et al. [32] optimized the cluster structure of sensor nodes to minimize the transmission distance and reduce the energy consumption of the WSNs by an improved PSO, in which the inertial weight of PSO is adjusted to avoid particles being trapped in local optima. To address the problem where the probability of selecting an optimal sensor node as rendezvous point are very low, Tabibi and Ghaffari [33] proposed a particle swarm optimization based selection (PSOBS) method to select the optimal rendezvous points which were responsible for gathering the data of other sensor nodes. To balance the load of CHs so as to prolong the network lifetime, Ray and De [34] proposed an energy efficient clustering protocol based on K-means algorithm (EECPK-means). The EECPK-means method can balance the number of sensor nodes between clusters. To maximize the lifetime of the network, Lata et al. [35] proposed a LEACH-Fuzzy clustering method which realized CH selection and cluster formation based on fuzzy logic. To effectively cluster the sensor nodes and select the optimal CH for each cluster, Fei et al. [36] proposed a hybrid clustering method based on fuzzy c means (FCM) and moth-flame optimization method (MFO) to improve the network quality (FCMMFO). In FCMMFO, FCM was used to form energy-efficient clusters and MFO was used to select the optimal CH for each cluster. To address uneven clustering and poor energy consumption, an inter-cluster multi-hop hierarchical routing algorithm based on iterative self-organizing data analysis techniques algorithm (IICMH) is proposed by Yang et al. [37]. In addition, Elhabyan et al. [38] proposed a single multi-objective problem to find the best network configuration formulation. The proposed formulation took into account the number of CHs, the number of clustered nodes, the link quality between the cluster members and CHs, and the link quality of the constructed routing tree. Furthermore, Wang et al. [39] proposed an improved artificial bee colony algorithm to select the CHs, which took into account the energy of CHs, the density of CHs, the location of CHs, and other similar factors.

The existing sensor node clustering methods based on computational intelligence have proved the effectiveness of computational intelligence technologies in providing effective solutions for WSNs. However, sensor nodes are usually randomly scattered in the sensing region. Most existing sensor node clustering methods do not balance the number of sensor nodes between clusters, which causes unbalanced energy consumption. In addition, in different applications, the distribution and number of sensor nodes are different, it is critical to automatically determine the number of clusters according to the number and distribution of sensor nodes. Furthermore, existing technologies of computational intelligence still have some shortcomings. For some optimization methods of computational intelligence, it is easy to converge into the local optimum when dealing with high-dimensional and complex problems. Additionally, in order to cope with different WSN problems, technologies of computational intelligence should be improved accordingly. To prolong the network lifetime and make WSN systems effective data collection [40], the sensor node clustering strategy and the performance of computational intelligence technologies need to be further studied.

Therefore, this work proposes an energy-balanced cluster-routing protocol using PSO with five mutation operators for WSN, which can automatically determine the number of clusters and balance the number of sensor nodes between clusters. In addition, five mutation operators are specially proposed to improve the performance of PSO in optimizing sensor node clustering by increasing the search diversity. The energy-balanced cluster-routing protocol can effectively balance the energy consumption of sensor nodes and thus prolonging the network lifetime.

3. Network Model

In this study, the WSN system comprising N static sensor nodes and one sink node. The sink is placed in the center of the sensing region R, and sensor nodes are randomly scattered over R to monitor the environment. Sensor nodes can be grouped into several clusters, with one sensor node being appointed as the CH to aggregate data from the remaining cluster members within the same cluster. The set of sensor nodes is defined as $\mathbf{S} = \{s_1, s_2, \ldots, s_N\}$, and the set of CHs is defined as $\mathbf{CH} = \{ch_1, ch_2, \ldots, ch_k\}$, where k is the number of clusters. Each CH creates a time division multiple access (TDMA) communication message for its cluster members, which is used to inform cluster members to send data to the corresponding CH according to the schedule. The data collection, processing, and communication functions of sensor nodes are determined by the architecture of the nodes (Figure 2).

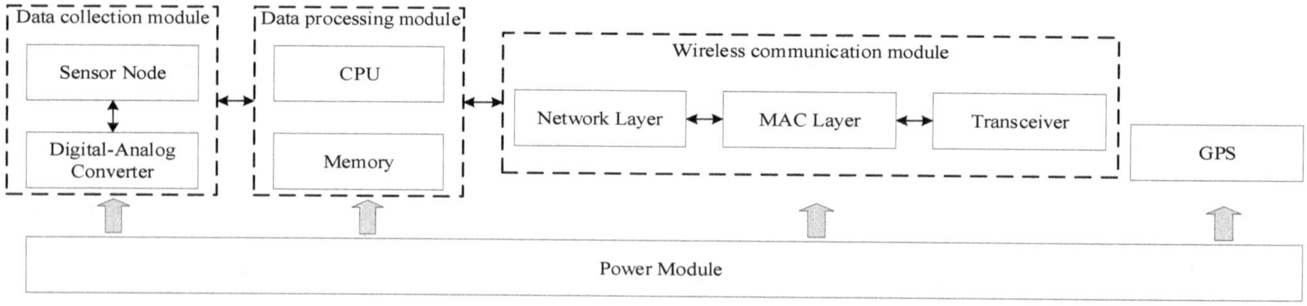

Figure 2. The main architecture of sensor nodes comprising four modules: data collection module, data processing module, wireless communication module, and power module.

Sensor nodes can convert their sensed information into digital signals in the data collection module. Then, the data processing module processes and stores these data. Sensor nodes can interact with each other via a wireless link within their communication range. The information they exchange includes collected data and some control information. Each Sensor node can locate its position through the global positioning system (GPS). The power module can provide energy to the above mentioned functional modules. Most of the available energy is consumed during data transmission. Unreasonable data transmission can lead to high and unbalanced energy consumption of sensor nodes, thereby reducing the lifetime of the network. The energy consumption model and network lifetime are described in the following subsections based on the following assumptions.

- All sensor nodes have the same communication range and equipment with the same initial energy.
- Sensor nodes can know their location via the GPS.
- The sink node has enough energy; the battery of sensor nodes cannot be replaced or recharged.

- Each sensor node has data of the same size to transmit in each round.

3.1. Energy Dissipation Model of Sensor Nodes

The energy consumed by sensor nodes is mainly used for data transmission. In this study, sensor nodes can be divided into two types: cluster members and CHs. Both of them can obtain data from the sensing region. Cluster members send the acquired data to their CH instead of directly to the sink. Each CH can aggregate data from its cluster members and send these data to the sink. The energy of cluster members is mainly used for sending data; the energy consumption of CHs includes both sending and receiving data. To calculate energy consumption, the channel model is employed based on the Euclidean distance d between the sender and the receiver. If d is smaller than a threshold d_0, the free channel model is selected; if $d > d_0$, the multipath model is adopted. The energy used by a sensor node to transmit or receive an L-bit packet is E_T and E_R respectively. E_T can be calculated as Equation (1):

$$E_T = \begin{cases} L \times E_e + L \times E_f \times d^2, & d < d_0 \\ L \times E_e + L \times E_m \times d^4, & d \geq d_0, \end{cases} \tag{1}$$

where E_e denotes the energy consumed to run the transmitter or receiver circuit; E_f and E_m depend on the used transmitter amplifier model (i.e., free channel or multipath model); d_0 denotes the threshold used to determine the model to be selected and can be defined as $\sqrt{E_f/E_m}$. E_R is defined as Equation (2):

$$E_R = L \times E_e. \tag{2}$$

It can be noticed from Equations (1) and (2) that the energy consumption of data transmission is greatly affected by the data size and distance between the sender and the receiver while the energy consumption of data reception is affected by the size of the received data. To reduce the energy consumption of sensor nodes and prolong the network lifetime, the data transmission distance and received data size should be reduced to lower the energy consumption of cluster members and CHs.

3.2. Lifetime Model of the Network

In this study, the lifetime of the network is considered to start from its initialization until it cannot work effectively. Generally, some existing studies define the network lifetime as a period from network initialization until the energy of at least one sensor node is depleted, that is, one sensor node fails [8,41,42]. The region covered by sensor nodes may not be monitored continuously after the energy of some sensor nodes is depleted, especially when a small number of sensor nodes are deployed in the sensed region; this situation can result in data loss. For a WSN system with a large number of sensor nodes, the network may continue to work effectively until a certain percentage of sensor nodes fail; this percentage is then used to determine the network lifetime [32].

Balancing the energy consumption of sensor nodes can extend the network lifetime of both small and large WSNs systems, because the energy consumption of sensor nodes become similar in this case. This can avoid the phenomenon where some sensor nodes consume energy faster while other sensors have a lot of residual energy, which is helpful to prolong the lifetime of WSNs.

4. EBCRP: Energy-Balanced Cluster-Routing Protocol

Figure 3 shows the main procedure of the proposed EBCRP.

Based on the sensing region R with randomly deployed sensor nodes, the sensor nodes are first grouped into clusters evenly using PSO with five mutation operators. In this step, the numbers of clusters and their centroids are optimized, in which some sensor nodes are selected as the centroids of clusters and other sensor nodes are assigned to their respective nearest centroid. Additionally, the sensor node clustering scheme is optimized according to the following two objectives: minimizing the average distances between sensor nodes and their centroids, and maximizing the balance index of the number of sensor nodes between clusters. To increase the diversity of search, five mutation operators are proposed for optimizing the sensor node clustering scheme. Finally, the data collection task is performed based on the clustering results. In each round, the sensor node with the highest residual energy in each cluster is selected as a CH. Each CH is responsible for aggregating data from its cluster members and sending these data to the sink. The residual energy of all sensor nodes is also updated in each round until the end of the network lifetime.

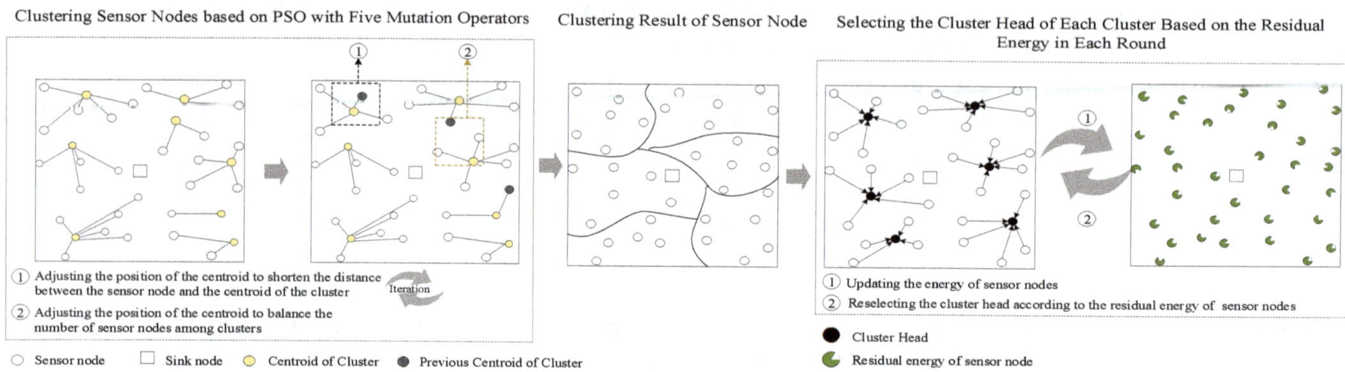

Figure 3. Main procedure of the proposed energy-balanced cluster-routing protocol (EBCRP).

4.1. *Clustering Sensor Nodes Based on Particle Swarm Optimization with Five Mutation Operators*

4.1.1. Particle Swarm Optimization

PSO is a population-based global optimization technology; the individuals in the population are called particles. In PSO, a swarm uses M particles to search for the best solution, with each particle representing a potential solution. Particles have two attributes: velocity v and position x. The solution (position) of the i-th particle can be expressed as $x_i = \{x_{i1}, x_{i2}, \ldots, x_{iD}\}$, where D is the dimension of particles. The velocity of the i-th particle, $v_i = \{v_{i1}, v_{i2}, \ldots, v_{iD}\}$, represents the speed of particle movement. In PSO, particles can cooperate and share information to find the best solution. All particles iteratively adjust their velocity and position according to their best experience $Pbest$ and the best experience $Gbest$ of the entire swarm. The scholars added the inertia weight w to the traditional PSO to change the convergence speed of the optimization algorithm, which is called standard PSO [43]. The velocity and position of the i-th particle are updated by Equations (3) and (4), respectively.

$$v_i = wv_i + rc_1(Pbest_i - x_i) + rc_2(Gbest - x_i), \tag{3}$$

$$x_i = x_i + v_i, \tag{4}$$

where w represents the inertia weight, c_1 and c_2 are acceleration constants, r is a random value in $[0,1]$, and $Pbest_i$ is the best experience of the i-th particle.

4.1.2. Topology of Particles in Sensor Node Clustering

In EBCRP, PSO is used to optimize the number of clusters and centroid of each cluster. Figure 4 illustrates the structure of particles utilized in sensor node clustering optimization.

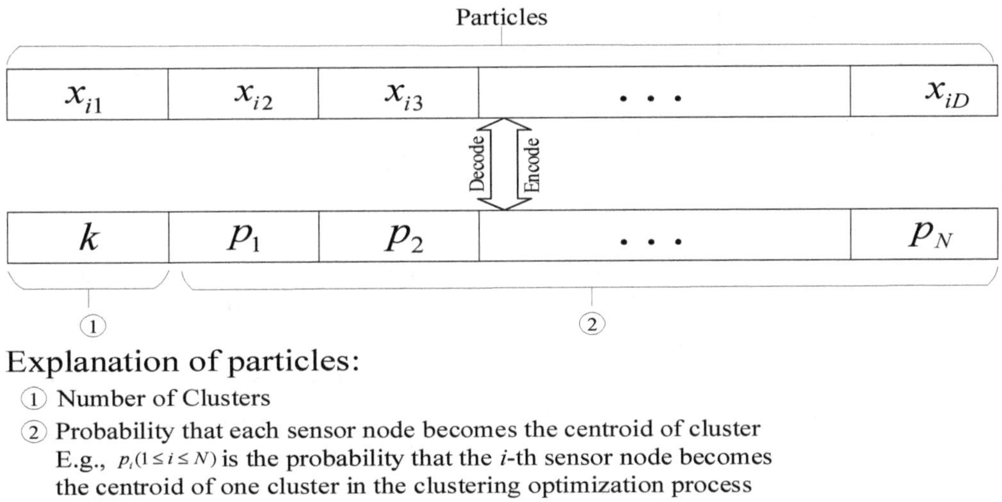

Figure 4. Structure of particles utilized in sensor node clustering optimization.

A total of $D = N + 1$-dimensional parameters are optimized, where N denotes the number of sensor nodes. In this structure of particles, the first dimension represents the number of clusters k, while the i-th ($i \in \{2, \ldots, N+1\}$) dimension represents the probability $P_i \in [0, 1]$ that the $(i-1)$-th sensor node is selected as the centroid of a cluster.

During the optimization process, for each particle (solution), k sensor nodes with the highest probability are selected as cluster centroids of clusters according to the first dimension parameter k in the particle. Other sensor nodes are assigned to their nearest centroid, which forms the sensor node clustering solution (scheme) of each particle in each generation.

4.1.3. Cost Function in Sensor Node Clustering

Each particle corresponds to a sensor node clustering solution in each generation. These solutions are expected to be searched towards the desired target (i.e., shorting the distance between sensor nodes and their centroids while balancing the number of sensor nodes between clusters). Therefore, the fitness of the clustering solutions need to be evaluated.

In this study, two goals are expected. Firstly, it is hoped that the nearest sensor nodes are grouped into a cluster instead of any distance. The average distance between the sensor nodes and the centroids they belong to is calculated to achieve the goal, which is expected to have a small value. Secondly, balancing the number of sensor nodes between clusters is expected, which helps to balance energy consumption of sensor nodes so as to prolong the lifetime of WSNs. Jain's fairness index [44] is used to evaluate the balance index of the number of sensor nodes between clusters. The corresponding balance index can be defined as Equation (5):

$$BI = \frac{(\sum_{i=1}^{k} n_i)^2}{k(\sum_{i=1}^{k} n_i^2)}, \tag{5}$$

where BI is the balance index of the number of sensor nodes between clusters, n_i is the number of sensor nodes in the i-th cluster. The value of BI ranges in the interval of [0, 1]. When BI is close to 1, the more balanced the number of sensor nodes between clusters. Therefore, the optimization objective can be defined to maximize the Equation (6):

$$max\ \ fit = \frac{w_{ad}}{Sum_d / (\sum_{i=1}^{k} n_i)} + BI, \tag{6}$$

where Sum_d represents the sum of the distances between all sensor nodes and their centroids, $\sum_{i=1}^{k} n_i$ means the number of sensor nodes in all clusters, that is, $\sum_{i=1}^{k} n_i = N$, Sum_d/N represents the average distance, and w_{ad} is the weight of the average distance in fitness fit ($w_{ad} = 10$). Since N is a positive integer, Equation (6) is equivalent to Equation (7).

$$max\ \ fit = \frac{10}{Sum_d} + \frac{N}{k(\sum_{i=1}^{k} n_i^2)}. \tag{7}$$

4.1.4. Five Mutation Operators

In the sensor node clustering optimization process, k (the value of the first dimension of a particle) sensor nodes with the highest probability are selected as cluster centroids of clusters. This means that it is necessary to not only simply optimize the value of each dimension of particles, but also adjust the position (dimension) of the value corresponding to the highest k probabilities. In addition, the clustering solution is significantly affected by the first dimension of particles representing the number of clusters. To increase the search diversity, the following five mutation operators are proposed for optimizing the sensor node clustering scheme.

- Reversing probability operator: it is used to reverse a part of the probabilities of sensor nodes. The detail process is shown in Algorithm 1.
- Decreasing probability operator: it is used to decrease the probability of some sensor nodes becoming cluster centroids. The detail process is shown in Algorithm 2.
- Increasing probability operator: it is used to increase the probability of some sensor nodes becoming the cluster centroids. The detail process is shown in Algorithm 3.
- Swapping probability operator: it is used to swap the probabilities of two sensor nodes in a particle. The detail process is shown in Algorithm 4.
- Transforming number operator: it is used to transform the number of clusters in the first dimension of particles. The detail process is shown in Algorithm 5.

Algorithm 1: Reversing probability operator

Input: Solution of the i-th particle $x_i = \{x_{i1}, x_{i2}, \ldots, x_{iD}\}$.
Output: The new solution x_i.

Randomly generate a value r_1 between 2 and D;
Randomly generate a value r_2 between 2 and D;
$tempX = x_i$;
$d_1 = min\{r_1, r_2\}$;
$d_2 = max\{r_1, r_2\}$;
for $j = d_1$ *to* d_2 **do**
    $h = d_2 - (j - d_1)$;
    $x_{ij} = tempX_h$;
end
Return the solution x_i.

Algorithm 2: Decreasing probability operator

Input: Solution of the i-th particle $x_i = \{x_{i1}, x_{i2}, \ldots, x_{iD}\}$.
Output: The new solution x_i.

for $j = 2$ *to* D **do**
    if $x_{ij} >= 0.8$ **then**
        $x_{ij} = \frac{x_{ij}}{2}$;
    end
end
Return the solution x_i.

Algorithm 3: Increasing probability operator

```
Input: Solution of the i-th particle x_i = {x_i1, x_i2, ..., x_iD}.
Output: The new solution x_i.
for j = 2 to D do
    if x_ij < 0.5 then
        x_ij = x_ij × 2;
    end
end
Return the solution x_i.
```

Algorithm 4: Swapping probability operator

```
Input: Solution of the i-th particle x_i = {x_i1, x_i2, ..., x_iD}.
Output: The new solution x_i.
Randomly generate a value r_1 between 2 and D;
Randomly generate a value r_2 between 2 and D;
tempX = x_i;
x_ir1 = tempX_r2;
x_ir2 = tempX_r1;
Return the solution x_i.
```

Algorithm 5: Transforming number operator

```
Input: Solution of the i-th particle x_i = {x_i1, x_i2, ..., x_iD}, maximum number of clusters c_max.
Output: The new solution x_i.
x_i1 = c_max × r; // r is a random value in [0, 1]
Return the solution x_i.
```

4.1.5. Optimization Process of Sensor Node Clustering

This section outlines the process of clustering sensor nodes based on PSO with the introduced five mutation operators. The parameters of PSO (position x, velocity v, and fitness) are first initialized for all particles. The best experience of particles ($Pbest$) and the best experience of entire swarm ($Gbest$) are also initialized. Then, particles begin to search for the best sensor node clustering solution through iterations. In each iteration, each particle updates its velocity and position; each particle also mutates when a random value $r > 0.5$ that is used to judge whether the current i-th particle is mutated. During the mutation process, five mutation operators are randomly selected to apply to the position for the i-th particle. Next, the fitness fit_i is evaluate. The best experiences $Pbest_i$ and $Gbest$ are updated if fit_i is higher than them. Until the stopping criterion is satisfied, the best clustering solution is obtained. The entire optimization process is shown in Algorithm 6.

Algorithm 6: Clustering sensor nodes based on PSO with five mutation operators

Input: Number of sensor nodes N, location distribution of sensor nodes S, maximum number of clusters $clustMax$.
Output: The best clustering solution $Gbest$.

Initialize inertia weight w, acceleration constants c_1 and c_2, number of particles $numP$.
Initialize position x and velocity v for all particles.
Initialize the fitness of each particle according to the clusters obtained by decoding its particle topology and Equation (7).
Initialize the best experience $Pbest_i(i \in [1, numP])$ for all particles by their current fitness.
Initialize the best experience $Gbest$ for whole swarm according to the current fitness of all particles.
while $Fes < MaxFes$ **do**
  for $i = 1$ ***to*** *the number of particles* $numP$ **do**
    Update the velocity of i-th particle by Equation (3).
    Update the position of i-th particle by Equation (4).
    if *generate a random value* $r > 0.5$ **then**
      //mutation
      Randomly select a number op from $\{1, 2, 3, 4, 5\}$;
      Switch op **do**
        case 1:
          Perform reversing probability operator on x_i by Algorithm 1;
        case 2:
          Perform decreasing probability operator on x_i by Algorithm 2;
        case 3:
          Perform increasing probability operator on x_i by Algorithm 3;
        case 4:
          Perform swapping probability operator on x_i by Algorithm 4;
        case 5:
          Perform transforming number operator on x_i by Algorithm 5;
    end
    Evaluate the fitness fit_i of i-th particle according to the clusters obtained by decoding its particle topology and Equation (7).
    if fit_i *is bigger than the fitness of* $Pbest_i$ **then**
      Update the personal best $Pbest_i = x_i$;
    end
    if fit_i *is bigger than the fitness of* $Gbest$ **then**
      Update the global best $Gbest = x_i$;
    end
  end
end
Return the best clustering solution $Gbest$.

4.2. Selecting the Cluster Heads Based on Residual Energy

Sensor nodes are grouped into k clusters based on PSO with five mutation operators as described in the above subsection, in which the number of clusters k is optimized by PSO. The set of clusters $\mathbf{CS}$ is defined as $\mathbf{CS} = \{cs_1, cs_2, \ldots, cs_k\}$. cs_i has a set of sensor nodes, $cs_i = \{s_i^1, \ldots, s_i^{n_i}\}$, where s_i^j denotes the j-th sensor node in cs_i and n_i is the number of sensor nodes in cs_i.

In this study, a rotation CH selection scheme based on the highest residual energy is used to balance the energy consumption of sensor nodes, thereby prolonging the network lifetime. In each round, the sensor node with the highest residual energy in each cluster is selected as a CH to aggregate data from its cluster members and send data to the sink. During the lifetime of the network system, the residual energy of sensor nodes in cs_i is defined as $e_i = \{e_i^1, \ldots, e_i^{n_i}\}$. The set of CHs in each round is defined as $\mathbf{CH} = \{ch_1, ch_2, \ldots, ch_k\}$ and calculated as Equation (8).

$$ch_i = \{j \in [1, n_i] \mid e_i^j \geq e_i^k, k \in [1, n_i]\}. \tag{8}$$

5. Simulations and Results

This section presents several simulations performed using MATLAB to evaluate the effectiveness and performance of the proposed EBCRP method. The effectiveness of EBCRP is first validated in terms of the clustering performance and balanced energy consumption of sensor nodes. Further, simulations are also performed to compare EBCRP with some existing methods, namely, LEACH [9], SEP [29], and IICMH [37] in terms of the network lifetime and balancing the energy consumption. For the simulations, the sensing region is set to 100 m × 100 m; the number of sensor nodes N is varied from 50 to 200. Some parameters of the WSN system are listed in Table 1.

Table 1. Parameters and their values of the wireless sensor network (WSN).

Parameter	Value
Initial Energy of Sensor Nodes (E_0)	0.1 J
Size of Data	4000 bit
E_e	50 nJ/bit
E_f	10 pJ/bit^2
E_m	1.3×10^{-3} pJ/bit^4

5.1. The Effectiveness of EBCRP

In EBCRP, an adaptive sensor node clustering scheme based on PSO is proposed to determine the number of clusters and group sensor nodes into clusters evenly. Additionally, five mutation operators are specially proposed to improve the performance of PSO in optimizing sensor node clustering. This subsection examines the effectiveness of the proposed EBCRP method.

The effectiveness of PSO and five mutation operators in EBCRP are verified. PSO with five mutation operators (PSO-FMO) and without five mutation operators (SPSO) are compared. The same parameters are used for both PSO-FMO and SPSO. The population size is set to 50. Acceleration constants c_1 and c_2 are set to 1.5 and 2.0, respectively. The number of fitness evaluations (FEs) is set to 10,000. Figure 5 shows the sensor node clustering results based on SPSO and PSO-FMO for different numbers of sensor nodes (50, 100, 150 and 200).

It can be noticed from the figure that while both PSO-FMO and SPSO group the sensor nodes scattered into clusters in the sensing region R. However, the allocation of cluster centroids by SPSO is not reasonable. For example, the clusters highlighted by circles have extremely asymmetrical centroids, which results in an increased distance between sensor nodes and their centroids and thus a deceased fitness fit. Furthermore, the result of SPSO also shows the imbalance of the number of sensor nodes between clusters, which is shown in the clusters highlighted by square. These results indicate the weak performance of SPSO, the method that does not use the five mutation operators in optimizing sensor node clustering. In contrast, PSO-FMO achieves more reasonable cluster centroids and a better balance of the number of sensor nodes between clusters. This is because the use of five mutation operators increases the search diversity, thus improving the probability of finding a better sensor node clustering solution. Therefore, the use of PSO-FMO justifies the good performance of EBCRP in optimizing sensor node clustering.

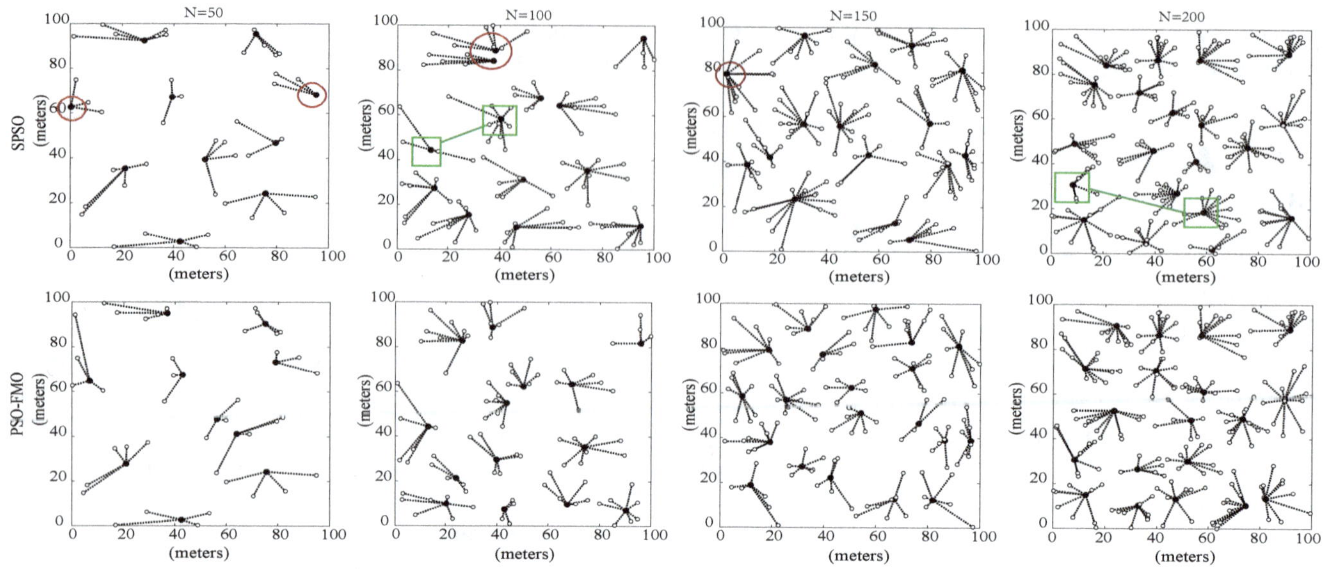

Figure 5. Clustering results of sensor nodes based on SPSO and PSO-FMO in terms of different number of sensor nodes 50, 100, 150 and 200. The clusters highlighted by circles have extremely asymmetrical centroids; while the clusters highlighted by square show the imbalance of the number of sensor nodes between clusters, which is the result of SPSO.

Figure 6 illustrates the mean convergence curves of SPSO and PSO-FMO in optimizing sensor nodes clustering.

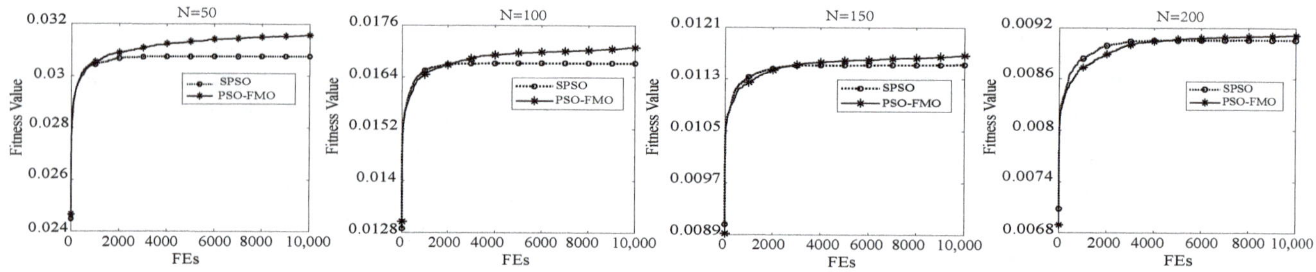

Figure 6. Mean convergence curves of SPSO and PSO-FMO in optimizing sensor node clustering.

To reduce the statistical error, both methods are run 30 times under the same conditions. It can be noticed from the figure that PSO-FMO achieves a higher average fitness for different numbers of sensor nodes compared to SPSO. SPSO converges quickly and falls into a local optima. In contrast, PSO-FMO continuously searches the best solution and avoids falling into local optima. This is because the five mutation operators continuously provide the diversity of particles for the entire swarm, which helps particles find a better solution. Hence, the mean convergence curves also verify the effectiveness of PSO-FMO in EBCRP.

To verify the effectiveness of PSO-FMO in balancing the number of sensor nodes between clusters, Table 2 also gives the comparison of balance index of the number of sensor nodes between clusters. It can be noticed from the table that PSO-FMO achieves a better balance for different numbers of sensor nodes, which also indicates the effectiveness of PSO-FMO in EBCRP.

Table 2. Balance Index of the Number of Sensor Nodes between Clusters.

Total Number of Sensor Nodes	$N = 50$	$N = 100$	$N = 150$	$N = 200$
Balance index based on SPSO	0.9615	0.9568	0.9527	0.9542
Balance index based on PSO-FMO	0.9766	0.9731	0.9715	0.9747

To verify the effectiveness of EBCRP method in balancing the energy consumption of sensor nodes, the residual energy of sensor nodes during the network lifetime is analyzed. Figure 7 shows the residual energy of each sensor node in different rounds for different numbers of sensor nodes (50, 100, 150 and 200).

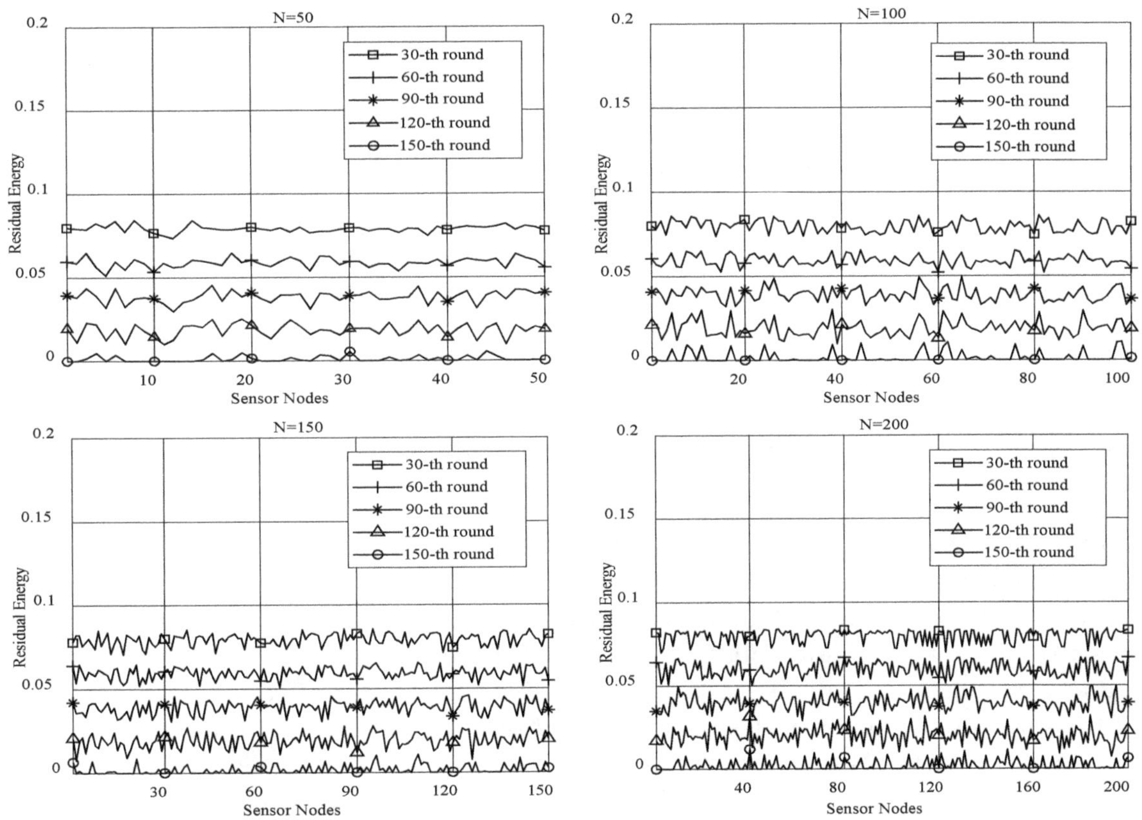

Figure 7. Residual energies of sensor nodes in different rounds.

While the residual energies of sensor nodes fluctuate up and down, it still demonstrates balance within a small energy range. In sensor node clusters, some sensor nodes are dynamically selected as CHs, while others are selected as cluster members. The rates of energy consumption of CHs and cluster members are different. Therefore, there are some sensor nodes with a higher energy consumption. Furthermore, all sensor nodes cannot consume energy completely synchronously. Small range fluctuations of the residual energy of sensor nodes prove the effectiveness of EBCRP in balancing the energy consumption of sensor nodes.

5.2. Comparison of EBCRP with Other Methods

To validate the performance of the proposed EBCRP method, it was compared to other methods, namely, LEACH [9], SEP[29], and IICMH [37]. The LEACH protocol selects some sensor nodes as CHs according to a random probability of each node. Other sensor nodes were assigned to corresponding CHs based on the received signal strength. SEP is similar to LEACH, except that a certain proportion of sensor nodes in SEP are equipped with more energy, and CHs are selected based on the residual energy of each sensor node. IICMH adopts iterative self-organizing data analysis techniques algorithm to select clustering centers; then, CHs are selected based on the distance between sensor nodes and the clustering centers. When the residual energy of CHs is lower than a set threshold, CHs are reselected based on the residual energy of sensor nodes.

To fully compare the performance of each method, simulations were performed for four different numbers of sensor nodes: 50, 100, 150 and 200. For LEACH and SEP, 5% of sensor nodes were selected as CHs. For the IICMH and EBCRP methods, 5% of sensor nodes are set as the expected number of clusters and a maximum number of clusters , respectively.

Figure 8 shows the number of dead sensor nodes for the different methods in different rounds.

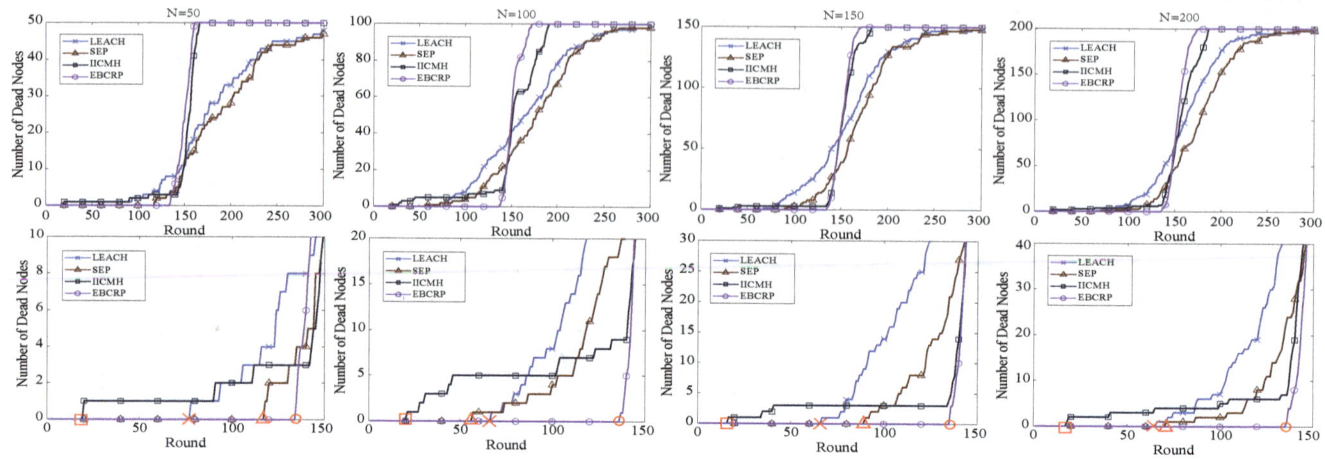

Figure 8. Number of dead nodes of different methods in different rounds, and the bottom subfigures detail the the first dead node.

The first row displays the number of dead sensor nodes from 0 to 300 rounds, while the second row shows the round corresponding to the first dead sensor node for each method. It can be noticed from the figure that the number of dead sensor nodes in LEACH and SEP increases slowly. It is obvious that some sensor nodes consume energy too fast, while others consume energy slowly due to the unbalanced energy consumption of sensor nodes in LEACH and SEP. In contrast, IICMH considers the balance of the number of sensor nodes between clusters when clustering sensor nodes, which helps balance the energy consumption of sensor nodes. Therefore, there is a period of a sharp increase in the number of dead sensor nodes. Nevertheless, IICMH still does not deal with the energy balance of sensor nodes well as some sensor nodes ran out of energy prematurely. EBCRP only shows a period of a sharp rise in the number of dead sensor nodes, whereas there is no problem of sensor nodes consuming energy prematurely. This is because EBCRP balances the energy consumption of sensor nodes well by balancing the number of sensor nodes between clusters and employing the rotation CH selection scheme based on the highest residual energy. It can be noticed that the first dead sensor node appears in a later round for EBCRP compared to the other methods, which also indicates that the energy balance of sensor nodes in EBCRP is better. Overall, EBCRP presents a good performance in balancing the energy consumption of sensor nodes.

To further verify the balance of energy consumption between sensor nodes in EBCRP, the energy consumption balance index (ECBI) is calculated based on the fairness index as shown in Equation (9).

$$ECBI = \frac{(\sum_{i=1}^{N} ec_i)^2}{N(\sum_{i=1}^{N} ec_i^2)}, \tag{9}$$

where ec_i is the energy consumption of the i-th sensor node. Figure 9 compares the curve of ECBI of different methods during rounds.

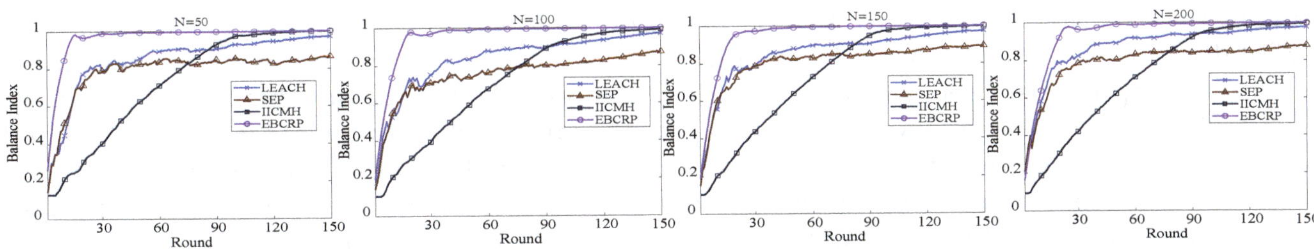

Figure 9. The curve of energy consumption balance index of different methods.

In addition, Table 3 also quantitatively compares the ECBI between the proposed EBCRP method and other methods in 30-th, 60-th, 90-th, 120-th, and 150-th rounds under the different number of sensor nodes (50, 100, 150 and 200), in which the best EBCI value is bolded for each round.

Table 3. The comparison of energy consumption balance index of different methods.

Number of Sensor Nodes	Round	LEACH	SEP	IICMH	EBCRP
	30-th	0.7883	0.7918	0.4026	**0.9906**
	60-th	0.8939	0.8362	0.7096	**0.9954**
$N = 50$	90-th	0.9056	0.8485	0.9282	**0.9967**
	120-th	0.9459	0.8270	0.9899	**0.9976**
	150-th	0.9712	0.8645	0.9992	**0.9997**
	30-th	0.7530	0.7094	0.3935	**0.9691**
	60-th	0.8784	0.7623	0.6768	**0.9937**
$N = 100$	90-th	0.8956	0.8005	0.8010	**0.9956**
	120-th	0.9346	0.8270	0.9712	**0.9965**
	150-th	0.9692	0.8719	0.9911	**0.9980**
	30-th	0.7894	0.7861	0.4364	**0.9722**
	60-th	0.8951	0.8276	0.7299	**0.9934**
$N = 150$	90-th	0.9042	0.8478	0.9464	**0.9960**
	120-th	0.9470	0.8683	0.9897	**0.9970**
	150-th	0.9725	0.8919	0.9973	**0.9994**
	30-th	0.8251	0.7858	0.4193	**0.9650**
	60-th	0.9185	0.8322	0.7107	**0.9890**
$N = 200$	90-th	0.9384	0.8402	0.9219	**0.9948**
	120-th	0.9584	0.8479	0.9843	**0.9966**
	150-th	0.9758	0.8785	0.9951	**0.9991**

The best results are highlighted in bold.

From Figure 9 and Table 3, it can be observed that the ECBI of all the methods increases gradually with the increase of rounds from 1 to 150 for all the case with different numbers of sensor nodes. This is because more and more sensor nodes consume a lot or even all energy with the increase of rounds, which will raise the ECBI. Also, the proposed EBCRP method and these comparison methods dynamically select CH for each cluster, which can balance the energy consumption between sensor nodes. However, the ECBI of the proposed EBCRP method is significantly higher than that of the other methods in different rounds. This is because LEACH and SEP randomly select some sensor nodes as CHs, and then other sensor nodes are assigned to their respective nearest CHs, which ignores the balance of the number of sensor nodes between clusters and increases the imbalance of energy consumption between sensor nodes. In contrast, during sensor node clustering, IICMH and the proposed EBCRP take into account the balance of the number of sensor nodes between clusters, thereby helping to balance the energy consumption of sensor nodes. Nevertheless, in the IICMH method, the CH of each cluster is reselected after the energy of the previous CH is lower than a threshold. This makes the energy of sensor nodes selected as CHs for the first time consume more energy. On the contrary, the used rotation CH selection based on the highest residual energy in the proposed EBCRP method can effectively balance the energy consumption between sensor nodes.

So the ECBI of IICMH is lower than that of the proposed EBCRP method. The significantly higher ECBI of the proposed EBCRP than other methods prove the effectiveness of EBCRP in balancing the energy consumption between sensor nodes. Furthermore, EBCRP shows a high ECBI in different rounds, whereas the other methods show significantly smaller ECBI in the early rounds than in the later rounds. This also proves the advantages of EBCRP over other methods. Therefore, the proposed EBCRP performs well in balancing the energy consumption of sensor nodes. However, the proposed EBCRP method dynamically selects CH for each cluster in each round, which reduces the stability of the network system.

The network lifetime of EBCRP is also compared with other methods. Figure 10 illustrates the network lifetime of different methods for different numbers of sensor nodes.

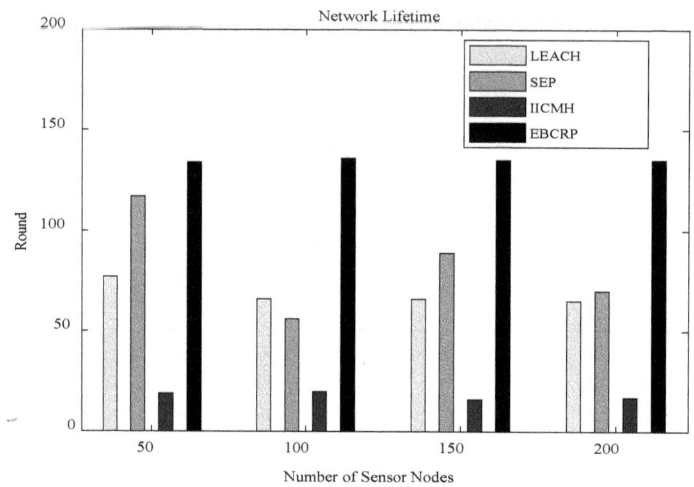

Figure 10. Network lifetime of different methods for different numbers of sensor nodes.

EBCRP achieves the longest lifetime compared to LEACH, SEP, and IICMH. There are two main factors that prolong the network lifetime of EBCRP. First, EBCRP employs an adaptive sensor node clustering scheme to determine the number of clusters and group sensor nodes into clusters evenly. This helps to balance the energy consumption between clusters. Second, EBCRP adopts a rotation CH selection scheme based on the highest residual energy to dynamically select CHs for each cluster. This approach can balance the energy consumption of sensor nodes within each cluster. Balancing the energy consumption between clusters and within clusters in EBCRP prevents some sensor nodes from consuming energy prematurely, thereby prolonging the lifetime of the WSN. However, LEACH and SEP randomly select some sensor nodes as CHs, while other sensor nodes are assigned to their respective nearest CHs, which cannot balance the number of sensor nodes between clusters and energy consumption of sensor nodes. At the same time, a certain proportion of sensor nodes in SEP are equipped with more energy, and CHs selection in SEP takes into account the residual energy of sensor nodes. Thus, the network lifetime of SEP is higher than that of LEACH, but it is still lower than the proposed EBCRP method. Although the IICMH also considers the balance of the number of sensor nodes between clusters, the CH of each cluster is reselected after the energy of the previous CH is lower than a threshold. This makes the energy of sensor nodes selected as CHs for the first time to quickly depleted. In addition, sensor nodes close to the sink in each cluster are preferentially selected as CHs in IICMH, which increases the energy consumption of some sensor nodes (i.e., cluster members far away from their CHs) when they send data to their CHs. Thus, the network lifetime of IICMH is lower compared to the other methods. Overall, the proposed EBCRP method performs better in prolonging the network lifetime by performing balanced clustering and employing the rotation CH selection scheme based on the highest residual energy.

6. Conclusions

This study proposes an energy balanced cluster-routing protocol based on particle swarm optimization with five mutation operators for wireless sensor networks to balance energy consumption of sensor nodes and prolong the network lifetime. An adaptive sensor node clustering scheme based on the PSO is proposed to determine the number of clusters and group sensor nodes into clusters evenly, which is helpful to balance the energy consumption between sensor nodes. To increase the search diversity, five mutation operators are specially proposed in EBCRP to improve the performance of PSO in optimizing sensor node clustering. In addition, a rotation CH selection scheme based on the highest residual energy is used in EBCRP to dynamically select a CH for each cluster. In each round, the sensor node with the highest residual energy in each cluster is selected as a CH responsible for aggregating data from the remaining sensor nodes within each cluster and sending these data to the sink. This can balance the energy consumption of sensor nodes within each cluster. Comprehensive simulations are performed to investigate the effectiveness of the proposed EBCRP method and compare it to some other methods. The simulation results demonstrate that the proposed EBCRP method can effectively balance the number of sensor nodes between clusters and energy consumption of sensor nodes, thereby prolonging the network lifetime.

Some limitations still exist in the EBCRP method and become important research work in the future. The PSO-based sensor node clustering scheme is greatly affected by the performance of the PSO. For the multimodal, high-dimensional, complex, and nonlinear optimization problems in WSNs, the PSO is easy to prematurely converges to local optima. To find an optimal or near-optimal solutions, PSO and other optimization algorithms could be studied in the future to further improve the performance of EBCRP. In addition, the EBCRP selects CH for each cluster in each round, which reduces the stability of the network system. A more efficient CH selection mechanism could be further studied, e.g, each cluster selects multiple CHs to cooperate with each other to aggregate the data from other sensor nodes within their cluster. When the energy of the current CHs is lower than a threshold, the new CHs are reselected. Another important future work is about the application of the EBCRP to complex WSNs problems with large sensing region, some relay nodes can be selected to assist the communication between CHs.

Author Contributions: Conceptualization, Y.H. and H.B.; methodology, Y.H., H.B. and L.Z.; software, Y.H. and L.Z.; validation, Y.H., H.B. and L.Z.; writing, Y.H., H.B. and L.Z.; funding acquisition, H.B. All authors have read and agreed to the published version of the manuscript.

References

1. Khan, I.; Belqasmi, F.; Glitho, R.; Crespi, N.; Morrow, M.; Polakos, P. Wireless sensor network virtualization: A survey. *IEEE Commun. Surv. Tutor.* **2016**, *18*, 553–576. [CrossRef]
2. Mutiara, G.A.; Herman, N.S.; Mohd, O. Using Long-Range Wireless Sensor Network to Track the Illegal Cutting Log. *Appl. Sci.* **2020**, *10*, 6992. [CrossRef]
3. Pandya, S.; Ghayvat, H.; Sur, A.; Awais, M.; Kotecha, K.; Saxena, S.; Jassal, N.; Pingale, G. Pollution Weather Prediction System: Smart Outdoor Pollution Monitoring and Prediction for Healthy Breathing and Living. *Sensors* **2020**, *20*, 5448. [CrossRef] [PubMed]
4. Fattah, S.; Gani, A.; Ahmedy, I.; Idris, M.Y.I.; Targio Hashem, I.A. A Survey on Underwater Wireless Sensor Networks: Requirements, Taxonomy, Recent Advances, and Open Research Challenges. *Sensors* **2020**, *20*, 5393. [CrossRef]
5. Raghunathan, V.; Ganeriwal, S.; Srivastava, M. Emerging techniques for long lived wireless sensor networks. *IEEE Commun. Mag.* **2006**, *44*, 108–114. [CrossRef]
6. Anastasi, G.; Conti, M.; Di Francesco, M.; Passarella, A. Energy conservation in wireless sensor networks: A survey. *Ad Hoc Netw.* **2009**, *7*, 537–568. [CrossRef]

7. Alnuaimi, M.; Shuaib, K.; Alnuaimi, K.; Abdel-Hafez, M. Data gathering in delay tolerant wireless sensor networks using a ferry. *Sensors* **2015**, *15*, 25809–25830. [CrossRef]
8. Zhao, M.; Yang, Y.; Wang, C. Mobile Data Gathering with Load Balanced Clustering and Dual Data Uploading in Wireless Sensor Networks. *IEEE Trans. Mob. Comput.* **2015**, *14*, 770–785. [CrossRef]
9. Heinzelman, W.R.; Chandrakasan, A.; Balakrishnan, H. Energy-efficient communication protocol for wireless microsensor networks. In Proceedings of the 33rd Annual Hawaii International Conference on System Sciences, Maui, HI, USA, 7 January 2000; Volume 2, p. 10.
10. Khan, A.W.; Abdullah, A.H.; Razzaque, M.A.; Bangash, J.I. VGDRA: A Virtual Grid-Based Dynamic Routes Adjustment Scheme for Mobile Sink-Based Wireless Sensor Networks. *IEEE Sens. J.* **2015**, *15*, 526–534. [CrossRef]
11. Kareem, M.M.; Ismail, M.; Altahrawi, M.A.; Arsad, N.; Mansor, M.F.; Ali, A.H. Grid Based Clustering Technique in Wireless Sensor Network using Hierarchical Routing Protocol. In Proceedings of the 2018 IEEE 4th International Symposium on Telecommunication Technologies (ISTT), Selangor, Malaysia, 26–28 November 2018; pp. 1–5.
12. Padmanaban, Y.; Muthukumarasamy, M. Scalable Grid-Based Data Gathering Algorithm for Environmental Monitoring Wireless Sensor Networks. *IEEE Access* **2020**, *8*, 79357–79367. [CrossRef]
13. Fanian, F.; Rafsanjani, M.K. Cluster-based routing protocols in wireless sensor networks: A survey based on methodology. *J. Netw. Comput. Appl.* **2019**, *142*, 111–142. [CrossRef]
14. Shahraki, A.; Taherkordi, A.; Haugen, Ø.; Eliassen, F. Clustering objectives in wireless sensor networks: A survey and research direction analysis. *Comput. Netw.* **2020**, *180*, 107376. [CrossRef]
15. Bhadeshiya, J.; Vora, S. A Reformed Cluster-Head of LEACH Protocol and Performance Analysis with Conventional Routing Protocol for WSN. *J. Inf. Knowl. Res. Electron. Commun. Eng.* **2012**, *2*, 812–816.
16. Sharma, P.; Yadav, A. Enhanced parameters incorporated in LEACH for wireless sensor network. *Int. J. New Innov. Eng. Technol.* **2013**, *2*, 5–8.
17. Iqbal, A.; Akbar, M.; Javaid, N.; Bouk, S.H.; Ilahi, M.; Khan, R. Advanced LEACH: A static clustering-based heteroneous routin protocol for WSNs. *J. Basic Appl. Sci. Res.* **2013**, *3*, 864–872.
18. Luan, W.; Zhu, C.; Su, B.; Pei, C. An improved routing algorithm on LEACH by combining node degree and residual energy for WSNs. In *Internet of Things*; Springer: Berlin/Heidelberg, Germany, 2012; pp. 104–109.
19. Muruganathan, S.D.; Ma, D.C.; Bhasin, R.I.; Fapojuwo, A.O. A centralized energy-efficient routing protocol for wireless sensor networks. *IEEE Commun. Mag.* **2005**, *43*, S8–S13. [CrossRef]
20. Taneja, H.; Bhalla, P. An improved version of leach: Three levels hierarchical clustering leach protocol (TLHCLP) for homogeneous WSN. *Int. J. Adv. Res. Comput. Commun. Eng.* **2013**, *2*, 3610–3615.
21. Kaur, R.; Sharma, D. Improvement of Leach Protocol with K Angle Optimization using an Optimized Algorithm in Wireless Sensor Networks. *Int. J. Comput. Appl.* **2013**, *70*, 37–41. [CrossRef]
22. Singh, B.; Lobiyal, D.K. A novel energy-aware cluster head selection based on particle swarm optimization for wireless sensor networks. *Hum.-Centric Comput. Inf. Sci.* **2012**, *1*, 1–18. [CrossRef]
23. Ma, D.; Ma, J.; Xu, P. An adaptive assistant-aided clustering protocol for WSNs using niching particle swarm optimization. In Proceedings of the 2013 IEEE 4th International Conference on Software Engineering and Service Science, Beijing, China, 23–25 May 2013; pp. 648–651.
24. Sasikumar, P.; Khara, S. K-Means Clustering in Wireless Sensor Networks. In Proceedings of the 2012 Fourth International Conference on Computational Intelligence and Communication Networks, Mathura, India, 3–5 November 2012; pp. 140–144.
25. Primeau, N.; Falcon, R.; Abielmona, R.; Petriu, E.M. A review of computational intelligence techniques in wireless sensor and actuator networks. *IEEE Commun. Surv. Tutor.* **2018**, *20*, 2822–2854. [CrossRef]
26. Khalifeh, A.; Rajendiran, K.; Darabkh, K.A.; Khasawneh, A.M.; AlMomani, O.; Zinonos, Z. On the Potential of Fuzzy Logic for Solving the Challenges of Cooperative Multi-Robotic Wireless Sensor Networks. *Electronics* **2019**, *8*, 1513. [CrossRef]
27. Kulkarni, R.V.; Förster, A.; Venayagamoorthy, G.K. Computational Intelligence in Wireless Sensor Networks: A Survey. *IEEE Commun. Surv. Tutor.* **2011**, *13*, 68–96. [CrossRef]
28. Tang, W.; Wu, Q. Evolutionary computation. In *Condition Monitoring and Assessment of Power Transformers Using Computational Intelligence*; Springer:London, UK, 2011; pp. 15–36.

29. Smaragdakis, G.; Matta, I.; Bestavros, A. SEP: A Stable Election Protocol for Clustered Heterogeneous Wireless Sensor Networks. In Proceedings of the Second International Workshop on Sensor and Actor Network Protocols and Applications (SANPA 2004), Boston, MA, USA, 22 August 2004.
30. Manzoor, B.; Javaid, N.; Rehman, O.; Akbar, M.; Ishfaq, M. Q-LEACH: A New Routing Protocol for WSNs. *Procedia Comput. Sci.* **2013**, *19*, 926–931. [CrossRef]
31. Marappan, P.; Rodrigues, P. An energy efficient routing protocol for correlated data using CL-LEACH in WSN. *Wirel. Netw.* **2016**, *22*, 1415–1423. [CrossRef]
32. Zhou, Y.; Wang, N.; Xiang, W. Clustering Hierarchy Protocol in Wireless Sensor Networks Using an Improved PSO Algorithm. *IEEE Access* **2017**, *5*, 2241–2253. [CrossRef]
33. Tabibi, S.; Ghaffari, A. Energy-efficient routing mechanism for mobile sink in wireless sensor networks using particle swarm optimization algorithm. *Wirel. Pers. Commun.* **2019**, *104*, 199–216. [CrossRef]
34. Ray, A.; De, D. Energy efficient clustering protocol based on K-means (EECPK-means)-midpoint algorithm for enhanced network lifetime in wireless sensor network. *IET Wirel. Sens. Syst.* **2016**, *6*, 181–191. [CrossRef]
35. Lata, S.; Mehfuz, S.; Urooj, S.; Alrowais, F. Fuzzy Clustering Algorithm for Enhancing Reliability and Network Lifetime of Wireless Sensor Networks. *IEEE Access* **2020**, *8*, 66013–66024. [CrossRef]
36. Fei, W.; Hexiang, B.; Deyu, L.; Jianjun, W. Energy-Efficient Clustering Algorithm in Underwater Sensor Networks Based on Fuzzy C Means and Moth-Flame Optimization Method. *IEEE Access* **2020**, *8*, 97474–97484. [CrossRef]
37. Yang, X.; Gao, L.; Wang, X. Inter-cluster multi-hop routing algorithm for wireless sensor networks based on ISODATA clustering. In Proceedings of the 2019 IEEE 4th Advanced Information Technology, Electronic and Automation Control Conference (IAEAC), Chengdu, China, 20–22 December 2019; Volume 1, pp. 2521–2525.
38. Elhabyan, R.; Shi, W.; St-Hilaire, M. A Pareto optimization-based approach to clustering and routing in Wireless Sensor Networks. *J. Netw. Comput. Appl.* **2018**, *114*, 57–69. [CrossRef]
39. Wang, Z.; Ding, H.; Li, B.; Bao, L.; Yang, Z. An Energy Efficient Routing Protocol Based on Improved Artificial Bee Colony Algorithm for Wireless Sensor Networks. *IEEE Access* **2020**, *8*, 133577–133596. [CrossRef]
40. Fortino, G.; Fotia, L.; Messina, F.; Rosaci, D. Trust and Reputation in the Internet of Things: State-of-the-Art and Research Challenges. *IEEE Access* **2020**, *8*, 60117–60125. [CrossRef]
41. Akbas, A.; Yildiz, H.U.; Tavli, B.; Uludag, S. Joint Optimization of Transmission Power Level and Packet Size for WSN Lifetime Maximization. *IEEE Sens. J.* **2016**, *16*, 5084–5094. [CrossRef]
42. Cheng, Z.; Perillo, M.; Heinzelman, W.B. General Network Lifetime and Cost Models for Evaluating Sensor Network Deployment Strategies. *IEEE Trans. Mob. Comput.* **2008**, *7*, 484–497. [CrossRef]
43. Shi, Y.; Eberhart, R. A modified particle swarm optimizer. In Proceedings of the 1998 IEEE International Conference on Evolutionary Computation Proceedings, Anchorage, AK, USA, 4–9 May 1998; pp. 69–73.
44. Liu, C.H.; Chen, Z.; Tang, J.; Xu, J.; Piao, C. Energy-Efficient UAV Control for Effective and Fair Communication Coverage: A Deep Reinforcement Learning Approach. *IEEE J. Sel. Areas Commun.* **2018**, *36*, 2059–2070. [CrossRef]

3

An SVM-Based Method for Classification of External Interference in Industrial Wireless Sensor and Actuator Networks

Simone Grimaldi *, Aamir Mahmood and Mikael Gidlund

Department of Information Systems and Technology, Mid Sweden University, 851 70 Sundsvall, Sweden; aamir.mahmood@miun.se (A.M.); mikael.gidlund@miun.se (M.G.)

* Correspondence: simone.grimaldi@miun.se

Academic Editor: Mário Alves

Abstract: In recent years, the adoption of industrial wireless sensor and actuator networks (IWSANs) has greatly increased. However, the time-critical performance of IWSANs is considerably affected by external sources of interference. In particular, when an IEEE 802.11 network is coexisting in the same environment, a significant drop in communication reliability is observed. This, in turn, represents one of the main challenges for a wide-scale adoption of IWSAN. Interference classification through spectrum sensing is a possible step towards interference mitigation, but the long sampling window required by many of the approaches in the literature undermines their run-time applicability in time-slotted channel hopping (TSCH)-based IWSAN. Aiming at minimizing both the sensing time and the memory footprint of the collected samples, a centralized interference classifier based on support vector machines (SVMs) is introduced in this article. The proposed mechanism, tested with sample traces collected in industrial scenarios, enables the classification of interference from IEEE 802.11 networks and microwave ovens, while ensuring high classification accuracy with a sensing duration below 300 ms. In addition, the obtained results show that the fast classification together with a contained sampling frequency ensure the suitability of the method for TSCH-based IWSAN.

Keywords: industrial wireless sensor and actuator networks; support vector machine; interference classification; spectrum-sensing; wireless LAN; microwave oven

---

1. Introduction

The use of wireless sensor networks (WSNs) is a growing trend in a myriad of application domains, including building-health monitoring [1], military applications [2], health monitoring systems [3] and disaster and emergency management [4], to mention a few. A common denominator for many of these networks is the underlying radio technology, which is based on the IEEE 802.15.4 standard [5]. However, depending on the application, different requirements are set regarding the quality of service (QoS). In particular, differently from common implementations of WSN, the requirements found in those deployed in industrial settings, also known as industrial wireless and actuator networks (IWSANs), are considerably more challenging. Furthermore, the inclusion of actuators allows the IWSAN to cover more specific applications, such as closed-loop control, in which bi-directional data-traffic is needed.

IWSANs are characterized by having star or few hops mesh topology with a small number of devices and for presenting stringent requirements on the end-to-end communication delay and reliability. These requirements commonly include downlink and uplink transmission of process data with refresh rates in the order of tens of milliseconds and a network uptime greater than 99.999%, which corresponds to a downtime of less than 5.26 min per year [6]. Fulfilling such communication requirements is critically important in order to enable the adoption of IWSAN as a replacement of traditional wired implementations, such as Fieldbus-based solutions [7]. A failure to meet the QoS requirements can result in unwanted and costly production halts, corruption of the industrial product or even physical damage to production devices and human harm.

The two main factors that hamper the performance of IWSANs are the harsh radio-propagation conditions of most industrial environments, with pronounced effects of multipath fading and attenuation (MFA), and the interference originated from RF emissions in the 2.4-GHz unlicensed industrial, scientific, and medical (ISM)-band. The combined effect of these phenomena can cause severe degradation of the IWSAN radio links, potentially generating prolonged communication outages in some sectors of the wireless network. The RF interference that affects IEEE 802.15.4-based WSNs is mainly generated by wireless systems sharing the same ISM-band and microwave ovens (MWO), while the RF emissions of other devices (e.g., electric motors or switches) is mainly confined to the sub-GHz region of the spectrum, as shown in [8] and the references therein. Nevertheless, while some industrial plants can employ MWO in their production process (e.g., industrial material drying or food processing [9]), the wireless systems that reside in the 2.4-GHz band are much more frequent. The most widespread technologies that operate in this band are the IEEE 802.11 and IEEE 802.15.1 standards, under the commercial name of Wi-Fi and Bluetooth, respectively. IEEE 802.11-based WLANs are generally acknowledged as the most severe cause of interference for a number of reasons. Primarily, IEEE 802.11 networks are now ubiquitous in both office and production areas due to the widespread diffusion of WiFi-enabled terminals, such as smartphones or laptops. Moreover, in order to achieve full coverage, numerous access points are deployed, which can represent an obstacle for coexistence with IWSANs. Additionally, the IEEE 802.11 standard defines a physical layer (PHY), which enables transmission powers ten-times higher than IEEE 802.15.4 devices and a 5–8-times wider channel bandwidth, as shown in Figure 1. As a result, a coexisting IEEE 802.11 network can cause a packet error rate (PER) up to 70% [10–12] for a WSN receiver under the worst-case scenarios, such as prolonged use of overlapping channels, proximity of an IEEE 802.11 access point and sustained utilization rate of the interfering network. While devices implementing the IEEE 802.15.1 standard can also be found in industrial settings, thanks to the limited channel bandwidth and the implemented frequency-hopping scheme, their impact on the performance of IEEE 802.15.4-based networks is limited compared to MWO and IEEE 802.11 interference, as reported in [13]. For this reason, the classification of IEEE 802.15.1 interference is not considered in this paper.

Time-slotted channel-hopping (TSCH) is a well-known technique implemented in IWSAN standards, including WirelessHART [14], ISA100.11a [15] and WIA-PA [16], to mitigate the effects of external interference. Nevertheless, none of these standards employs intelligent methods for classifying the source of interference and adopting ad hoc strategies for interference mitigation. Since the first release of the IEEE 802.15.4 standard in 2003, a consistent number of research works has been carried out addressing interference-awareness in WSN. This matter can be separated into two different, but tightly-related aspects: interference classification and interference mitigation. In the terminology of cognitive-radio systems [17], the secondary-users (i.e., WSN-devices) are required to gain a certain level of spectrum awareness in order to utilize the unused resources opportunistically. A common approach for spectrum sensing methods in the literature is to adopt a relatively high sampling frequency and a sensing-time in the order of seconds, in order to maximize classification accuracy or make inference on the inflicted PER [18]. However, this is not suitable in the context of the time-critical IWSANs, where a long spectrum-sensing time implies slow network reactivity to the variations of the interference-scenario and waste of network resources due to the need for reserving numerous silent timeslots for channel sensing.

In this article, an interference detection and classification method is proposed and analyzed, with particular focus on minimizing the time required for channel sensing and the complexity of feature selection, while ensuring a good level of in-channel detection accuracy. For this purpose, a distributed spectrum sensing strategy and a centralized classification algorithm are employed to generate a space-frequency map of interference-free channels (IFCs). The IFC map is valuable information in the context of interference-aware resource scheduling for interference mitigation. The proposed interference classifier uses a three-step classification strategy, comprising a lightweight feature extraction stage, a set of four support vector machines (SVMs) performing preliminary binary classifications and a final stage composed of a logic decisor. The introduced mechanism is able to discriminate among interference from IEEE 802.11 networks, even when no terminal is associated with the access point, RF leakage from MWO and an IFC. Differently from other methods in the literature, such as [19–22], the proposed method does not rely on features based on the periodicity of IEEE 802.11 beacons. This fact, in conjunction with the novel classification scheme based on multiple SVMs, helps to ensure good classification performance while requiring an extremely limited sensing time.

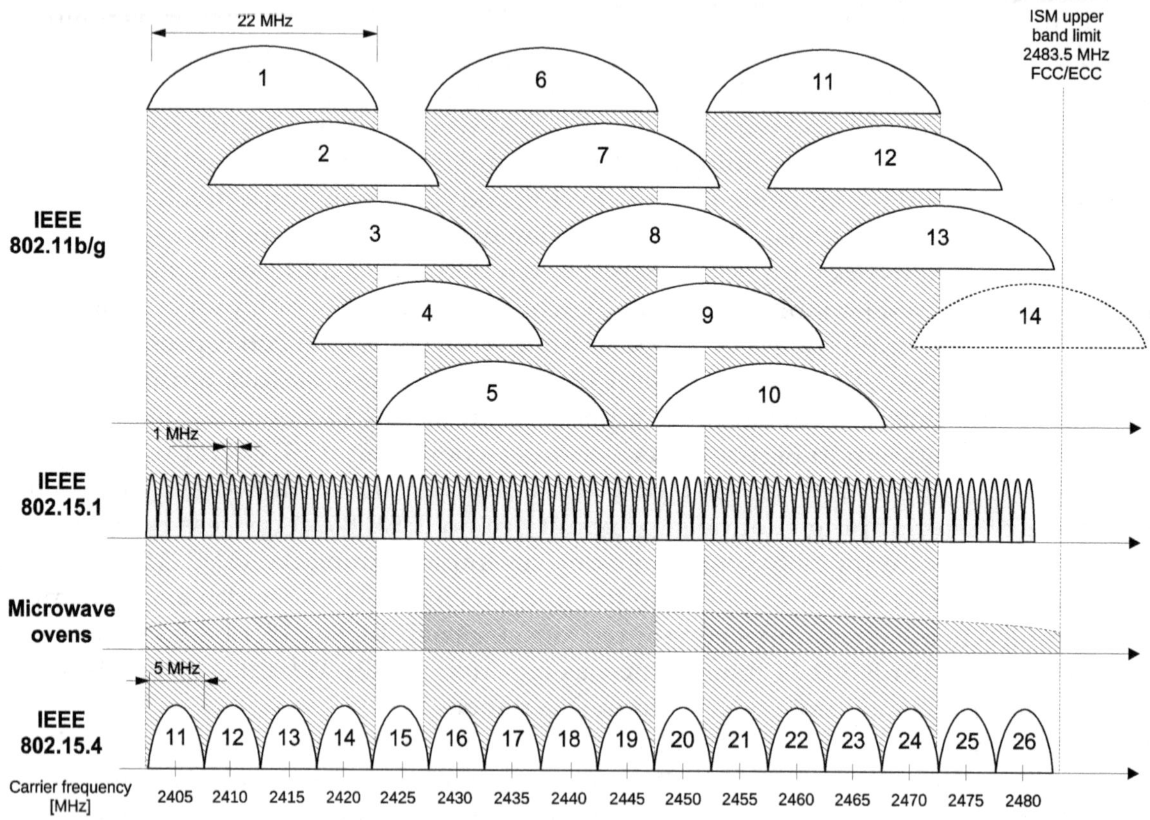

Figure 1. The 2.4-GHz industrial, scientific and medical (ISM) spectrum. Channel allocation for heterogeneous technologies with RF emissions within the band: IEEE 802.11, Bluetooth, microwave oven (MWO), IEEE 802.15.4.

The main contributions of this work are as follows:

- This is the first study that employs an SVM classifier to process signal features extracted from received signal strength indicator (RSSI) traces to identify the source of external interference. The proposed method employs four lightweight signal features, designed considering hardware constraints of commercial off-the-shelf (COTS) WSN devices.
- It is shown that, in order to ensure good detection performance, the proposed classifier requires a time window for spectrum sensing consistently below 300 ms, which, to the best knowledge of the authors, places the proposed solution amongst the quickest and most reliable methods reported in the literature.
- The performance of the proposed solution is validated by using an RSSI dataset collected in different industrial environments. Both the controlled and uncontrolled interferences from IEEE 802.11 networks are taken into account.

- The often overlooked influence of device calibration on spectrum sensing-based interference classification is analyzed, showing that the classifier accuracy is subject to the intrinsic hardware variations of the employed devices. However, we show that this factor can be easily corrected by means of a straightforward calibration process.

The remainder of this article is structured as follows. In Section 2, relevant work available in the literature about interference classification and mitigation in WSNs is presented. Section 3 provides a general background of the topic, discussing the various sources of cross-technology interference, with specific interest in the IEEE 802.11 standard. In Section 4, the basic concepts and mathematical formulation for SVMs are explained. In Section 5, a detailed description of the proposed solution is given, highlighting feature selection and the structure of the proposed classifier. In Section 6 and Section 7, the experimental setup and the results from experiments are described. In Section 8, the achieved results are discussed, and lastly, conclusions and final considerations are drawn in Section 9.

2. Related Works

The unrestricted and widespread usage of the unlicensed 2.4-GHz ISM bands, coupled with the asymmetric transmit power and medium access rules, results in harmful mutual interference among coexisting wireless systems. The most affected are the low-power systems, such as IEEE 802.15.4-based WSNs. Various experimental and theoretical studies have highlighted WSNs' susceptibility to the external interference, especially from high transmit power IEEE 802.11-based WLANs. Many experimental studies (e.g., [10,23,24]) show that an IEEE 802.15.4 link operating on a channel overlapped by an IEEE 802.11 network can experience packet losses of up to 50–70%. In light of these performance studies, it is evident that without an interference detection and avoidance mechanism, WSNs cannot satisfy any reliability or dependability conditions required by the aforementioned industrial applications. The most common interference detection technique, also recommended by the ZigBee standard [25], is to utilize energy detection-based spectrum sensing and avoid the channels with an energy level above a certain threshold. However, in order to design an intelligent interference avoidance technique, the type of interference and its behavior in the time and frequency domains need to be identified first. As interference scenarios may evolve in time, adaptive mitigation approaches with an individual strategy are efficient and are recommended [11]. There exist two main approaches to interference classification, where the distinction is made based on the information source used to extract the features to analyze: (i) raw channel energy measurements (i.e., RSSI samples), and (ii) bit error patterns in a corrupted packet. The existing methods for interference classification available in the literature are shown in Figure 2 and further discussed in the following sections.

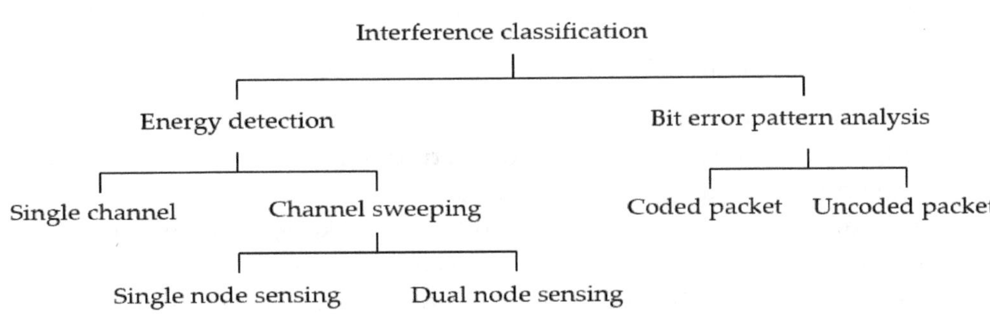

Figure 2. An overview of interference classification methods.

2.1. *Energy Detection-Based Interference Classification*

In this approach, a node actively collects energy samples on one or more IEEE 802.15.4 channels when the WSN devices are not transmitting. Signal processing techniques are then applied to the stored samples in order to extract a number of signal features, according to the implemented method.

Hard conditioning or machine learning techniques are then employed in order to map these features to a class of interference, such as IEEE 802.11, Bluetooth and MWO. The advantage of this approach is that no packet transmission is required, since there is no feature extraction from received packets. On the other hand, these methods require a certain time window in order to collect the required energy samples, meaning that specific idle-periods have to be reserved for channel sensing, potentially reducing the availability of network resources for data transmission.

In [19,26], Zacharias et al. propose a lightweight interference classification method in which a series of conditions are tested to identify the dominant source of interference. In these works, a node collects the RSSI samples on a single channel over a duration of one second at a sampling frequency of 8 kHz. The samples are then binarized using a fixed threshold of −85 dBm. Based on the binary data, the temporal features such as channel idle, busy time and signal periodicity are extracted. The classification conditions are then applied to the extracted features to identify the type of interference, achieving a classification time between 600 ms and 700 ms.

The detection of multiple sources of interference is studied by the authors in [27]. In this study, a clustering algorithm is applied to RSSI samples (collected by a node at a sampling frequency of 21 kHz) to distinguish the RSSI bursts from different interferers. In addition, a classifier identifies the channel activity patterns as periodic, bursty or a combination of both to determine channel suitability. The identification of periodic signals such as IEEE 802.11 beacons is also considered, achieving a classification accuracy of over 90% for sampling windows greater than or equal to 3 s.

The detection of the IEEE 802.11 beacons for the discovery of an IEEE 802.11 network has also been the subject of investigation in [20–22]. The collection of RSSI measurements over available channel sets (channel sweeping) is considered in [28], which employs two IEEE 802.15.4 radios to achieve pair-wise synchronized channel sensing. The objective of collecting samples over multiple channels is to identify the interference by matching the observed spectral pattern with the stored reference shape. The work targets only IEEE 802.11 interference, achieving a classification rate of 96% with sensing time in the order of 300 ms. In [29], instead, a single node is used for channel sweeping, and an interference classification method targeting IEEE 802.11 and MWOs is proposed in the context of an interferer-aware transmission adaptation mechanism.

Based on the high-resolution scanning feature of Atheros-based WLAN cards, the authors in [30] were able to extract detailed timing and frequency information of the interfering signal, at the cost of using additional hardware. In this context, a decision-tree classifier for interference identification was implemented yielding 91–96% detection accuracy. In [31], Weng et al. have developed two algorithms for the identification of MWO, IEEE 802.11 and Bluetooth signals based on 20 MHz I/Q data sampling performed by means of additional spectrum sensing hardware. However, these approaches are beyond the scope of the hardware capability of commonly-used resource-constrained sensor nodes.

2.2. Bit Error Pattern-Based Interference Classification

This class of methods does not require the active collection of RSSI samples; rather, the interference classification is based on the analysis of bit error patterns in the packets exchanged during the normal operation of the network. In [32,33], the authors show that the different interferers, such as IEEE 802.11, Bluetooth and ZigBee, corrupt IEEE 802.15.4 packets, leaving specific error footprints. In addition, the bit error pattern can also be used to reveal the presence of weak links. In particular, Hermans et al. [32] propose identification of the interference source by combining (i) the signal strength variations during packet reception, (ii) the link quality indicator (LQI) associated with a packet and (iii) the position of corrupted bytes in the payload. The classification accuracy of the proposed method is 72%, while this result is also IEEE 802.15.4 packet size dependent, since packets with a small payload size are partially overlapped with the interferer, thus carrying a small interference fingerprint. In [34] Barac et al. use forward error correction (FEC) in order to identify the source of bit errors in a received packet (i.e., multipath fading and attenuation, as well as the IEEE 802.11 b/g interference). Therefore, instead of packet retransmissions (which is used in [32]), the FEC method in [34] emerges as an energy-efficient alternative for interference classification, yielding more than 91% classification rate with just one received packet.

3. Background

3.1. Cross-Technology Interference Sources

In this section, we discuss the salient features of the cross-technology sources of interference targeted in the current work, namely IEEE 802.11-based WLAN and MWO.

3.1.1. IEEE 802.11

The prevalent WLAN networks in the 2.4-GHz band are based on the IEEE 802.11 b/g/n specifications. The IEEE 802.11 b/g PHY supports up to 14 channels, 20 MHz wide each. On the other hand, the IEEE 802.11 n can support both the 20 MHz- or 40 MHz-wide channels. There are only three non-overlapping usable channels in the U.S. and other countries with similar regulations (Channels 1, 6, 11, with 25-MHz separation) and four in Europe (Channels 1, 5, 9, 13, with 20-MHz separation). The transmit power of WLAN devices ranges from 15 dBm–20 dBm, and depending on the underlying standard, different modulation schemes and data-rates are available. However, the maximum air-time of an IEEE 802.11 packet remains below 600 μs. The standard specifies a carrier-sense multiple access with collision-avoidance (CSMA/CA)-based MAC with certain timing rules between the consecutive packets. In commonly-used infrastructure mode, an access point advertises the network by sending the periodic beacon frames. For compatibility reasons and in order to increase the network detection range, the beacons are usually sent at the lowest data rate (1 or 2 Mb/s). The default beacon frequency period is 100 time units, which is equal to 102.4 ms [26]. The above-stated heterogeneous medium access rules and PHY specifications for WLAN networks render ZigBee systems vulnerable. Firstly, WLAN networks deployed on non-overlapping channel allocation, such as the typical $\{1, 6, 11\}$ configuration, leave a small number of IFCs for ZigBee. Secondly, the high-power concurrent transmission on an overlapping channel from a WLAN device will likely cause severe packet corruption in an IEEE 802.15.4 packet. Thirdly, the duration of an IEEE 802.15.4 clear-channel assessment (CCA) is 128 μs, while 192 μs [14] are required to switch from CCA to transmit mode. Conversely, the IEEE 802.11 CCA procedure takes 28 μs, while the switching time is negligible. As a result, the chances of the corruption of a ZigBee packet are very high, as the WLAN transmission can disrupt the ZigBee transmission during the switching mode.

3.1.2. Microwave Ovens

The energy leakage from the residential MWOs usually affects the whole 2.4-GHz band. However, as depicted by various studies [29,35], the RF emissions from MWOs peak at about a 2.45-GHz frequency, while the number and center frequencies of peaks may vary slightly according to the specific model, as shown in [36]. As a result, the IEEE 802.15.4 Channels 20 and 21 have a high probability of being strongly affected by the MWO operation. A prominent feature of MWO is the periodicity of *on* and *off* phases during the heating process, where the time from one *on* phase to the next is $\frac{1}{2f}$ s, with f the frequency of the power supply (i.e., 50/60 Hz).

4. Support Vector Machines

In this section, we outline the basic formulation of the mathematical problem for an SVM, focusing on the training and classification tasks, while we leave more in-depth analysis to specific machine-learning literature, such as [37], and to [38] for details about the related convex optimization methods.

4.1. The Standard Model for SVM

An SVM is a supervised classification algorithm that allows a binary decision to be performed, assigning an M-dimensional feature-vector to one of two classes. Being a supervised approach, an SVM needs to be trained using an appropriate dataset, which should be sufficiently large and representative of the two classes, with respect to the selected features. A training phase is then needed to determine a subset of the training vectors (called support vectors), which will actually be used for solving the classification problem. One important advantage of SVMs resides in the fact that the number of support vectors is generally much smaller than the cardinality of the training dataset. Hence, while the training of the SVM can be a resource-intensive task, the actual classification algorithm can be very slender. The standard formulation for a two-class classification problem is:

$$y(\mathbf{x}) = \mathbf{w}^T \phi(\mathbf{x}) + b \tag{1}$$

which is a linear model where $\mathbf{x}$ is the M-dimensional input vector, M is the size of the feature space, $\mathbf{w} = \{w_1, w_2, \ldots, w_M\}$ is the vector of coefficients for the linear model, ϕ is a general feature-space transformation function (which can eventually be non-linear) and b represents the bias of the model.

Hence, the training set for the SVM is composed of a set of N training feature-vectors $\mathbf{x}_1, \ldots, \mathbf{x}_N$ where each vector is associated with one of the two classes (C_1, C_2) via the parameters $t_n = \{-1, 1\}$, which are the class labels for the training vectors. The decision logic is then the following: an unknown vector $\mathbf{x}^*$ belongs to class C_1, if $y(\mathbf{x}^*) < 0$ and to class C_2 if $y(\mathbf{x}^*) > 0$. The implicit assumption is that the training data are linearly separable, so that the coefficient vector $\mathbf{w}$ and the parameter b can be determined (i.e., there exists at least one feasible combination of $\mathbf{w}$ and b).

4.2. SVM: Training and Classification

The training of an SVM can be seen geometrically as the problem of maximizing the minimum Euclidean distance between the decision hyperplane and the points of the training set. This problem can be formulated in an equivalent fashion, observing that since $t_n = \{-1, 1\}$ are the target values for the two classes, the following is verified for any correctly-labeled input vector $\mathbf{x}$:

$$t_n y(\mathbf{x}) > 0 \tag{2}$$

It can be easily shown that the optimization problem can be expressed as:

$$\begin{aligned} &\text{minimize} \quad \|\mathbf{w}\|_2 \\ &\text{subject to:} \quad t_n(\mathbf{w}^T \phi(\mathbf{x}_n) + b) - 1 > 0 \end{aligned} \tag{3}$$

with $n \in [1, N]$. Hence, due to the definition of two-norm, the function to minimize in (3) is a quadratic cost function with M variables. The optimization problem that arises is then a quadratic program (QP) with M variables (size of the feature space) and N inequality constraints (size of the RSSI input vector).

Once the model is trained, the solution of the decision problem for a generic input vector $\mathbf{x}^*$ can be obtained by simply evaluating the sign of $y(\mathbf{x})$ in the original linear model $y(\mathbf{x}) = \mathbf{w}^T \phi(\mathbf{x}) + b$, with the coefficient vector $\mathbf{w}$ populated using the results from the minimization of the cost function in (3), hence calculating:

$$y(\mathbf{x}) = \sum_{n=1}^{N} t_n \alpha_n k(x, x_n) \tag{4}$$

where α_n are the Lagrange multipliers of the dual problem. Equation (4) is subject to the Karush-Kuhn-Tucker (KKT) conditions:

$$\alpha_n \geq 0$$

$$t_n y(x_n) \geq 0$$

$$\alpha_n (t_n y(x_n) - 1) = 0$$

An important result is that each point of the cost function for which the respective Lagrange multiplier $\alpha_n = 0$ can be discharged, since it will not influence the calculation, yields a consistent reduction of the dataset size, which is one of the key advantages of SVMs.

5. The Proposed Solution

5.1. Classifier Setup

The proposed interference detection method employs an SVM-based classifier, which processes input data composed of observations of the background RF noise on a specific IEEE 802.15.4 radio channel. The method is based on the basic assumption that when there is no transmission on a certain channel (and thus, there is an absence of intra-network interference), the devices can collect samples of the RF radiation and process the data to detect and classify eventual interferers, as well as assessing the eventuality of an IFC. This assumption nicely fits with the time-division multiple-access (TDMA) approach employed in ISWANs, since in these networks, the allocation of frequency-time resources for data transmission is known a priori; thus, a contiguous set of time slots on a specific channel can be reserved for spectrum sensing. The common hypothesis for spectrum sensing is that the classification can be done with a certain level of accuracy if the time window is sufficiently long for specific signal features to emerge. The proposed solution is designed to keep this detection time as short as possible. As shown in Figure 3, the first stage of the classifier employs a process of signal feature extraction, in which data are processed in order to extrapolate a number of signal features in the time and amplitude domains.

The second stage of the classifier is composed of four SVMs, which perform a first decision stage, outputting single binary partial hard decisions with respect to the related interference scenarios. The different SVMs are hereby described:

1. SVM-free channel: this SVM is trained to detect the presence of an IFC.
2. SVM-active network: targets the presence of an active IEEE 802.11 network occupying the related IEEE 802.15.4 PHY channel (i.e., an IEEE 802.11 access point with at least one associated terminal, generating uplink/downlink traffic).
3. SVM-silent network: targets a silent IEEE 802.11 network overlapping the specific channel. This is the case of an IEEE 802.11 access point with no associated terminal or an access point with associated terminals that are not generating data traffic in the observation time window.
4. SVM-microwave oven: detects the presence of RF leakage from a microwave-oven operating in close proximity to the radio node.

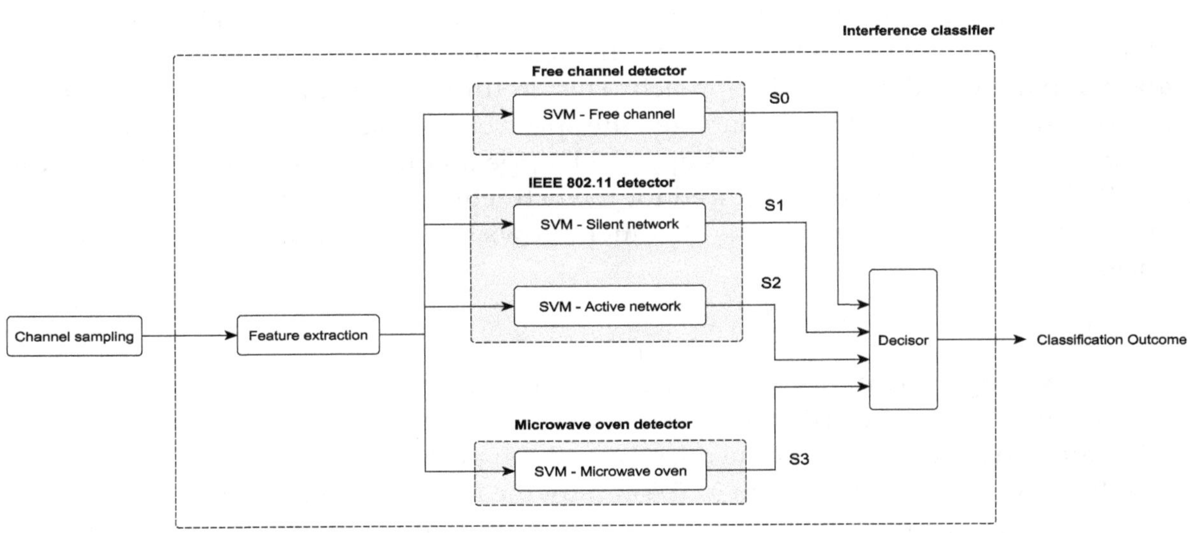

Figure 3. Setup of the support vector machine (SVM)-based interference classifier.

The outputs of the four SVM are represented by binary signals, S_1, S_2, S_3 and S_4, which have the value 1 (0) if the related decision is positive (negative). The binary decisions preformed by the SVMs are then processed by the logic decisor shown in Figure 4. The logic function of the decisor has been synthesized considering the cross-detection resilience of the single SVM. The final decision is composed of the four different classes listed in Table 1.

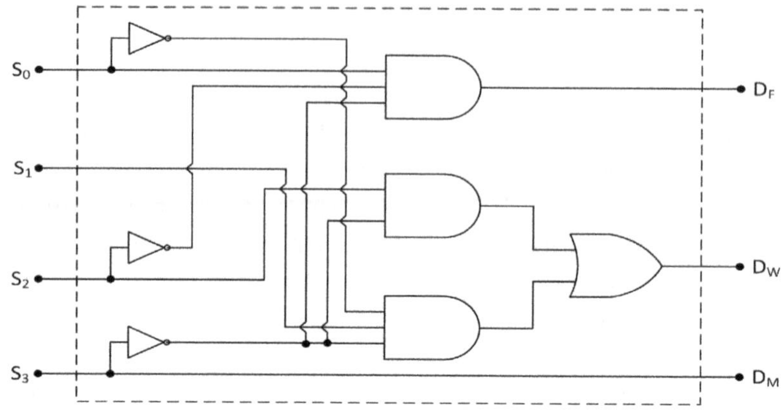

Figure 4. Details of the decisor for the proposed detection algorithm. The logic input signals are generated by the four SVMs in Figure 3.

Table 1. The four interference classes in the analysis.

D_F	D_W	D_M	**Classification Outcome**
1	0	0	The channel is free from the interference sources in the analysis.
0	0	1	A MWO was active during the sensing period.
0	1	0	An IEEE 802.11 network was overlapping the channel in the analysis.
0	0	0	The source of interference is unknown.

It must be highlighted that the classification performed based on the observation of a single radio node of the network only has local validity. This is because radio devices located in different locations of an industrial plant may be subjected to different interference conditions. In this context, the proposed method allows the interference scenario to be captured for each of the deployed nodes, opening the possibility of mapping the different sources of interference in the space-frequency domain. Nevertheless, since the aforementioned classification scheme exhibits a computational complexity that is beyond the capabilities of COTS WSN nodes, the proposed implementation relies on a centralized classification in place of a distributed approach. This, in turn, means that while the classifier can be implemented in the IWSAN network manager, the spectrum sensing and feature extraction process can be carried out by IWSAN nodes. This approach appears rather convenient, since, as described in Section 5.2, the signal processing required for the extraction of the selected signal features is kept to a minimum, while the efficiency of the classifier allows the radio nodes to work with small RSSI sample traces.

5.2. Signal Features

We select four signal features, belonging to two main classes: time domain and amplitude domain. The logic behind the selection is related to the properties of the interfering signals, such as the transmission airtime of IEEE 802.11 transmission and the periodicity of time domain pattern of the MWO RF leakage, as discussed in Section 3. To simplify the feasibility of the whole spectrum sampling and feature extraction processes on COTS WSN devices, our approach is to minimize the size of the RSSI trace, as well as the complexity of feature calculation methods.

5.2.1. Number and Length of Signal Bursts

The first time domain feature includes information about the burstiness of the observed signal, employing a threshold-based burst detection. The feature is an M-element vector in which each element represents the number of bursts of a certain sample length. Hence, we define:

$$\mathbf{F}_B = \{F_1, F_2, F_3, F_4, .., F_M\} \tag{5}$$

where $F_n \in \mathbf{N}$ represents the the number of bursts of length n found in the RSSI trace in the analysis, while with F_M, we mark all of the bursts with sample length $L \geq M$. In particular, we require a certain number of samples under the selected threshold to identify the end of a burst. This is to avoid the case where a single or a few incorrect readings of the RSSI register will lead to a misclassification of long signal bursts into shorter ones. As will be discussed in Section 7, the detection of longer (i.e., $>$ 5 ms) bursts is extremely important because it is a specific feature of the RF emissions of microwave ovens. The choice of a proper value of the threshold with respect to the calibration of the radio nodes will be discussed in Section 6.

5.2.2. Mean, Variance and Cardinality of Over-Threshold Samples

The second feature belongs to the amplitude domain and is defined as the mean value of the RSSI samples over the selected threshold θ. We define the vector containing all of the RSSI samples collected during the continuous observation window as $\mathbf{S} = \{s_0, s_1, s_2, ..., s_{N_S}\}$. Then, indicating with $\mathbf{S}^{(\mathbf{OT})}$ the subset of $\mathbf{S}$, such that $\mathbf{S}^{(\mathbf{OT})} = \{s_n \in \mathbf{S} \mid s_n > \theta\}$:

$$F_M = \frac{1}{N_{OT}} \sum_{i=1}^{N_{OT}} s_i^{(OT)} \tag{6}$$

with N_{OT} representing the cardinality of $\mathbf{S}^{(\mathbf{OT})}$, hence the number of above-threshold samples in the set. The third feature F_V follows directly from the definition of sample variance, hence using the same notation employed for F_M:

$$F_V = \frac{1}{N_{OT}} \sum_{i=1}^{N_{OT}} (s_i^{(OT)} - F_M)^2 \tag{7}$$

The last feature F_C counts the occurrences of RSSI samples above the threshold and hence is simply the cardinality of the set $\mathbf{S}^{(\mathbf{OT})}$. This feature nicely complements the previous two, adding information about the activity level of the interference source. It must be noted that while the signal features F_M, F_V and F_C are scalars, the feature F_B is an M-element vector; hence the SVM feature-space will be $M + 3$-dimensional, even using only four features.

6. Experimental Setup

6.1. Hardware Setup

The WSN devices selected for the experiments are Crossbow's TelosB motes CA2400 [39], equipped with Texas Instrument CC2420 transceiver [40]. The devices are programmed to collect a continuous set of RSSI samples with a sampling frequency of 2 kHz, over a sampling window that is selected according to the specific experiment. The RSSI value for each sample is fetched from the first 8 bits of register 0x16 of the CC2420 transceiver and represents the incident RF power in the selected 5 MHz-wide channel averaged over 128 μs (hence, eight IEEE 802.15.4 O-QPSK symbols). The RSSI data, fetched from the register in the form of an 8-bit signed integer, are buffered in the RAM and periodically saved to the internal flash memory. At the end of the sampling process, the content of the flash memory is sent over the USB port to a laptop, which logs the received data. The choice of this method for collecting RSSI data, in place of the direct sample-and-send over USB approach, was due to the insufficient bitrate (i.e., 115,200 baud including serial message overhead) available at the serial interface. In order to validate the performed measurements, we time stamp all of the observations and measure the delay of the instructions and task implemented in Tiny OS. This aspect will be further discussed in Section 8.2.2. Since Chen et al. [41] reported a consistent offset among RSSI readings performed with CC2420-equipped radio devices, we have also profiled our devices by means of a simple calibration process. We will discuss the effect of node calibration as well as the effect of this process on the performance of the proposed solution in Section 8.2.1.

6.2. Test Environments

The collection of experimental data has been carried out in three locations. Location A is a three-storey production plant employed for mineral processing. The environment is an open space cluttered with metal tanks, production machinery and a radio-controlled crane, while the three storeys are separated by metal grate flooring. A resident IEEE 802.11 WLAN covering the whole production plant was running at the time of the experiments.

Location B is a small mechanical workshop with an abundance of metal cluttering and soldering tools. A total of fourteen IEEE 802.11 access points with overlapping spectrum allocation on Channels {1, 6, 7, 11} is detectable in this environment. In Figure 5, we show the position of three of the nearest access points, which we label with AP1, AP2 and AP3, while the remaining devices were placed outside the range of the map, or on the upper floors of the building.

Location C is an office area, with nine IEEE 802.11 access points and residential MWO. We use this location to perform experiments on the classification of microwave interference, since neither industrial, nor residential MWO were present in the other two selected sites.

(a) (b)

Figure 5. Two of the selected experimental environments. (**a**) Location A: industrial plant; (**b**) Location B: mechanical workshop.

6.3. The Collection of Training Data for SVM

As described in Section 4, the availability of a representative training dataset is fundamental for supervised-learning classification algorithms such as SVMs. Hence, particular attention has been put into building the dataset from both controlled and uncontrolled sources of interference and covering all 16 IEEE 802.15.4 channels.

6.3.1. Training Data from Uncontrolled IEEE 802.11 Networks

A preliminary set of measurements is collected in Location A, by means of Metageek channel analyzer [42], in order to determine the ground truth on the spectrum allocation of the resident IEEE 802.11 access points present at the industrial site. The network was composed of three IEEE 802.11 b/g/n access points statically allocated on IEEE 802.11 Channels 1, 6, and 11.

The training set is collected by means of a TelosB mote deployed in a fixed location of the industrial plant, programmed to sense each IEEE 802.15.4 channel for 10 min, collecting traces with over 1 M-sample per channel. Subsequently, the traces collected from the sampling of IEEE 802.11 Channels {15, 20, 25, 26} were assigned to the IFC class, since the IEEE 802.11 network did not overlap these channels (as shown in Figure 1), while the remaining traces were assigned to class IEEE 802.11 interference.

6.3.2. Training Data from Controlled Sources

The dynamics of an IEEE 802.11 network can vary greatly according to several factors (e.g., the number of connected devices and the traffic data-rate), and this in turn reflects the characteristics of the observed RSSI sample trace. While different methods for generating controlled interference are available in the literature, such as the one presented in [43], we use IEEE 802.11 hardware and a server-client architecture in order to have full control over the traffic distribution and transmission parameters. Following this approach, a controlled IEEE 802.11 network has been deployed at Location A. The structure of the network is represented in Figure 6 and is composed of a Linksys WRT610N IEEE 802.11 access point connected by an Ethernet cable to a Linux laptop running a traffic generator application generating the user datagram protocol (UDP) traffic with uniform, exponential and Pareto distributions. A second laptop, employing a Wi-Fi interface and running a Linux client, was used to receive and monitor the IEEE 802.11 packets. The access point was set on Channel 3, in order to overlap IEEE 802.15.4 Channel 15 (which was not affected by the resident industrial network), in order to isolate the observation from the effects of the resident network and capture only the effects of the custom IEEE 802.11 network.

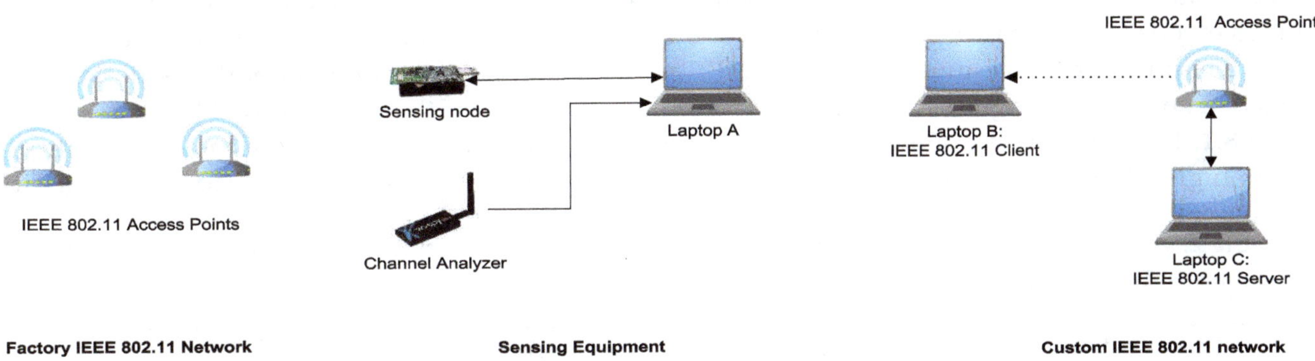

Figure 6. Experimental setup for the collection of training data for IEEE 802.11 interference detection.

6.3.3. Training Data from Microwave Oven

A set of measurements was collected in proximity (1 m) of a consumer Samsung MW82Y MWO set at the maximum heating power, achieving an active-passive heating phase with a duty cycle of 50%. The training data were collected along all of the IEEE 802.15.4 channels, since the temporal features of RF emissions from MWO can vary considerably moving within the 2.4-GHz ISM band due to the employed technology, as discussed in Section 3.

6.3.4. Test Data

We collected an extensive test dataset in the three described locations in order to thoroughly test the proposed algorithm.

At Location A, multiple RSSI traces from both the resident IEEE 802.11 network and the access point employed in the experiments were collected over all of the IEEE 802.15.4 channels. The traces were collected at several points of the three floors of the factory, taking care of including both line-of-sight (LOS) and non-line-of-sight NLOS propagation scenarios between the access points and the sensing node.

At Location B, we instead deployed the sensing node at one fixed point of the workshop (Point CS in Figure 5) and sensed each of the 16 channels for 5 min. At Location C, we deployed the radio node in the proximity of the active MWO, taking care of collecting measurements for all of the channels, randomizing the node position in the range 0.5 m–2 m from the oven. In all of the data collection points of the selected locations, the spectrum analyzer was used to determine the actual interference status of the sensed channels (similarly to the training data collection process), in order to determine the ground-truth for assessing the performance of the off-line classifier.

7. Results

7.1. Global Classification Accuracy

We tested the performance of the proposed algorithm by splitting each of the RSSI traces into several data chunks, with a length varying according to the tested sampling window, in the range of 50–500 ms. The data chunks were then processed in MATLAB, where we implemented the proposed classifier, including the feature extraction process and the four SVMs using the standard MATLAB SVM implementation with the Gaussian kernel, as well as the final decisor stage. For each test set, we calculated a detection accuracy metric by analyzing the outcome of the predicted interference source (according to Table 1) and comparing to the actual source determined during experiments. The detection accuracy was then simply calculated by dividing the number of correct classifications by the total number of classification rounds. In Table 2, we show the classification rates calculated for Locations A, B and C, including test data for all 16 IEEE 802.15.4 channels when the sampling widow is 250 ms; hence, chunks of 500 eight-bit RSSI samples are analyzed in each round of the test. It should be noted that the validity of the presented results is expected to be quite broad in nature as our dataset includes extensive traces from a broad range of scenarios including both controlled and uncontrolled IEEE 802.11 interference, spanning all of the IEEE 802.15.4 channels. A more in-depth discussion of the effects of a shorter or longer sampling window on the accuracy of the solution will be carried out in Section 8.1. In the following tables, we also include data about the distribution of misclassification in order to highlight which sources of interference were most likely to be misinterpreted by the classifier.

Table 2. Average classification accuracy for the 250 ms sampling window calculated over all of the scenarios. IFC, interference-free channel.

Channel Status	Detected Interference Source			
	IFC	**IEEE 802.11**	**Microwave Oven**	**Unknown**
IFC	91.2%	6.6%	2.1%	0.1%
IEEE 802.11	12.4%	83.9%	1.4%	2.3%
Microwave Oven	0.8%	16.3%	82.8%	0.1%

As shown in Table 2, the classifier was able to determine the presence of a free IEEE 802.15.4 channel 91.2% of the time, and the primary source of misclassification was the IEEE 802.11 network, which was detected 6.6% of the time. This fact is mainly due to the similarity of RSSI traces originating from an IEEE 802.11 network with low data traffic or even no associated terminal with RSSI originated from background noise. The similarity becomes more prominent when the signals originating from an IEEE 802.11 network and received by the WSN node are weak, due to attenuation effects. The same effect can also be used to explain the IFC misclassification rate of 12.4% when the interference comes from an IEEE 802.11 network. Nevertheless, in both cases, the introduction of the second support vector machine targeting silent IEEE 802.11 networks together with the employed decision logic helped to ensure a full-spectrum average detection accuracy of 83.9% for IEEE 802.11, even with a 250 ms sampling window, which is significantly shorter than other approaches presented in the literature (e.g., [21,26,28]). The classifier shows a detection accuracy of 82.8% when the source of interference was an MWO, where the most likely misclassification output was IEEE 802.11 interference, due to the similarities between temporal features of IEEE 802.11 and RF leakage from MWO. In Section 7.2, we point out that the detection accuracy appears to be significantly higher than the average for a specific contiguous set of IEEE 802.15.4 channels. This gives insight about dynamic channel-sensing strategies for maximizing the classification rate for this class of interference.

In Tables 3 and 4, we give more details about the full-spectrum detection accuracy from the datasets collected at Location A and Location B.

Table 3. Average classification accuracy for the 250 ms sampling window for Location A.

Channel Status	Detected Interference Source			
	IFC	**IEEE 802.11**	**Microwave Oven**	**Unknown**
IFC	98.2%	1.7%	0.1%	0.0%
IEEE 802.11	0.1%	98.9%	0.3%	0.7%

Table 4. Average classification accuracy for the 250 ms sampling window for Location B.

Channel Status	Detected Interference Source			
	IFC	**IEEE 802.11**	**Microwave Oven**	**Unknown**
IFC	84.9%	11.2%	3.8%	0.1%
IEEE 802.11	10.7%	77.9%	5.2%	6.1%

The classifier showed notable performances at Location A, being able to determine the correct source of interference 98.2% of the time when the channel was free and 98.9% when the RF emissions from the IEEE 802.11 network were overlapping the sensed channel. The average detection accuracy appeared lower at Location B, down to 77.9% for IEEE 802.11 interference. This is because the channel allocation of the resident IEEE 802.11 networks present at Location B was more challenging, including multiple overlapping networks with weaker signals and thus complicating the task of correctly classifying the interference on some WSN channels. This fact will be further analyzed in Section 7.2, where we provide in-channel detection accuracy analysis.

7.2. Channel-Specific Accuracy

In this section, the in-channel classification accuracy will be analyzed for all of the locations included in the tests. In Figure 7, we show the detection accuracy for Location A for a sampling window of 150 ms, together with information collected by means of the spectrum analyzer, showing the energy density in the 2.4-GHz ISM band at the industrial site. We chose to show the results for a shorter sampling window (150 ms) with respect to the previous section in order to highlight the impact of this aspect on the classification accuracy. As can be seen, even with this short sampling window, the detection rate ranged around 90% for the channels overlapped by the $\{1, 6, 11\}$ configuration of the IEEE 802.11 network, while the IEEE 802.15.4 Channels $\{15, 20, 25, 26\}$ were accurately reported free from interference.

In Figure 8, we show the classification outcome for Location B. In this test scenario, there are multiple IEEE 802.11 networks occupying IEEE 802.11 Channels 1, 6 and 11, while a distant access point with an average RSSI level < -80 dBm at the data collection point was present on Channel 7. As expected, while IEEE 802.15.4 Channels $\{15, 25, 26\}$ were reported free in more then 85% of the tests, the decision for Channels 19 and 20 was uncertain since the IEEE 802.11 network was reported in only around 40–50% of the tests. We select this scenario to stress the performance of the proposed solution in the presence of an access point, which is barely detectable at the channel sensing location.

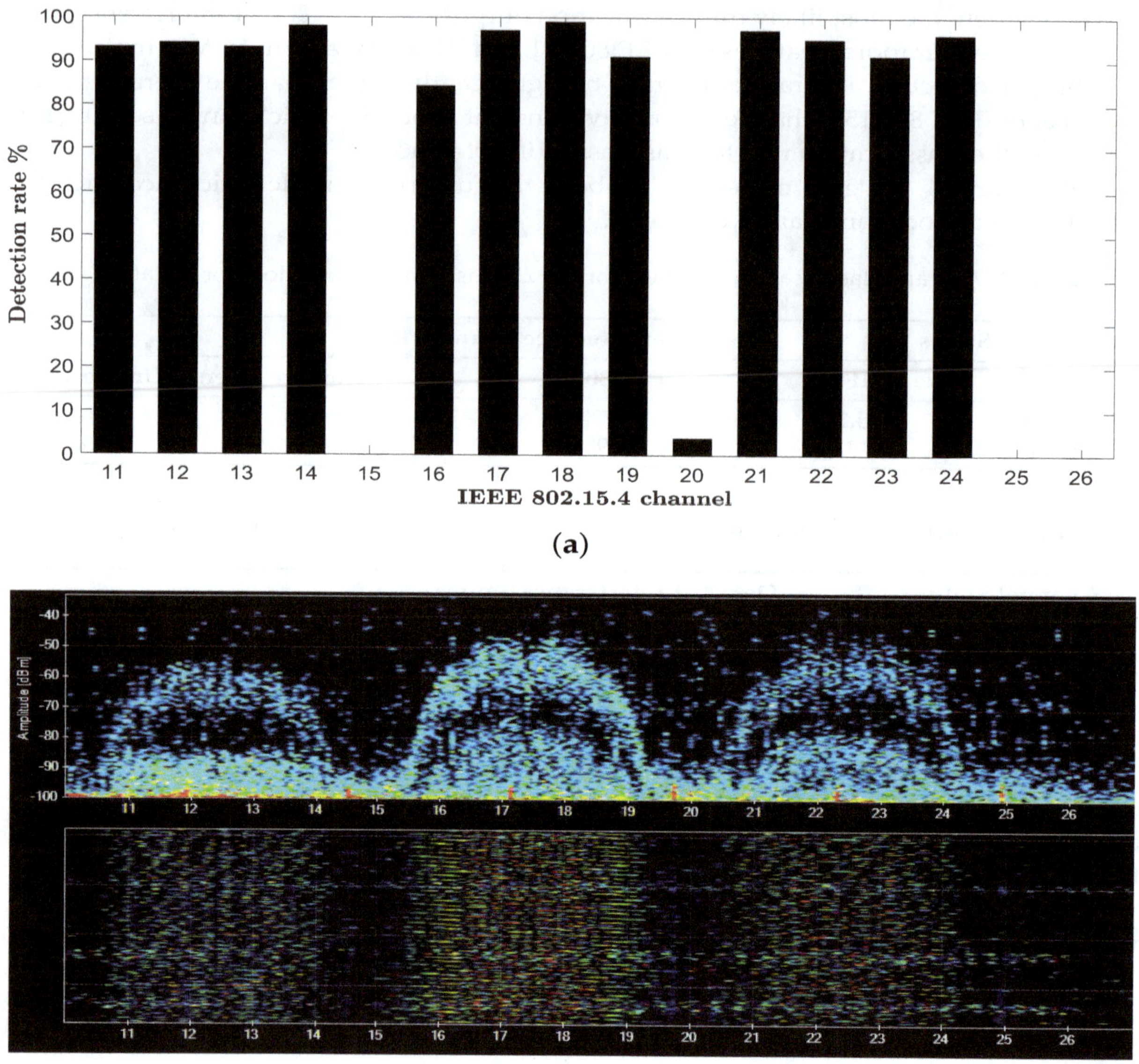

(b)

Figure 7. Location A, three IEEE 802.11 access point operating on IEEE 802.11 Channels {1, 6, 11}. (**a**) Channel-specific detection rate at Location A for the sampling window length of 150 ms; (**b**) IEEE 802.11 power spectral density (PSD) as observed by the channel analyzer.

In Figure 9, we show both the classification rate at Location C for a MWO and the average RSSI value and the standard deviation for the collected test data. As mentioned beforehand, while the RF leakage from MWO spans all along the 2.4-GHz ISM band, the detection accuracy presents considerable variations along the 16 IEEE 802.15.4 channels. In particular, the eight channels 16–23 seem to offer the best chance for microwave detection, while Channel 21 shows the maximum classification accuracy. This is because, as reported in [36] and the references therein, the residential MWOs have an emission peak frequency around 2.45 GHz, which corresponds to Channel 20 in the IEEE 802.15.4 mapping. In this case, we are likely to be experiencing an MWO with center emission frequency at 2.455 GHz, which consequently triggers a very high detection rate on Channel 21. Nevertheless, since the emission pattern may vary from model to model, the channel-specific performance is expected to vary from the one shown in Figure 9. In any case, for any model, the average detection accuracy is expected to remain consistent, since there will always be a region of maximum emission inside the ISM band of interest.

For the aforementioned reasons, a reasonable approach for channel sensing could be to perform the sensing on Channel 20 or on the adjacent channels in order to maximize the classification accuracy.

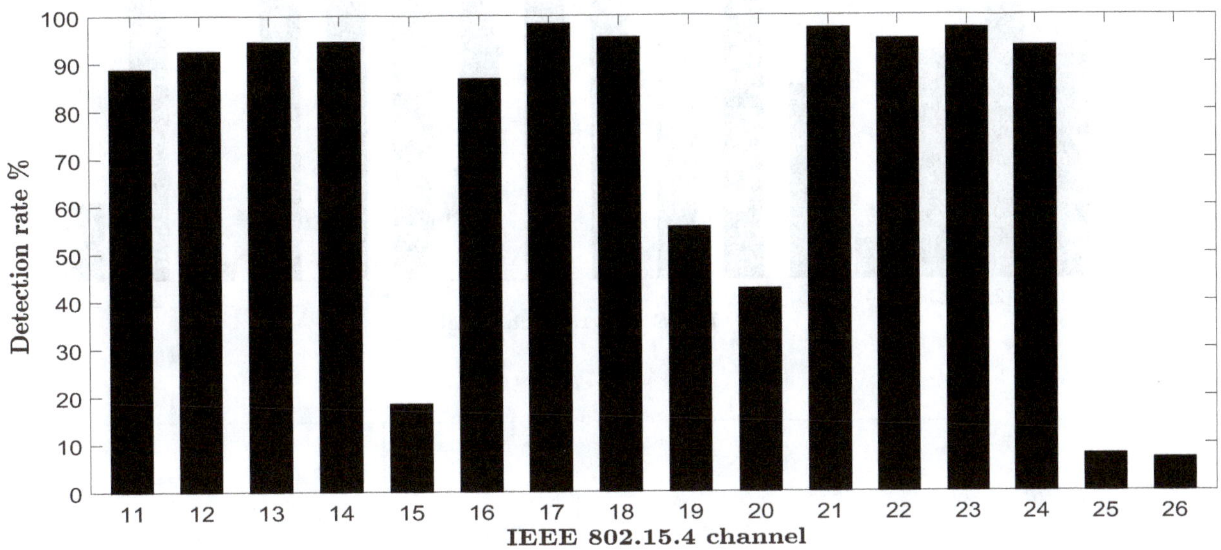

(a)

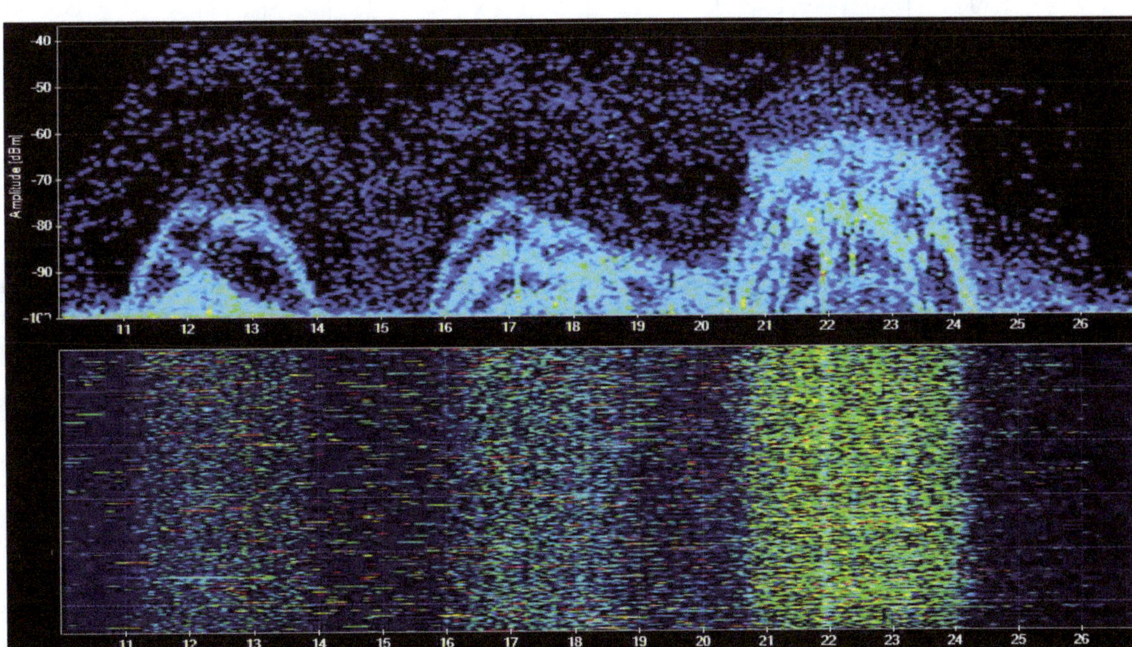

(b)

Figure 8. Location B, multiple IEEE 802.11 access points operating on Channels {1, 6, 7, 11}. (**a**) Channel-specific detection rate at Location B for a sampling window length of 150 ms.; (**b**) IEEE 802.11 power spectral density (PSD) as observed by the channel analyzer.

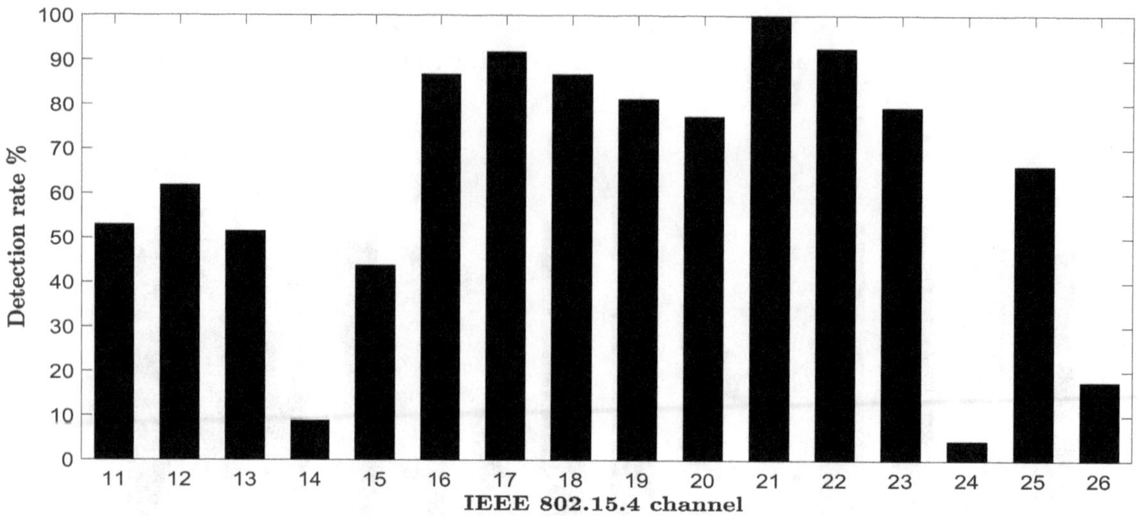

(a)

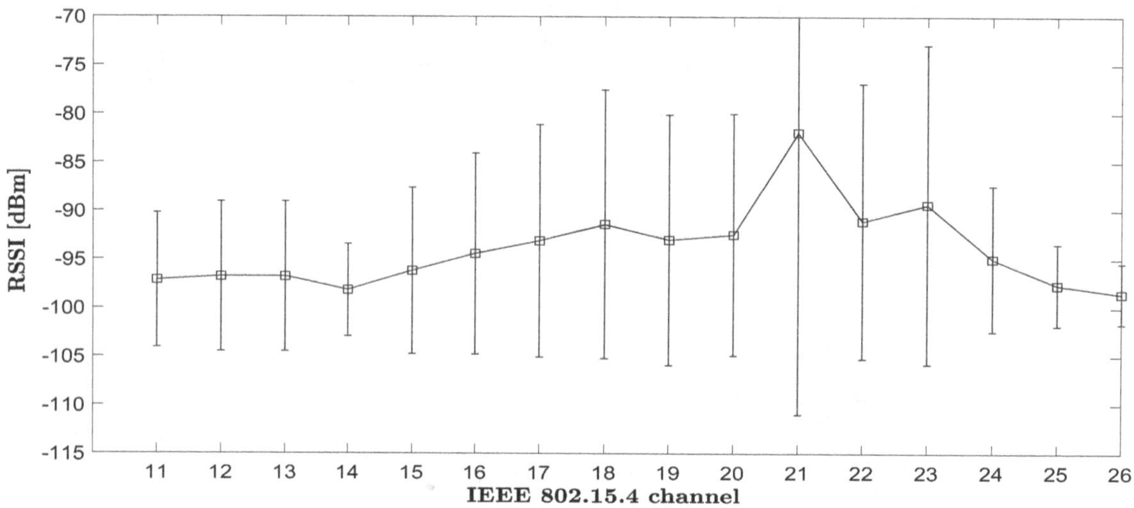

(b)

Figure 9. Location C, classification results for a sampling window length of 300 ms. (**a**) Channel-specific microwave oven detection rate; (**b**) average and standard deviation of RSSI traces collected in proximity of the tested microwave oven.

8. Discussion

8.1. The Influence of Sampling Window Length

In this section, we analyze the impact of different sampling window lengths on the classification accuracy of the three channel status classes of interest. The different sampling windows are tested by varying the number of samples included in the feature extraction process, as described in Section 5. In Figure 10, we show the curves for the average full-spectrum classification accuracy of IEEE 802.11 interference for sampling windows spanning from 50 ms–400 ms. We additionally show the curves representing the misclassification rate in order to highlight how the separation between classes is influenced by the sampling window.

It is interesting to note that the proposed classifier was not able to ensure proper separation between the classes IEEE 802.11 and IFC when the sampling window is $T_{SW} = 50$ ms, while the accuracy increases rapidly as T_{SW} approaches 200 ms, stabilizing around 84%. This behavior is mainly driven by the dynamics of IEEE 802.11 silent networks. Since a silent network shows by definition a low or null rate of exchanged data packets, due to a limited number of associated terminals, the on-air transmission is mainly due to the beacons emitted by the access point. Since the beacon period for all of the networks in experiments was set to the default value of 102.4 ms, a short sampling window can result in an increased possibility of missing the sensing of the beacon, which in turn reflects an insufficient separation between the vectors representing the IFC class and the IEEE 802.11 class in the employed $M + 3$-dimensional feature space. Nevertheless, thanks to the supervised-learning structure of the classifier, the proposed method allows the detection an IEEE 802.11 network with good accuracy in less than two beacon periods, while in concurrent approaches (e.g., [26]), the channel should be sensed for the time of several beacon periods in order to maximize the detection rate. The curves representing the classification rate for the IFCs are shown in Figure 11.

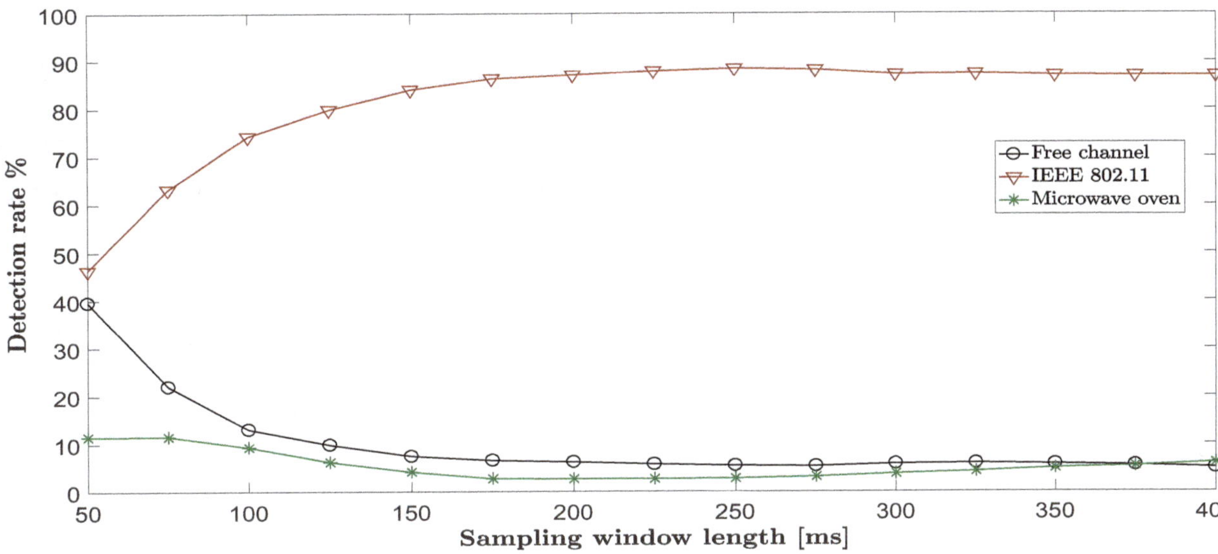

Figure 10. Average full-spectrum detection rate for IEEE 802.11 scenarios for different sampling window lengths.

In the case of IFC, the classification accuracy trend is opposite with respect to the IEEE 802.11 interference classification. This can be simply explained by the fact that shorter observation windows will in turn mean a lower probability of encountering amplitude fluctuations of the background noise, which can potentially drift the feature vector in the decisional zone of IEEE 802.11 and MWO classes. Despite this fact, the IFC classification rate was reported consistently above 90%, even for sampling windows greater than 200 ms, while we observed an increase of MWO and IEEE 802.11 misclassification, for the reason just described.

In Figure 12, we show the full-spectrum classification accuracy for MWO interference.

The figure shows insufficient separation between the classes MWO and IEEE 802.11 for $T_{SW} = 50$ ms, while increasing the sampling time improves the classification accuracy, even if the improvement is significantly slower with respect to the case of IEEE 802.11 interference. This in turn means that in order to maximize the separation between classes, a sampling window of $T_{SW} \geq 250$ ms is required so that the selected features can emerge with sufficient clarity and ensure a full spectrum classification accuracy greater than 82%. As discussed in Section 7, this behavior is due to the similarity of the temporal features of IEEE 802.11 signals in the case of active networks and the RF leakage of MWO.

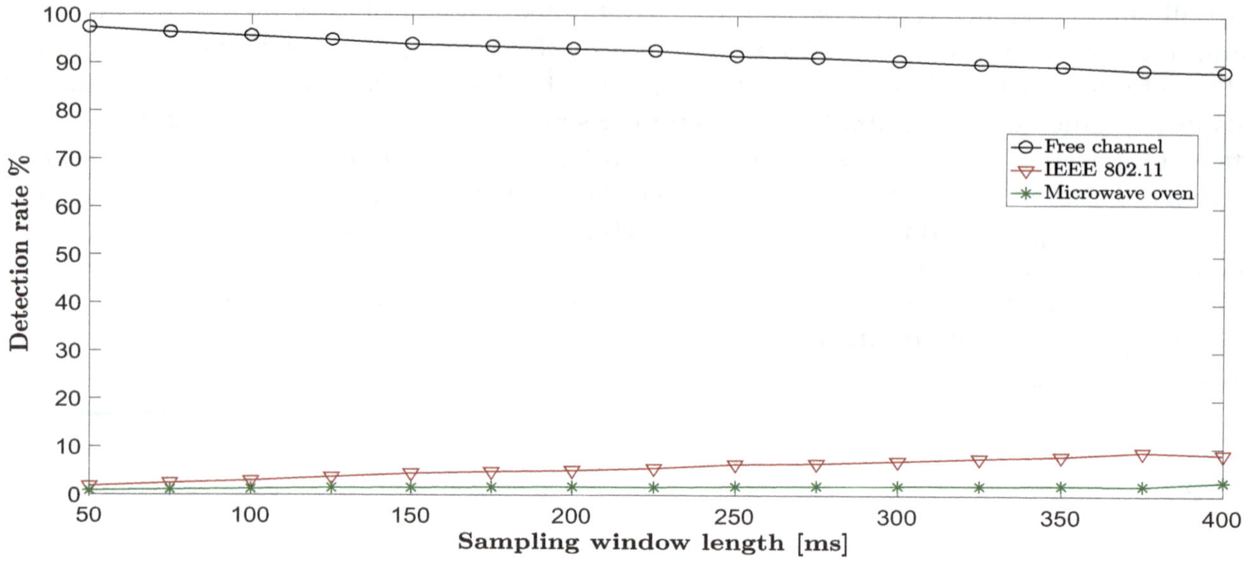

Figure 11. Average full-spectrum detection rate for interference-free scenarios for different sampling window lengths.

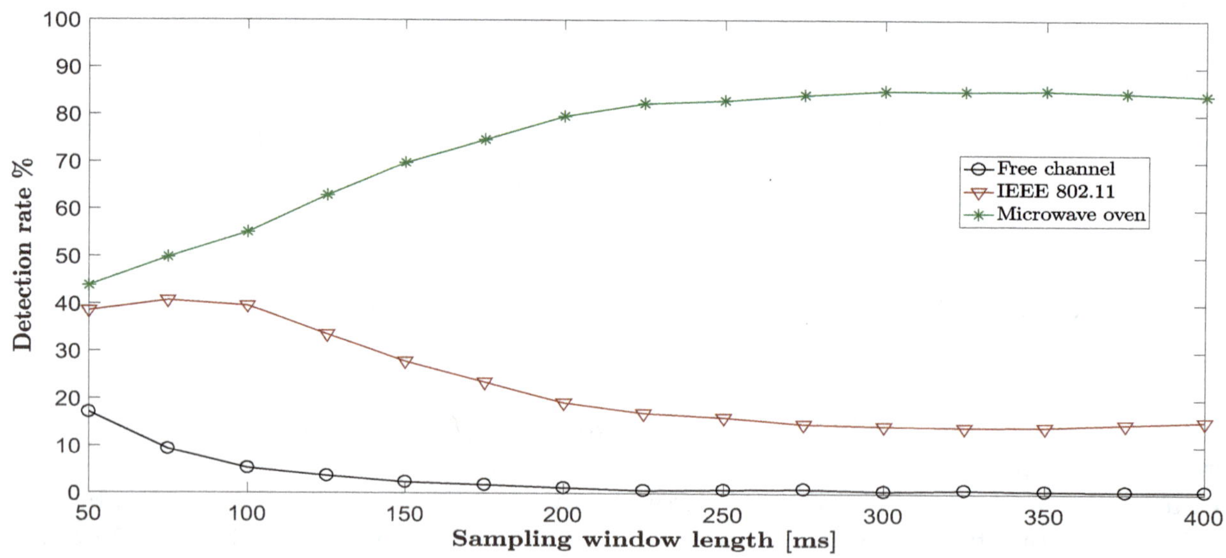

Figure 12. Average full-spectrum detection rate for interference from MWO for different sampling window lengths.

In Figure 13, we finally show the effects of different sampling windows on the in-channel detection accuracy for MWO at Location C.

From the plot, we observe that the classification accuracy is monotonically increasing for all of the channels, meaning that longer sampling windows are always beneficial for MWO detection. It can also be noted that the classification accuracy dip on Channel 14 and on Channel 24 experiences a significant improvement when the sampling window approaches 350 ms, giving a hint about the bursty time distribution of the RSSI samples on the side spectrum of MWO leakage.

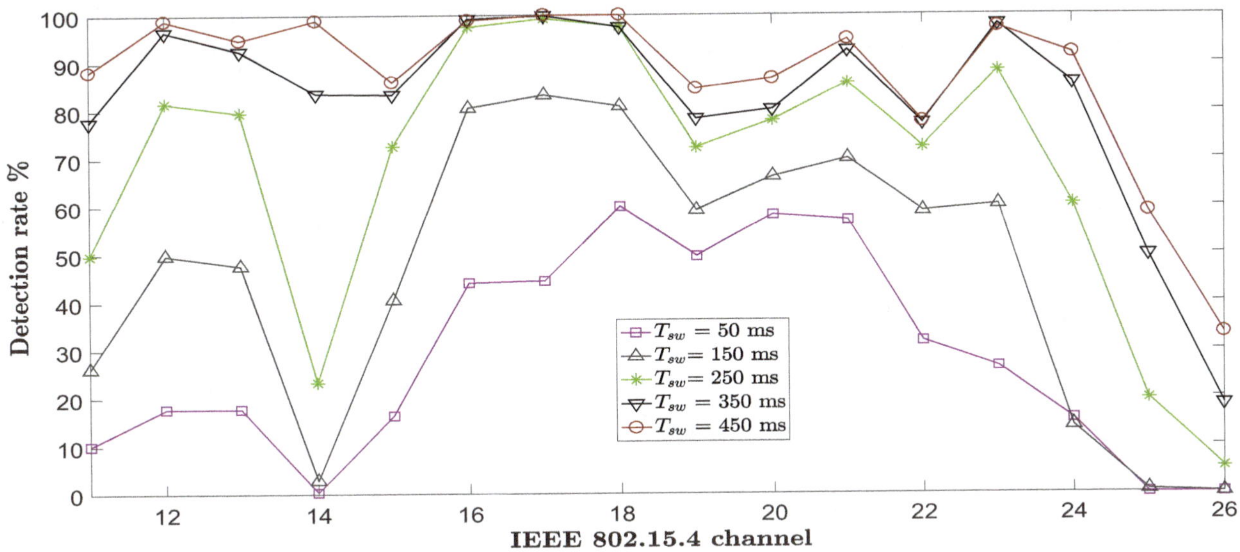

Figure 13. Detection rate for microwave oven at Location C for different sampling window lengths. Some curves have been removed for clarity.

8.2. Hardware-Related Considerations

Since COTS WSN nodes are low-power devices with resource-constrained hardware, particular attention has to be paid when implementing a complex methods on these platforms. In this section, we discuss how certain characteristics of the selected hardware (i.e., CC2420-equipped TelosB motes) influence the spectrum-sensing task and consequently the applicability and performance of the proposed classification method.

8.2.1. The Role of Node Calibration

It is a well-known fact that different CC2420-based devices can show variation in the nominal response of the RSSI curve. Since the core of the proposed method is based on RSSI sampling and threshold-based features, it is of primary importance to analyze if these variations can hamper the performance of the classifier. In their work, Chen et al. [41] showed that these variations are due to two different phenomena: a non-linearity in the CC2420 RSSI response curve and the presence of a node-dependent offset. While the first phenomenon is of minor relevance, since the non-linear and non-injective regions do not influence significantly the RSSI curve (which remains mostly linear), a consistent offset of ±6 dB is reported among different nodes.

We have tested several different TelosB nodes, sampling IEEE 802.15.4 Channel 26 in a radio-controlled environment to determine both the amplitude distribution and the mean of the collected RSSI traces for different nodes. As shown in Figure 14, we have observed a maximum RSSI offset of ±5 dB.

We carried out an analysis of the influence of the RSSI offset existing between the network device deployed for channel sensing and the device used for preliminary training set collection. In Figure 15, we show the impact of RSSI offset on the classification accuracy of the three targeted interference classes.

As can be observed, performing the channel sensing using a node with a consistent RSSI offset can greatly hamper the performance of the classifier, also considering that the offset between two nodes could theoretically span up to 12 dB. Even an offset of 5 dB, such as the one reported in our node set, can decrease the performance of both IEEE 802.11 andMWO classification up 15–20%, rendering a node calibration process a relevant step from the perspective of safeguarding the performance of the proposed approach. Fortunately, this process is straightforward, since as shown in [41], only a simple noise-floor-based RSSI offset calculation and compensation is needed. In the proposed approach, for example, once the RSSI offset is acquired, the offset compensation can be simply implemented by employing a software-based adaptation of the energy-threshold used for the feature-extraction task.

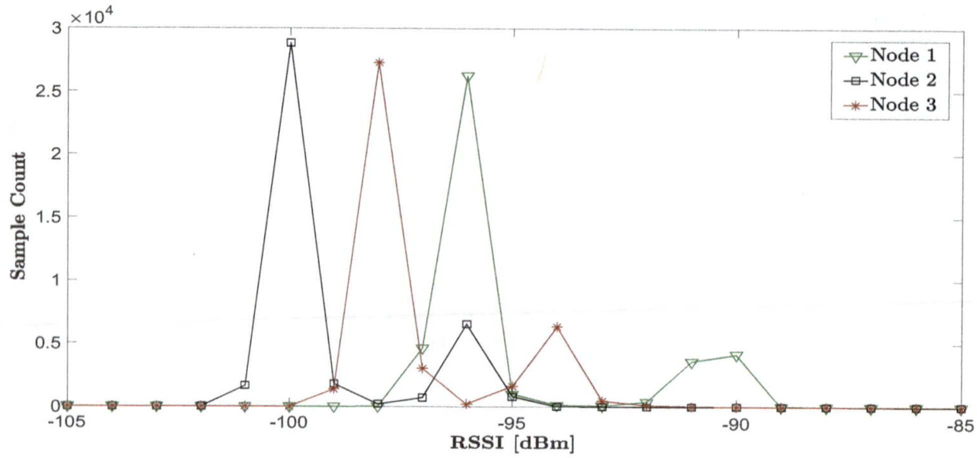

(a)

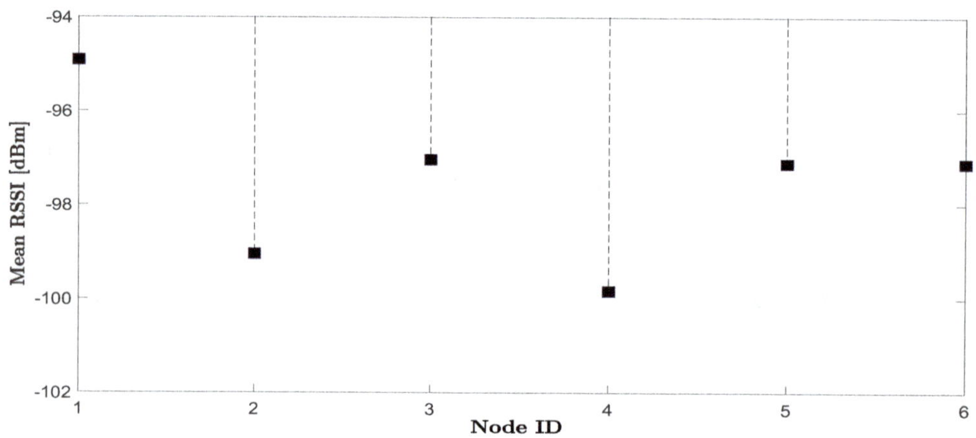

(b)

Figure 14. The RSSI profiling process for TelosB motes. (**a**) Amplitude distribution of RSSI traces from background noise sensing for different CC2420 nodes. Some curves have been removed for clarity. (**b**) Mean of recorded RSSI sample traces.

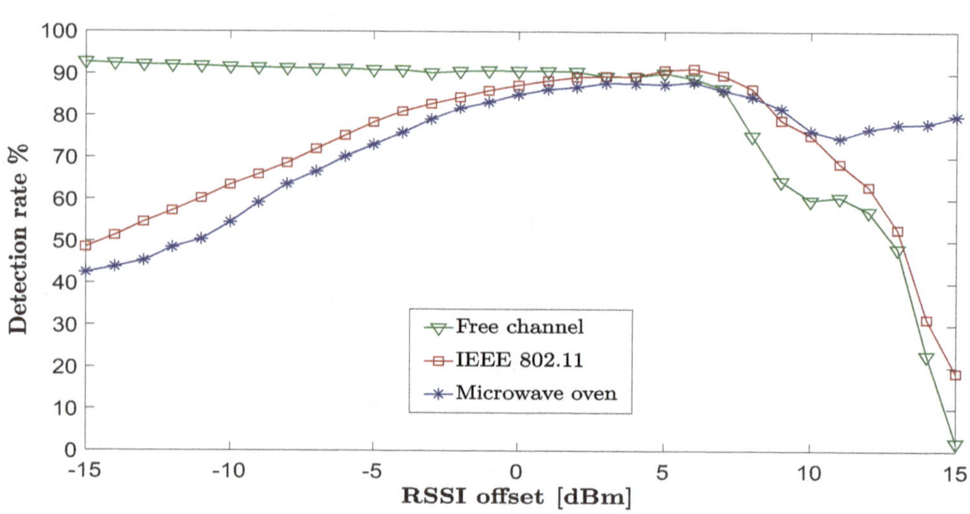

Figure 15. Detection accuracy with respect to RSSI offset between the node used for collecting training data and the actual sensing node.

8.2.2. Assessing the Timeliness of the Sampling Process

In order to discuss the feasibility of the proposed sensing scheme with respect to the employed COTS hardware platform, we monitor and analyze the delay generated by the various operations needed to perform the in-node channel sensing. In Figure 16, we show the partial duration of the tasks implemented in TelosB motes in order to acquire and store the RSSI samples. Two of the most demanding tasks in terms of delay are the request for accessing and releasing the I/O resources, requiring 212 µs and 74 µs, respectively. In addition, the tasks of setting the CSn (chip select) pin for reading the CC2420 RSSI register lasts 12 µs, while the actual operation of sampling the value of RSSI register takes 112 µs to be completed.

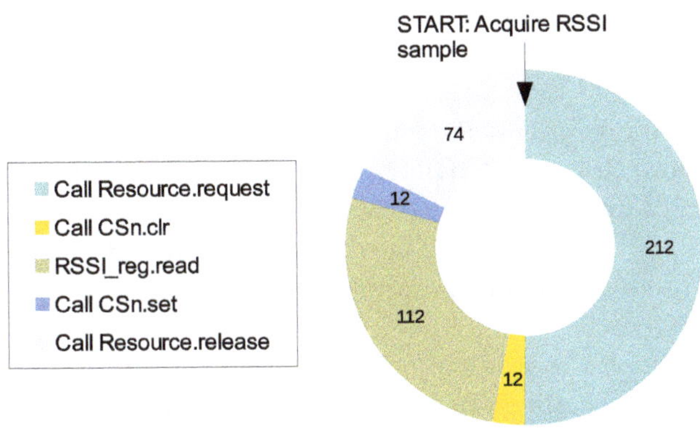

Figure 16. Operation delays in µs for the channel-sensing task implemented in TelosB motes.

The total delay for collecting and storing one sample is then 429 µs, while we use a sampling frequency of 2 kHz, corresponding to a 500 µs sampling period. With the current approach, the sampling frequency could be theoretically pushed up to 4.6 kHz if the CC2420 resources are not released until the end of the whole sampling process. In other approaches (e.g., [26]), the implementation for the channel sensing in Contiki OS allows for a sampling rate up to 8.13 kHz. Nevertheless, a higher sampling frequency means more data to process, as well as a more stressful and energy-consuming sampling process. Therefore, in this work, we have employed a more relaxed sampling timing, while we rely on the approach of an advanced classification algorithm, in order to maintain high classification performance while ensuring a lower memory footprint.

9. Conclusions

In this paper, we present a novel scheme employing machine learning methods for cross-technology interference classification in IWSAN. The proposed method employs a three-step classifier composed of a lightweight feature-extraction process, a preliminary classification stage employing four SVMs and a final decisor, allowing for classification among interference from IEEE 802.11 networks, microwave ovens, as well as the presence of interference-free channels. The tests conducted in industrial environments, including a wide range of interference scenarios, show an average classification accuracy of 84% and up to 98% for IEEE 802.11 active networks, with a channel sensing time of 300 ms. The memory footprint of the samples collected in this sensing time remains below 600 bytes per channel thanks to the limited sampling frequency. The extremely short time required for sensing renders the developed solution a promising candidate for the adoption in superframe-based TSCH networks by means of spectrum-sensing-reserved timeslots. In this paper, we have also highlighted the fundamental influence of device calibration on the performance of spectrum-sensing-based methods using COTS WSN hardware, which is a matter often overlooked in related literature. In particular, it is shown that the classification accuracy of the proposed solution is significantly influenced by the intrinsic hardware variations.

We leave to future works further investigations on the potentialities of SVM-based-methods for interference classification in IWSAN. Other notable aspects of interest are the inclusion of the channel-sensing and classification mechanism in a TSCH network and a run-time assessment of the solution, as well as the development of interference mitigation strategies.

Acknowledgments: The authors would like to thank Imerys Mineral AB for the access to their production plant in Sundsvall and R.Rondón from Mid Sweden University for the helpful feedback.

Author Contributions: S.G. conceived and implemented the proposed method; S.G and A.M. designed and conducted the experiments; S.G. analyzed the data. M.G. supervised the overall work. S.G., A.M. wrote the paper. All authors contributed in discussing and revising the manuscript.

References

1. Prathap, U.; Shenoy, P.D.; Venugopal, K.R.; Patnaik, L.M. Wireless Sensor Networks Applications and Routing Protocols: Survey and Research Challenges. In Proceedings of the 2012 International Symposium on Cloud and Services Computing, Mangalore, India, 17–18 December 2012; pp. 49–56.
2. Đurišić, M.P.; Tafa, Z.; Dimić, G.; Milutinović, V. A survey of military applications of wireless sensor networks. In Proceedings of the 2012 Mediterranean Conference on Embedded Computing (MECO), Bar, Montenegro, 19–21 June 2012; pp. 196–199.
3. Mangali, N.K.; Kota, V.K. Health monitoring systems: An energy efficient data collection technique in wireless sensor networks. In Proceedings of the 2015 International Conference on Microwave, Optical and Communication Engineering (ICMOCE), Bhubaneswar, India, 18–20 December 2015; pp. 130–133.
4. Benkhelifa, I.; Nouali-Taboudjemat, N.; Moussaoui, S. Disaster Management Projects Using Wireless Sensor Networks: An Overview. In Proceedings of the 2014 28th International Conference on Advanced Information Networking and Applications Workshops, Victoria, BC, Canada, 13–16 May 2014; pp. 605–610.
5. *IEEE Standard for Information technology—Local and metropolitan area networks—Specific requirements—Part 15.4: Wireless Medium Access Control (MAC) and Physical Layer (PHY) Specifications for Low Rate Wireless Personal Area Networks (WPANs)*; IEEE Std 802.15.4-2006, IEEE: New York, NY, USA; 2006.
6. Åkerberg, J.; Gidlund, M.; Björkman, M. Future research challenges in wireless sensor and actuator networks targeting industrial automation. In Proceedings of the 2011 9th IEEE International Conference on Industrial Informatics, Lisbon, Portugal, 26–29 July 2011; pp. 410–415.
7. Galloway, B.; Hancke, G. Introduction to Industrial Control Networks. *IEEE Commun. Surv. Tutor.* **2013**, *15*, 860–880.
8. Rappaport, T.S. Indoor radio communications for factories of the future. *IEEE Commun. Mag.* **1989**, *27*, 15–24.
9. Gwarek, W.K.; Celuch-Marcysiak, M. A review of microwave power applications in industry and research. In Proceedings of the 15th International Conference on Microwaves, Radar and Wireless Communications (IEEE Cat. No.04EX824), Warsaw, Poland, 17–19 May 2004; Volume 3, pp. 843–848.
10. Sikora, A. *Wireless Personal and Local Area Networks*; Wiley: Chichester, UK; Hoboken, NJ, USA, 2003.
11. Liang, C.J.M.; Priyantha, N.B.; Liu, J.; Terzis, A. Surviving Wi-Fi Interference in Low Power ZigBee Networks. In Proceedings of the 8th ACM Conference on Embedded Networked Sensor Systems, Zürich, Switzerland, 3–5 November 2010; ACM: New York, NY, USA, 2010; pp. 309–322.
12. Yang, D.; Xu, Y.; Gidlund, M. Wireless Coexistence between IEEE 802.11- and IEEE 802.15.4-Based Networks: A Survey. *Int. J. Distrib. Sens. Netw.* **2011**, *7*, 912152.
13. Hermans, F.; Rensfelt, O.; Voigt, T.; Ngai, E.; Norden, L.A.; Gunningberg, P. SoNIC: Classifying Interference in 802.15.4 Sensor Networks. In Proceedings of the 12th International Conference on Information Processing in Sensor Networks (IPSN '13), Philadelphia, PA, USA, 8–11 April 2013; ACM: New York, NY, USA, 2013; pp. 55–66.
14. *HART Communication Protocol Specification, Revision 7.4*; Technical Report; HART Communication Foundation: Austin, TX, USA, 2012.
15. *Wireless Systems for Industrial Automation: Process Control and Related Applications*; ISA 100.11a-2011; International Society of Automation: Research Triangle Park, NC, USA, 2011.

16. *Industrial Communication Networks—Fieldbus Specifications—WIA-PA Communication Network and Communication Profile*; IEC 62601; International Electrotechnical Commission: Geneva, Switzerland, 2011.
17. Yucek, T.; Arslan, H. A survey of spectrum sensing algorithms for cognitive radio applications. *IEEE Commun. Surv. Tutor.* **2009**, *11*, 116–130.
18. Brown, J.; Roedig, U.; Boano, C.A.; Römer, K. Estimating packet reception rate in noisy environments. In Proceedings of the 39th Annual IEEE Conference on Local Computer Networks Workshops, Edmonton, AB, Canada, 8–11 September 2014; pp. 583–591.
19. Zacharias, S.; Newe, T.; O'Keeffe, S.; Lewis, E. A Lightweight Classification Algorithm for External Sources of Interference in IEEE 802.15.4-Based Wireless Sensor Networks Operating at the 2.4 GHz. *Int. J. Distrib. Sens. Netw.* **2014**, *10*, 265286.
20. Zhou, R.; Xiong, Y.; Xing, G.; Sun, L.; Ma, J. ZiFi: wireless LAN discovery via ZigBee interference signatures. In Proceedings of the sixteenth annual international conference on Mobile computing and networking, Chicago, IL, USA, 20–24 September 2010; ACM: New York, NY, USA, 2010; pp. 49–60.
21. Gao, Y.; Niu, J.; Zhou, R.; Xing, G. ZiFind: Exploiting cross-technology interference signatures for energy-efficient indoor localization. In Proceedings of the 2013 Proceedings IEEE INFOCOM, Turin, Italy, 14–19 April 2013; pp. 2940–2948.
22. Choi, J. WidthSense: Wi-Fi Discovery via Distance-based Correlation Analysis. *IEEE Commun. Lett.* **2016**, *21*, 422–425.
23. Petrova, M.; Wu, L.; Mahonen, P.; Riihijarvi, J. Interference Measurements on Performance Degradation between Colocated IEEE 802.11 g/n and IEEE 802.15.4 Networks. In Proceedings of the Sixth International Conference on Networking (2007. ICN '07), Sainte-Luce, France, 22–28 April 2007; p. 93.
24. Hossian, M.M.A.; Mahmood, A.; Jäntti, R. Channel ranking algorithms for cognitive coexistence of IEEE 802.15.4. In Proceedings of the 2009 IEEE 20th International Symposium on Personal, Indoor and Mobile Radio Communications, Tokyo, Japan, 13–16 September 2009; pp. 112–116.
25. ZigBee Standards Organization. *ZigBee Specifications*; ZigBee Standards Organization: San Ramon, CA, USA, 2012; pp. 1–622.
26. Zacharias, S.; Newe, T.; O'Keeffe, S.; Lewis, E. 2.4 GHz IEEE 802.15.4 channel interference classification algorithm running live on a sensor node. In Proceedings of the 2012 IEEE Sensors, Taipei, Taiwan, 28–31 Octorber 2012; pp. 1–4.
27. Iyer, V.; Hermans, F.; Voigt, T. Detecting and Avoiding Multiple Sources of Interference in the 2.4 GHz Spectrum. In Proceedings of the12th European Conference on Wireless Sensor Networks (EWSN), Porto, Portugal, 9–11 February 2015; Abdelzaher, T., Pereira, N., Tovar, E., Eds.; Springer International Publishing: Cham, Switzerland, 2015; pp. 35–51.
28. Ansari, J.; Ang, T.; Mähönen, P. WiSpot: fast and reliable detection of Wi-Fi networks using IEEE 802.15.4 radios. In Proceedings of the 9th ACM International Symposium on Mobility Management and Wireless Access, Miami, FL, USA, 31 October–4 November 2011; ACM: New York, NY, USA, 2011; pp. 35–44.
29. Chowdhury, K.R.; Akyildiz, I.F. Interferer Classification, Channel Selection and Transmission Adaptation for Wireless Sensor Networks. In Proceedings of the 9th IEEE International Conference on Communications (ICC '09), Dresden, Germany, 14–18 June 2009; pp. 1–5.
30. Rayanchu, S.; Patro, A.; Banerjee, S. Airshark: Detecting non-WiFi RF Devices Using Commodity WiFi Hardware. In Proceedings of the 2011 ACM SIGCOMM Conference on Internet Measurement Conference (IMC '11), Berlin, Germany, 2–4 November 2011; ACM: New York, NY, USA, 2011; pp. 137–154.
31. Weng, Z.; Orlik, P.; Kim, K.J. Classification of wireless interference on 2.4 GHz spectrum. In Proceedings of the 2014 IEEE Wireless Communications and Networking Conference (WCNC), Istanbul, Turkey, 6–9 April 2014; pp. 786–791.
32. Hermans, F.; Larzon, L.A.; Rensfelt, O.; Gunningberg, P. A Lightweight Approach to Online Detection and Classification of Interference in 802.15.4-based Sensor Networks. *SIGBED Rev.* **2012**, *9*, 11–20.
33. Nicolas, C.; Marot, M. Dynamic link adaptation based on coexistence-fingerprint detection for WSN. In Proceedings of the 2012 11th Annual Mediterranean Ad Hoc Networking Workshop (Med-Hoc-Net), Ayia Napa, Cyprus, 19–22 June 2012; pp. 90–97.
34. Barać, F.; Gidlund, M.; Zhang, T. Ubiquitous, Yet Deceptive: Hardware-Based Channel Metrics on Interfered WSN Links. *IEEE Trans. Veh. Technol.* **2015**, *64*, 1766–1778.

35. Zheng, X.; Cao, Z.; Wang, J.; He, Y.; Liu, Y. ZiSense: Towards Interference Resilient Duty Cycling in Wireless Sensor Networks. In Proceedings of the 12th ACM Conference on Embedded Network Sensor Systems, (SenSys '14), Memphis, TN, USA, 3–6 November 2014; ACM: New York, NY, USA, 2014; pp. 119–133.
36. Rondeau, T.W.; D'Souza, M.F.; Sweeney, D.G. Residential microwave oven interference on Bluetooth data performance. *IEEE Trans. Consum. Electron.* **2004**, *50*, 856–863.
37. Bishop, C.M. *Pattern Recognition and Machine Learning (Information Science and Statistics)*; Springer: Secaucus, NJ, USA, 2006.
38. Boyd, S.; Vandenberghe, L. *Convex Optimization*; Cambridge University Press: New York, NY, USA, 2004.
39. Crossbow TelosB Mote Plattform, Datasheet. Available online: http://www.willow.co.uk/TelosB_Datasheet.pdf (accessed on 16 June 2017).
40. Texas Instruments CC2420 - 2.4 GHz IEEE 802.15.4/ZigBee-ready RF Transceiver. Available online: http://www.ti.com/lit/ds/symlink/cc2420.pdf (accessed on 16 June 2017).
41. Chen, Y.; Terzis, A. On the mechanisms and effects of calibrating RSSI measurements for 802.15.4 radios. In Proceedings of the 7th European conference on Wireless Sensor Networks, Coimbra, Portugal, 17–19 February 2010; Springer: Berlin/Heidelberg, Germany, 2010; pp. 256–271.
42. Metageek Wi-Spy Chanalizer. Available online: http://files.metageek.net/marketing/data-sheets/MetaGeek_Wi-Spy-Chanalyzer_DataSheet.pdf (accessed on 16 June 2017).
43. Boano, C.A.; Voigt, T.; Noda, C.; Römer, K.; Zúñiga, M. JamLab: Augmenting sensornet testbeds with realistic and controlled interference generation. In Proceedings of the 10th ACM/IEEE International Conference on Information Processing in Sensor Networks, Chicago, IL, USA, 12–14 April 2011; pp. 175–186.

4

Prototyping and Validation of MEMS Accelerometers for Structural Health Monitoring—The Case Study of the Pietratagliata Cable-Stayed Bridge

Chiara Bedon *, Enrico Bergamo, Matteo Izzi and Salvatore Noè

Department of Engineering and Architecture, University of Trieste, Piazzale Europa 1, 34127 Trieste, Italy; enrico.bergamo@me.com (E.B.); izzimatteo@gmail.com (M.I.); noe@units.it (S.N.)

* Correspondence: chiara.bedon@dia.units.it

Abstract: In recent years, thanks to the simple and yet efficient design, Micro Electro-Mechanical Systems (MEMS) accelerometers have proven to offer a suitable solution for Structural Health Monitoring (SHM) in civil engineering applications. Such devices are typically characterised by high portability and durability, as well as limited cost, hence resulting in ideal tools for applications in buildings and infrastructure. In this paper, original self-made MEMS sensor prototypes are presented and validated on the basis of preliminary laboratory tests (shaking table experiments and noise level measurements). Based on the well promising preliminary outcomes, their possible application for the dynamic identification of existing, full-scale structural assemblies is then discussed, giving evidence of their potential via comparative calculations towards past literature results, inclusive of both on-site, Experimental Modal Analysis (EMA) and Finite Element Analytical estimations (FEA). The full-scale experimental validation of MEMS accelerometers, in particular, is performed using, as a case study, the cable-stayed bridge in Pietratagliata (Italy). Dynamic results summarised in the paper demonstrate the high capability of MEMS accelerometers, with evidence of rather stable and reliable predictions, and suggest their feasibility and potential for SHM purposes.

Keywords: Micro Electro-Mechanical Systems (MEMS) accelerometers; Structural Health Monitoring (SHM); prototyping and validation; dynamic identification; cable-stayed bridge; Experimental Modal Analysis (EMA); Finite Element Analytical (FEA) modelling

---

1. Introduction, State-of-the-Art and Objectives

Nowadays, buildings and infrastructure are designed to sustain ordinary or extreme dynamic loads (such as wind, traffic, earthquakes, impacts, etc.), whose magnitude is determined from probabilistic approaches (i.e., EN 1991 [1]). In most of the cases, simplified design methods and simulation techniques are conventionally used, to describe the mechanical features of different structural typologies. However, their actual structural behaviour (i.e., fundamental period, vibration shapes, etc.) is properly assessed for a limited number of cases only, i.e., for critical buildings and infrastructures whose integrity and serviceability is of high importance for public safety and civil protection. Only a few of these strategic constructional facilities are then equipped with continuous monitoring systems.

The information that is typically obtained from structural monitoring tools, in this regard, is of fundamental importance in view of the consequences associated to possible collapse phenomena. Those systems provide in fact the authorities with a careful evaluation of the damage evolution, supporting the planning of the restoration interventions (e.g., [2–5], etc.). Structural Health Monitoring (SHM) and non-destructive testing have key roles for structural systems in operational conditions, for monumental buildings ([6–8], etc.), industrial facilities, or aerospace components [9–13], tunnels, and underground environments [14,15].

Several research efforts have been devoted in the last decade to the development of reliable and cost-effective monitoring devices equipped with Micro Electro-Mechanical Systems (MEMS). MEMS technology has evolved considerably, leading to a general improvement of the sensors performance, as well as to a price minimization [16,17]. Comparative experimental studies of literature report a 1-to-10 cost ratio of MEMS, with respect to traditional piezoelectric accelerometers (i.e., [18,19], etc.). MEMS-based systems, in addition, proved to be efficient for several types of dynamic applications. Dynamic measurements of human body movements, for example, were carried out via MEMS accelerometers by Benevicius et al. [20]. Hand-arm and whole-body MEMS-based vibration records were critically discussed, aiming at investigating the reliability of MEMS techniques for biomedical applications. The so-called bioMEMS gave evidence of their potential for the medical field especially, in the last five years [21]. At the same time, MEMS accelerometers proved to be efficient also for vibration monitoring in industrial machines and rotors (e.g., [22–25], etc.).

Since the 1990s, major efforts and well-promising results were reported in the literature from the application of MEMS accelerometers in the SHM of civil engineering facilities, as well as in the early-bird monitoring of seismological hazards. In the first case, MEMS systems have been efficiently used for the monitoring of strong-motion events in rigid structures, but positive efforts have been also achieved from continuous MEMS measurements of flexible structures (such as vehicular and pedestrian bridges), as deeply discussed in several research papers. Bassoli et al. [26] reported on the dynamic identification of an ancient masonry bell tower in Italy, seriously damaged after the Emilia earthquake of 2012 and subjected to experimental tests after the retrofitting interventions. Dynamic tests were carried out based on a MEMS acquisition system, including comparative measurements and a critical discussion of experimental results, as derived from the installed MEMS-based system or from traditional analogue instruments. This study is in line with the investigation presented in [27], where numerical model updating is carried out for ancient masonry bell towers, based on continuous SHM via a wired piezoelectric sensor network (commercially available, mono-axial accelerometers).

Feng et al. [28] explored the potential use of smartphone accelerometers for measuring the structural vibrations in buildings, hence as active instruments for SHM and post-event damage diagnostics. The shake table tests discussed in [28] gave evidence of well promising MEMS performances and results, both for low-amplitude ambient vibrations and high-amplitude seismic responses. Wargantiwar et al. [29] gave further evidence of the high potential of MEMS accelerometers, when working as earthquake alarm tools for buildings and civil engineering infrastructures. Major benefits were found in their typical low cost, limited power consumption and relatively small size. Kok et al. [30] experimentally assessed the accuracy of MEMS accelerometers for modal analysis purposes, giving evidence of maximum expected frequency errors up to 5%, within their working range. In [31,32], experimental shaking table tests are discussed for tri-axis MEMS accelerometers. The collected vibration data showed close agreement with the experimental measurements derived from commercial devices for SHM purposes. Beskhyroun and Ma [33] also presented an application of MEMS accelerometers for the experimental modal analysis of a high rise, reinforced concrete building subjected to strong aftershocks. The experimental study highlighted the high accuracy of MEMS accelerometers for the prediction of the modal parameters of the monitored building, compared to traditional testing instruments. A list of additional positive MEMS applications for the SHM and dynamic identification of civil engineering constructions, including wireless options, can be found in the literature (see for example [30,34–37]). In [38], the use of MEMS devices is proposed for the SHM of a suspension bridge in Istanbul. Domaneschi et al. [39] also explored the seismic performance of the Shimotsui-Seto suspension bridge in Japan. In [39], two MEMS sensor families (with low- and high-density noise levels) were taken into account, giving evidence of the related effects and sensitivity of measurements for localised damage detection purposes. The same suspension bridge was further numerically investigated in [40] under wind excitation, exploring the

MEMS noise effects on the damage detection, for different scenarios of technical interest (i.e., damage location and severity).

A number of research projects aimed to assess the feasibility of MEMS applications in the form of seismological alarm systems can then be found in the literature. Dashti et al. [41], for example, explored the use of cellular phones as ground motion instruments, giving evidence of their accuracy as seismic monitoring devices via comparative shake table tests. Similar results are also reported in [42,43], etc.

In this context, the paper presents original self-made MEMS accelerometers, as a possible suitable tool for SHM of engineering systems and constructed facilities. Major features of the prototyped devices are first described in Section 2, including a preliminary experimental validation of the assembled sensors via shacking table tests and noise level measurements (see Section 3). The collected test measurements are compared with commercially available devices. The feasibility and potential of the proposed self-made MEMS sensors are then emphasized via a full-scale Experimental Modal Analysis (EMA) investigation, carried out on the cable-stayed bridge of Pietratagliata (Italy). Compared to existing literature efforts, the current study aims at further assessing the reliability of SHM via low-cost, portable MEMS sensors that could be used for the continuous, on-site monitoring of constructed facilities. The selected bridge was opened to traffic in 2008, and is of particular interest for SHM and diagnostic purposes, due to its intrinsic dynamic behaviour. In addition, the bridge is representative of a strategic infrastructure located in a high seismic region. During 2010 and 2012, moreover, the bridge was affected by localised damage in two of the cables-to-deck connections, hence resulting in partial modification of its actual boundary conditions and suggesting detailed investigations with continuous data acquisition. In Section 4, for comparative purposes, MEMS experimental results are hence post-processed and assessed towards past EMA predictions and Finite Element Analytical (FEA) data available in the literature for the same structural system [44].

2. Measuring Devices

The typical measuring device considered in this study is composed of a printed circuit (PC) board with two RJ45 connectors for in-and-out connections (see Figure 1a). The main components of the PC board are:

(i) a logic unit, programmed with the synchronisation and recording routines;
(ii) an accelerometer;
(iii) an Analogue-to-Digital Converter (ADC);
(iv) a micro SD memory card, to store the recorded data;
(v) a real-time clock, to keep the synchronisation between the devices consistent.

(a)

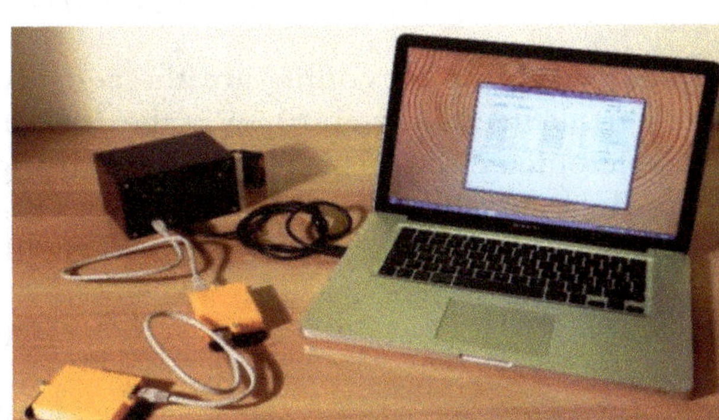

(b)

Figure 1. (**a**) PC board and (**b**) assembled system.

Each measuring device uses two CAT6 24AWG Ethernet cables with four couples of twisted wires: two couples of wires carry the power supply and the other two are used for the data transmission and the synchronisation signal.

The synchronisation of the devices is provided—prior to starting each registration—by a personal computer (Figure 1b), which sends data packets with the current date and time. Each data packet has a trigger, which activates the oscillators simultaneously. Furthermore, to ensure the consistence of the measurements when recording, a check square wave with 1 Hz frequency is sent from the personal computer and is recorded by each device. The sensors generate a square wave (1024 Hz), which allows collecting the input data at the sampling frequencies of 256 Hz, 128 Hz, and 64 Hz. The precision of the sensors strictly depends on the quality of the installed crystal, typically in the order of 20 ppm. Finally, in the post-processing phase, each synchronisation trace is compared with the reference trace and any small delay is corrected. Consequently, the alignment of the square wave recorded by each sensor with respect to the original signal is verified.

At the time of the prototyping, the accelerometer was chosen based on the most convenient trade-off between price and self-noise level. Several sensors were analysed before choosing the Kionix KXR94-2050 (Kionix, Inc.®, Ithaca, NY, USA), a tri-axis silicon micromachined accelerometer with a full-scale output range of ±2 g. The acceleration sensing is based on the principle of a differential capacitance arising from the acceleration-induced motion of the sensor. Furthermore, each board is equipped with an ADC Texas Instrument device, ADS1220 type (Texas Instruments, Dallas, TX, USA), which has a resolution of 24 bits and features two differential or four single-ended inputs through an input multiplexer. Table 1 lists the electrical properties of the chosen accelerometer and ADC.

Table 1. Electrical properties for the chosen accelerometer and ADC.

Accelerometer: Kionix KXR94-2050		**ADC: Texas Instrument ADS1220**	
Measurement axes	3	Type	Sigma-Delta
Measurement range	±2 g	Resolution	24 bit
Sensitivity	0.66 V/g	Channels	2 diff./4 single ended
Noise density	45 µg/$\sqrt{\text{Hz}}$	Data rate	2000 SPS
Supply voltage	3.3 V (typical)	Supply voltage	3.3 V (typical)
Temperature range	from −40 °C to 85 °C	Temperature range	from −40 °C to 125 °C

The resolution R of the prototyped MEMS accelerometer is rationally calculated as follows:

$$R = \frac{3.3\ \text{V}}{2^{23}\ \text{counts}\cdot 0.66\ \text{V/g}} = 0.596\ \mu\text{g/count} \tag{1}$$

where 0.66 V/g and 2^{23} counts are the sensitivity and the quantisation levels available in each accelerometer, respectively, and 3.3 V is the operating voltage of the ADC. The electro-mechanical noise of the accelerometer, conversely, is evaluated using the nominal specifications declared by the supplier in the product datasheet. The theoretical root-mean-square (*rms*) noise is evaluated by filtering the noise density with a first-order low-pass 20 Hz filter leading to:

$$rms = 45\frac{\mu\text{g}}{\sqrt{\text{Hz}}}\cdot\sqrt{20\ \text{Hz}\cdot 1.57} = 0.252\ \text{mg} \tag{2}$$

In this context, the result of Equation (2) is a theoretical value; the actual electro-mechanical noise might be even higher, being influenced by the final layout of the PC board, the production techniques, the frequency of the power supply, and the temperature.

3. Laboratory Testing and Validation

3.1. Shaking Table Testing

Preliminary hardware tests were carried out at the University of Trieste (Italy), Department of Engineering and Architecture, aimed at assessing the accuracy of the measuring devices with respect to commercial products available on the market. Tests compared the output response of the PC boards (Figure 2a) with a reference accelerometer used for laboratory measurements, the PCB 356A16 type of Figure 2b. To this aim, three boards were randomly selected from the full set of instruments (S#1, S#2 and S#3 in Table 2) and were simultaneously mounted on a vertical shaking table, together with the PCB 356A16 accelerometer. The shaking table operates at a frequency range of 5–50 Hz, while the PCB 356A16 has a sensitivity of 100 mV/g, an acceleration range of ±50 g and a frequency range of 0.5–5000 Hz.

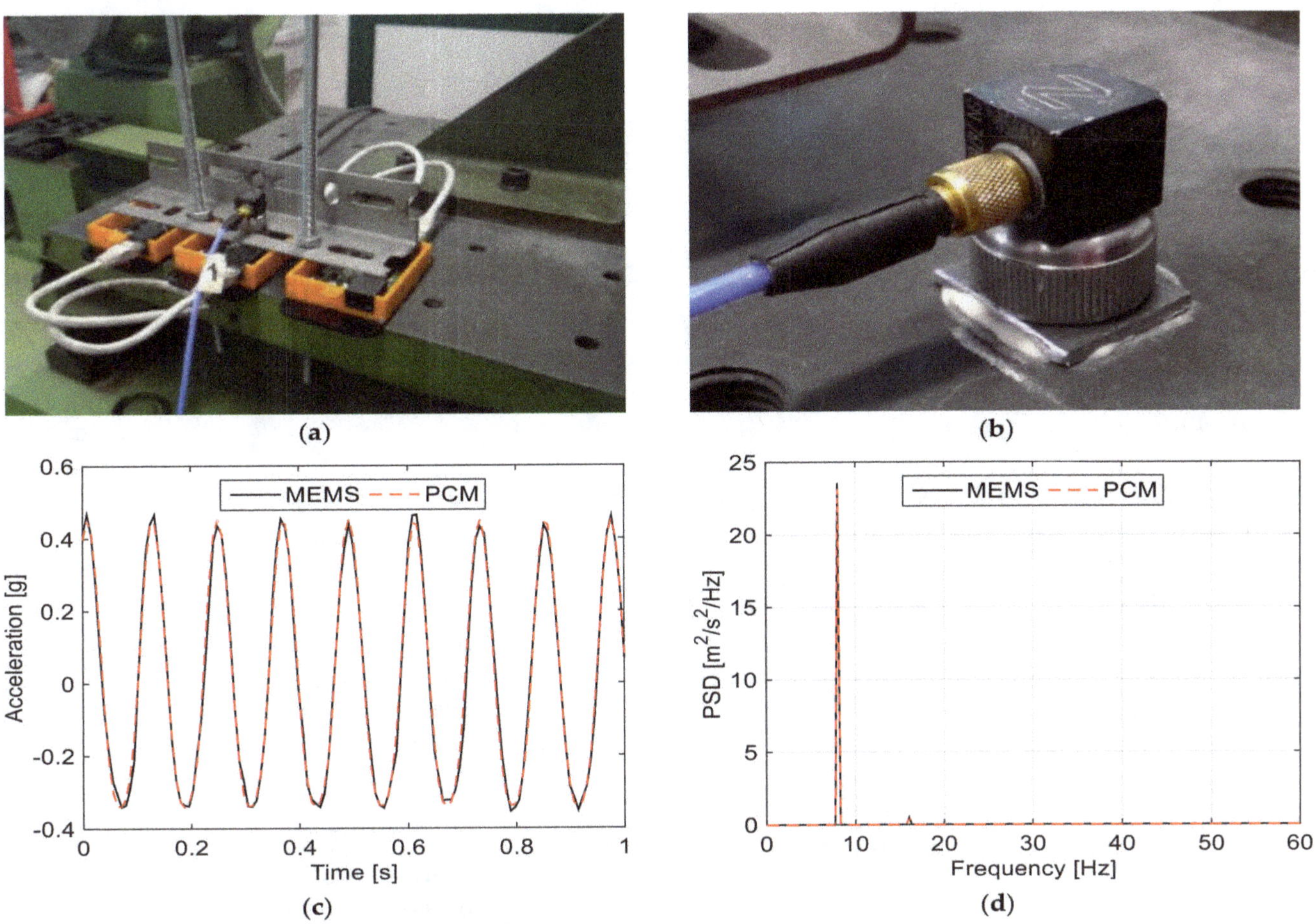

Figure 2. Laboratory shaking table tests: (**a**) MEMS and (**b**) comparative PCM sensors; (**c**,**d**) examples of test measurements in terms of acceleration-time plots and pseudo-spectral density (PSD).

The experimental tests investigated the response of three randomly-selected PC boards (see Table 2), at the frequency of 5 Hz, 8 Hz, and 11 Hz. Such an interval of tested frequencies was taken into account to assess the reliability of MEMS measurements, in a sufficiently wide frequency range of interest for the SHM and dynamic identification of buildings and civil engineering infrastructures.

For each one of these frequencies, a two-minute recording was carried out at a sampling rate of 128 Hz. The shaking table was activated at the desired frequency and a recording window of 45 s was selected for comparative purposes, 60 s after the activation of the shaking to avoid the occurrence of transient starting frequencies. To optimise the experimental output and to assess the response of the accelerometers, three test setups were considered (i.e., one for each axis), resulting in 27 measurements (three axes multiplied by three sensors, multiplied by three frequencies).

Comparative test calculations and a correlation assessment between the recorded signals from all sensors were carried out using the Pearson's correlation coefficients $\rho_{X,Y}$ and the ratio between the root-mean-squares $rms_{X,Y}$ of the prototypes (X) and the reference (Y) sensors, respectively. The $\rho_{X,Y}$ coefficient, defined in Equation (3), measures the correlation between two variables X and Y, giving a value in the range from 1 to -1 and allowing to quantify the linearity and phase distortion of the tested sensors. The $rms_{X,Y}$ (Equation (4)), conversely, is a statistical measure for the magnitude of a varying quantity and was used to quantify the difference in the amplitude response. In both the cases, a $\rho_{X,Y}$ or $rms_{X,Y}$ value equal to 1 means that two signals are identical (i.e., perfect match), while -1 denotes two opposite signals.

$$\rho_{X,Y} = \frac{\sum_i (x_i - \overline{x})(y_i - \overline{y})}{\sqrt{\sum_i (x_i - \overline{x})^2 \sum_k (x_k - \overline{y})^2}} = \frac{COV(X,Y)}{\sigma_x \sigma_y} \tag{3}$$

$$rms_{X,Y} = \sqrt{\frac{\frac{1}{n}\sum_i x_i^2}{\frac{1}{n}\sum_i y_i^2}} = \frac{rms(X)}{rms(Y)} \tag{4}$$

Table 2 lists the statistical coefficients derived from the measurements. A rather close correlation is observed for the Z-axis, while a major scatter is progressively perceived for the X- and Y-axes as far as the reference frequency f is increased. Such an effect could be partly justified by different internal production processes for the X- and Y-axes; however, most probably, it is due to alignment issues during the setup of the tests. Nevertheless, Table 2 suggests a rather good stability of the tested instruments for all the recorded frequencies, and a sufficient reliability of the test measurements under a repeated input.

Table 2. Statistical coefficients derived from laboratory test measurements (Equations (3) and (4)).

Reference Axis	f (Hz)	$\rho_{X,Y}$			$rms_{X,Y}$		
		S#1	S#2	S#3	S#1	S#2	S#3
Z	5	0.9960	0.9950	0.9951	0.9942	0.9983	0.9916
	8	0.9980	0.9971	0.9976	0.9938	0.9971	0.9920
	11	0.9989	0.9971	0.9987	0.9930	0.9960	0.9906
X	5	0.9797	0.9795	0.9799	0.9652	0.9737	0.9789
	8	0.9607	0.9609	0.9608	0.9615	0.9701	0.9755
	11	0.9515	0.9510	0.9519	0.9631	0.9714	0.9714
Y	5	0.9809	0.9809	0.9806	0.9955	0.9959	0.9974
	8	0.9628	0.9621	0.9625	0.9918	0.9924	0.9937
	11	0.9535	0.9533	0.9531	0.9888	0.9892	0.9905

3.2. Noise Level Assessment

The noise level of the prototyped devices was also preliminary assessed, due to its effects on the quality of measurements (see for example [39,40]). To this aim, all the MEMS sensors were installed on a rigid foundation block and additional records were collected (steady-state regime, Z-axis component only), at the sampling frequencies of 256 Hz, 128 Hz, and 64 Hz. The actual noise level was, hence, evaluated by filtering the noise density with a first-order low-pass 20 Hz filter.

Compared to the theoretical noise value expected from the sensors (0.252 mg, see Table 1 and Equation (1)), the experimentally-derived noise level was generally found to lie in the order of 0.317 mg (+25% the nominal value), suggesting a rather stable performance for the full set of prototyped sensors. Additional calculations were carried out by taking into account further MEMS sensors available in the literature, and in particular the wireless, three-axis MEMS devices designed by the University of Illinois, Urbana-Champaign, see [45] and Figure 3a. Such a solution was used for the SHM of the historical Basilica Santa Maria of Collemaggio in L'Aquila, Italy, after the 2009 seismic event [46–48]. The typical device—with a sampling range of DC-1500 Hz—consists of ISM400 sensor boards [49], and accelerometers (LIS344ALH type) produced by ST Microelectronics (Geneva, Switzerland). In [46–48], positive feedback was reported for the adopted wireless MEMS sensors, based on preliminary laboratory tests. At the same time, after one-year on-site data acquisition, the limited performance of LIS344ALH accelerometers was also highlighted, being responsible of major troubles for the identification of the dynamic parameters for the basilica object of study. In Figure 3b, the herein collected comparative results are proposed, in the form of the noise level as a function of the percentage of tested sensors. As shown, the prototyped devices generally proved to offer a more stable performance even compared to the ISM400 sensor board solution, with a significantly lower noise density, hence giving evidence of the potential of the proposed MEMS.

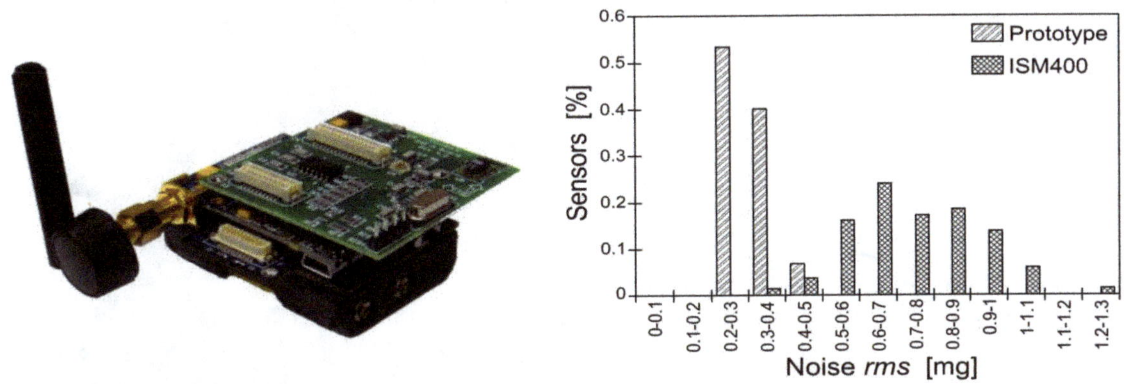

(a) (b)

Figure 3. Noise level assessment: (**a**) detail for the wireless MEMS designed by the University of Illionis, Urbana-Champaign, and (**b**) noise level comparisons (Z-axis), as obtained from laboratory testing and the literature [45,49].

4. Dynamic Identification of the Pietratagliata Cable-Stayed Bridge

4.1. The Case-Study Bridge

On-site experimental tests were then carried out and compared with earlier research efforts available in the literature, to validate the reliability of the assembled measuring devices when in use for SHM of existing structural systems. To this aim, the dynamic identification of the Pietratagliata Bridge (Italy) was taken into account, in accordance with [44].

The bridge consists of a steel-concrete composite deck simply supported at the ends, a system of double-plane cables supporting the deck, and an inclined steel tower (Figure 4). The total length of the deck is 67 m, while the bridge width is 11.1 m including two lanes and two lateral footways. The deck structure consists of "Predalles" concrete panels and a reinforced concrete (RC) slab supported by two lateral steel girders and a longitudinal central beam. The lateral longitudinal and transverse girders have double-T cross-section with height equal to 1.27 m and 1.2 m, respectively; the central longitudinal girder, conversely, is an H-shaped profile, with 0.5 m its height (HEB500 cross-section type, according to European standard, wide flange H steel beam specifications). The interaction between the RC slab and the upper flange of the longitudinal girders consists of welded steel stud connectors. The bridge deck is supported on a RC pier on the National Route (NR) n.13 side and on a cast-in-place RC foundation block on the Pietratagliata side (Figure 4a,b). On the NR n.13 side, two unidirectional bearing supports are used to sustain the lateral girders. On the Pietratagliata side, conversely, the lateral girders are restrained by means of spherical hinges. Three groups of forestays on the upstream and downstream side of the bridge provide additional support to the deck. Each group of cables consists of four Dywidag bars, which are connected to the main girders by means of special metal devices (see the detail of Figure 4c). Furthermore, the backstays connect the steel tower to a RC foundation block. The tower consists of two inclined columns having a thin-walled circular cross-section (1.1 m in diameter and 20 mm thickness). The connection between the inclined columns is given by two additional thin-walled tubes, 0.5 m in diameter (thickness 15 mm). Special steel restraints are located at the base of the steel tower, to reproduce the effect of spherical hinges.

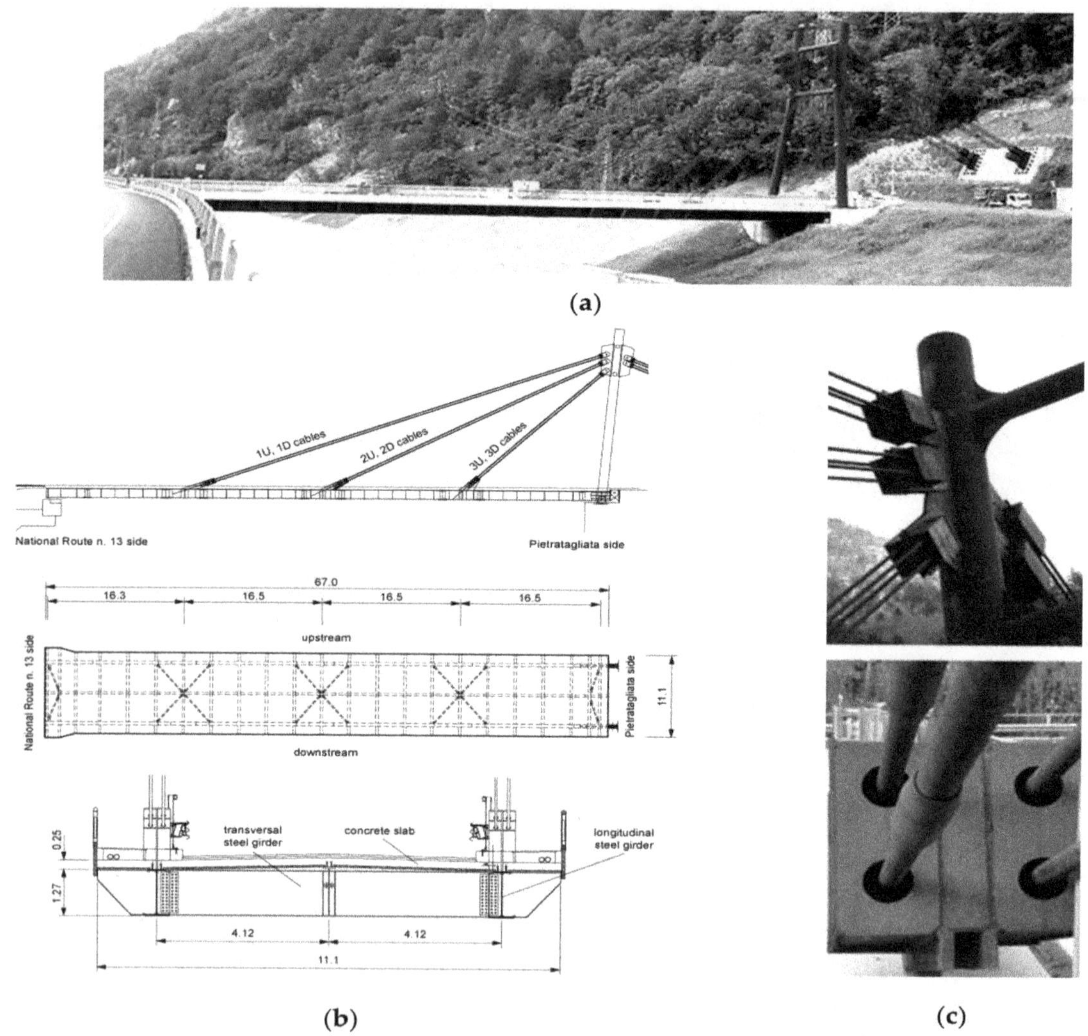

Figure 4. Pietratagliata Bridge (Italy): (**a**) general view; (**b**) technical drawings with lateral view, plan, and transversal cross-section; (**c**) stays-to-tower and stays-to-deck connection details (dimensions in meters).

4.2. On-Site Experimental Testing and Dynamic Identification

4.2.1. Summary of Past EMA and FEA Predictions

For comparative purposes, past EMA and FEA dynamic estimations reported in [44] were taken into account for the examined bridge. There, on-site vibration test measurements have been presented to assess the dynamic parameters of the cable-stayed bridge under investigation, including an advanced FEA analysis aimed at further exploring the experimental observations and at assessing the effects of some key input parameters on the overall performance of the bridge (i.e., boundaries, structural detailing, pre-stressing force in the stays, etc.).

More in detail, in terms of EMA measurements (herein referred as "TEST0"), an ambient vibration dynamic test has been carried out with the aim of identifying the low vibration modes of the bridge (see [44]). At the time of past experiments, no additional excitation due to traffic was accounted for, due to strict requirement of the Pietratagliata Municipal Authority. The instrumentation chain consisted of a 16-channel data acquisition system, connected to a remote personal computer, and 11 Sprengnether mono-axial servo-accelerometers sensors, operating in the frequency range of 0–25 Hz. Each sensor was provided with a pre-amplifier having variable gain controlled by the remote computer. The instruments were located at 20 selected points (16 on the deck and four on the tower), to capture the deformed shapes of both the deck and the tower. Regarding the numerical simulations, the here referred FEA model was implemented by means of the ABAQUS/Standard computer package [50], see Figure 5 and [44]. The geometrical description of the bridge components (deck, pylon, cables, and pier, see the A-to-E key details in Figure 5a), and the definition of their reciprocal mechanical interaction was, hence, carried out based on technical drawings and preliminary sensitivity studies. To this aim, additional FEA models representative of structural details were presented for a further assessment of boundary conditions effects on the dynamic parameters of the bridge. Refined calibration of major input features was, hence, carried out, by including fine-tuning towards available on-site measurements (see [44]).

(a)

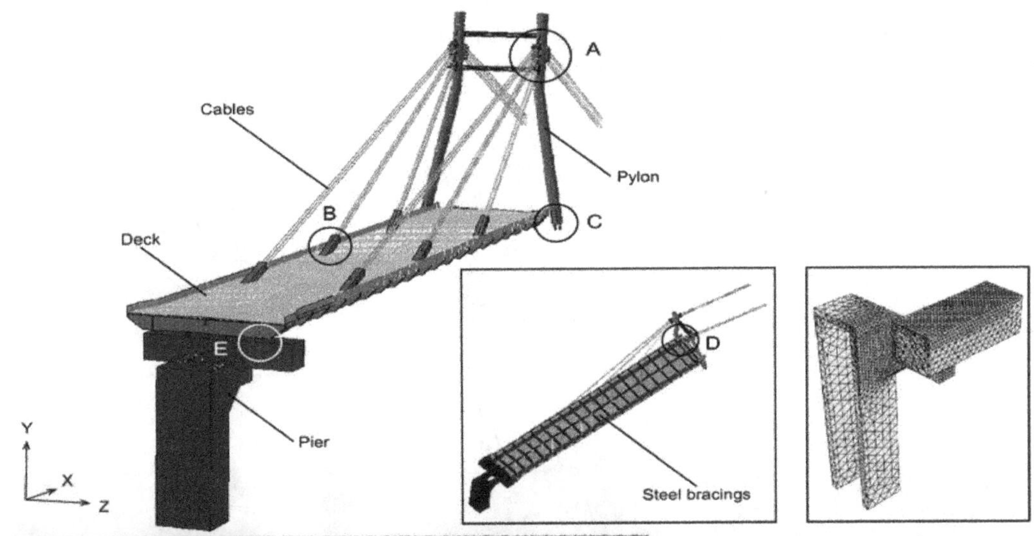

(b)

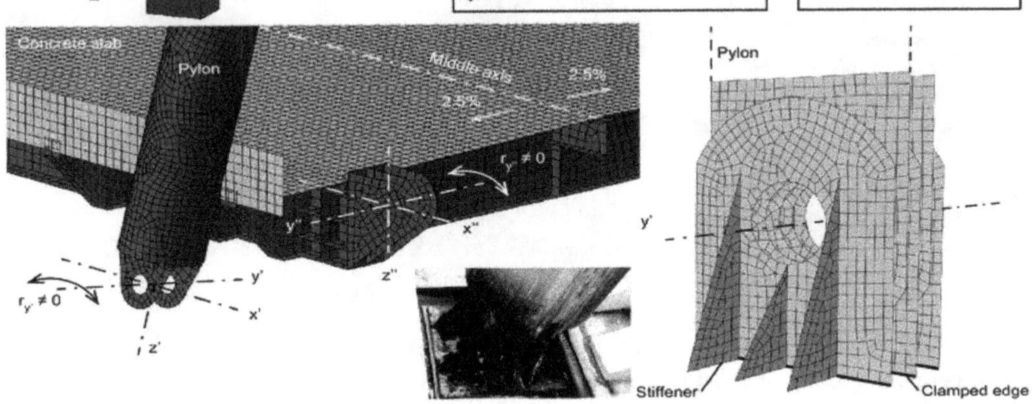

Figure 5. Refined FEA model for the dynamic identification of the Pietratagliata Bridge (ABAQUS), (**a**) global assembly and bottom/pier detailing, in accordance with [44], and (**b**) selected details.

Additional local EMA measurements for the natural frequencies of the stay cables were also reported in [44] from ambient vibration tests, and used to identify the axial force on the supporting cables. Based on combined parametric FEA simulations, it was shown that the vibration frequencies of the bridge are not particularly sensitive to these structural modifications, with an average reduction up to 0.5–1% the fundamental frequencies of the reference, undamaged configuration. A maximum scatter up to −5% was estimated for some torsional shapes only, when damage was imposed in the stays with the closest connection to the tower (i.e., with a key role for restraining the bridge deck for the modal shapes of interest). On the contrary, possible variation in the axial force amount, and/or damage in the cables-to-deck restraints was found to induce even important changes in the shape of the lower vibration modes (i.e., loss of symmetry of restraints for the deck and, hence, of the corresponding deformations, with respect to the longitudinal axis of the bridge), suggesting a potential use of such a kind of information for diagnostic purposes.

4.2.2. MEMS Experiments: Test Methods and Setup

The experimental investigation was carried out using ten sensors, aiming at acquiring and monitoring the slab deformations under the imposed input vibrations. In accordance with [44], three-component deformations of the deck were separately recorded for each control point, in accordance with the test setup reported in Figure 6a. Given the limited number of available instruments, the final setup of measuring devices was optimised based on preliminary investigations and past experimental findings summarized in Section 4.2.1, to capture the modal deformations of the deck. In this regard, the dynamic contribution of the pylon was not accounted for through the on-site investigation. Compared to [44], ambient vibration testing of the bridge was carried out under ordinary traffic loading.

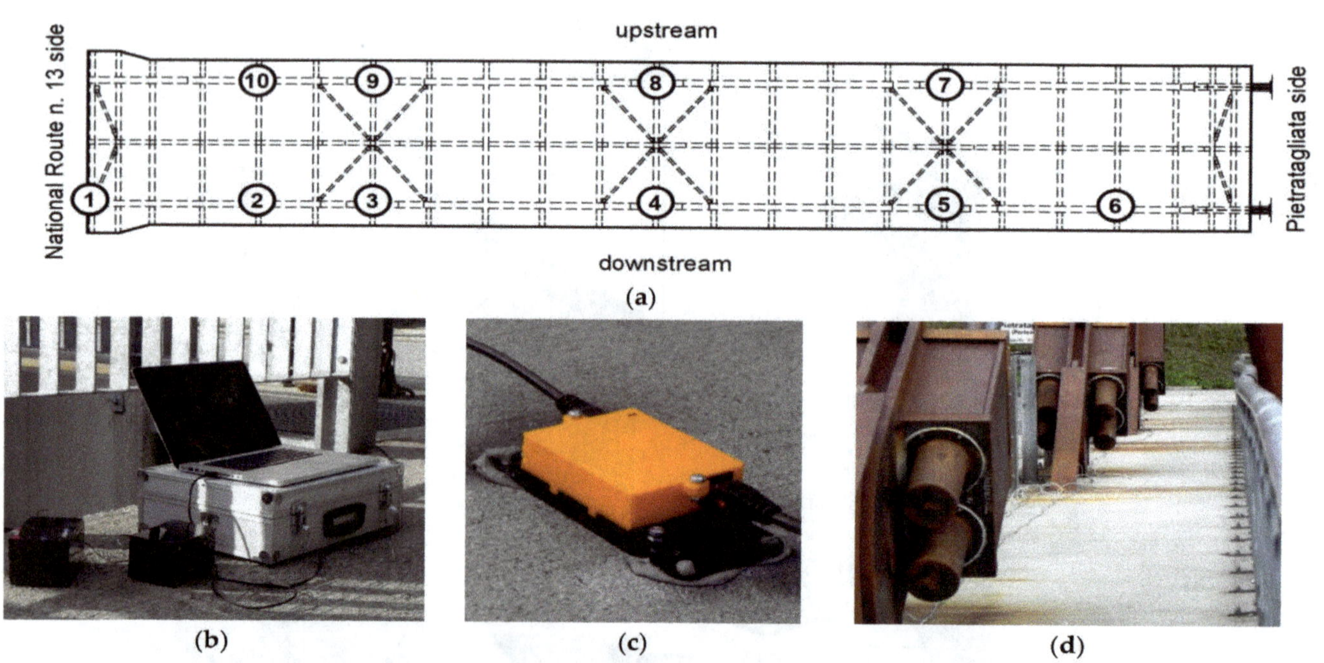

Figure 6. Experimental testing on the Pietratagliata bridge. (**a**) test setup (top view of the steel-concrete composite slab (dimensions in meters); in the circles, the instruments labels) and (**b–d**) instrumentation details.

4.2.3. Vibration Modes and Modal Correlation

Based on the available MEMS sensors and the collected measurements, the dynamic parameters of the bridge were estimated by means of the Structural Modal Identification Toolsuite software

(SMIT [51]). The ERA-OKID-OO approach [52,53], being representative of the extension of the simple ERA technique to vibrating systems whose initial conditions and dynamic external excitation are unknown, was used for natural frequencies, damping ratios, vibration shapes (see Figures 7 and 8, and Table 3). In general, the ERA-OKIDO-OO technique offers more stable identification results, compared to other approaches (see [52,53]).

In this regard, Figure 7a shows the typical test measurements for the examined bridge under ambient vibration, while Figure 7b gives evidence of six vibration modes—i.e., PSD peaks—emerging from the noise level.

Due to the test setup configuration and input vibrations, the post-processing of the collected experimental data proved to allow a clear detection of the first six modes of the bridge, especially the flexural ones (i.e., major peaks in Figure 7b, where the EMA modes #1, #3, and #5 are emphasized), but also giving evidence of the fundamental torsional modes for the deck (EMA #2, #4, and #6 in Figure 7b).

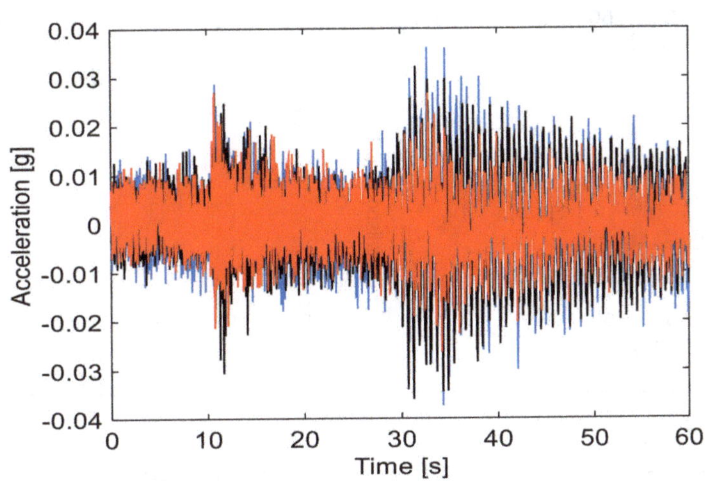

(a)

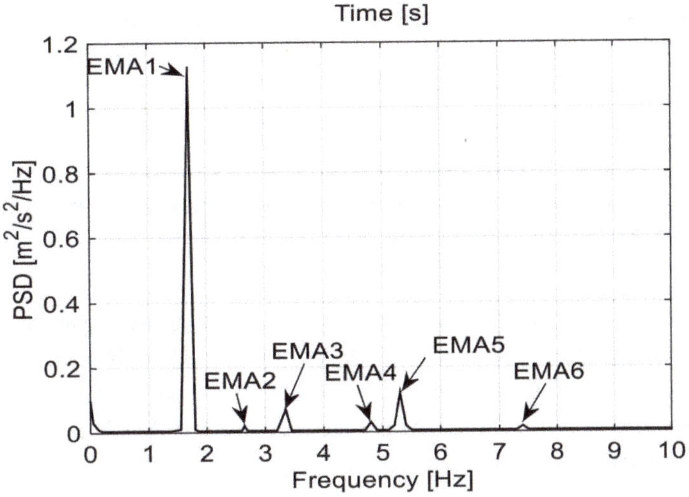

(b)

Figure 7. Dynamic identification of the Pietratagliata bridge via MEMS sensors: (**a**) example of test measurements (three sensors only are shown) and (**b**) pseudo spectral density with evidence of fundamental modes.

The experimentally-predicted vibration shapes are, in fact, reported in Figure 8 (lateral view of the bridge deck), while the corresponding vibration frequencies and damping ratios are listed in Tables 3 and 4.

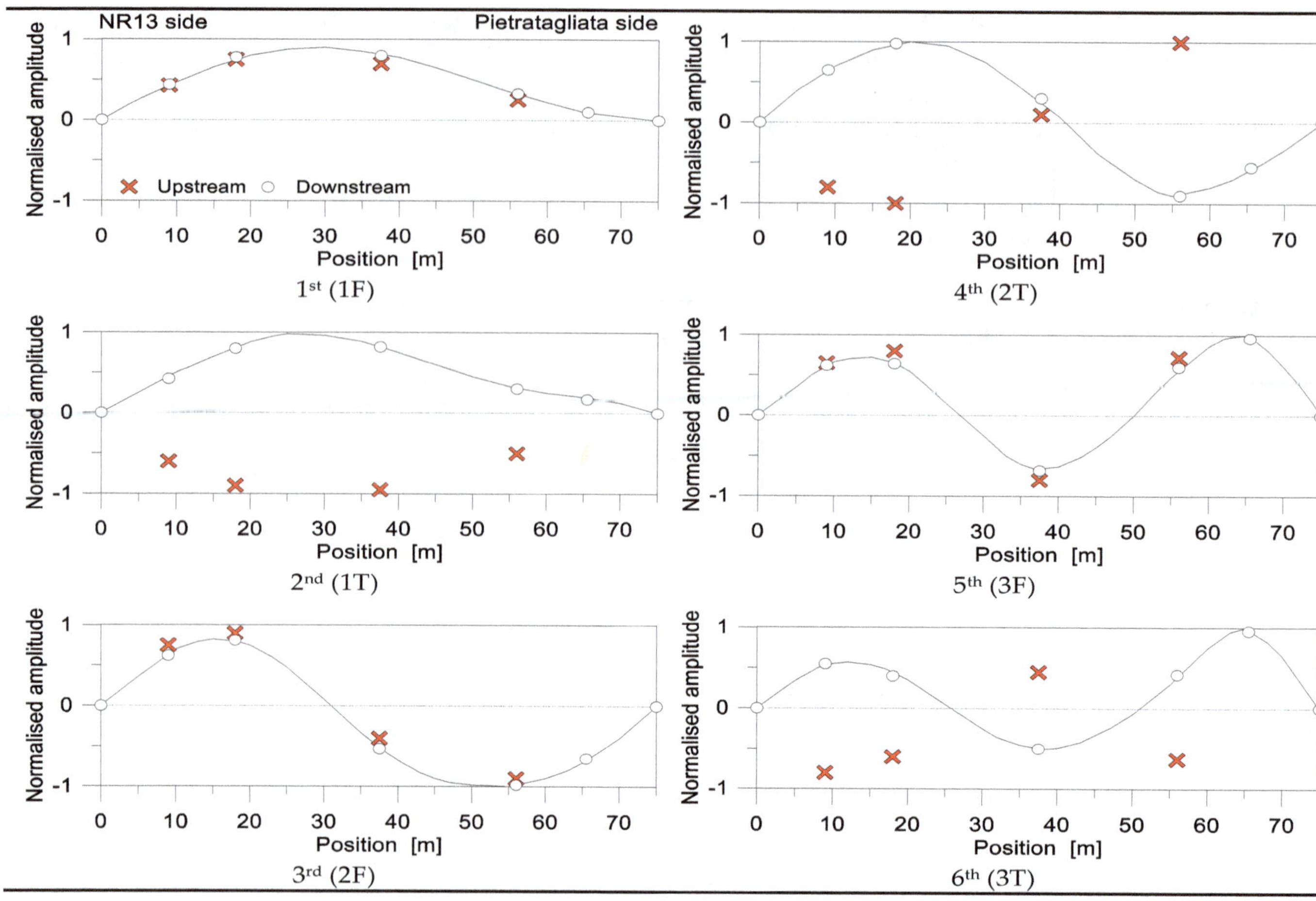

Figure 8. EMA vibration modes (normalised amplitudes), as obtained from MEMS measurements and SMIT post-processing [51].

Compared to past literature results, a close qualitative correlation was found for the detected modes. The fundamental mode of the bridge, see Figure 8, was found to be a first order flexural mode (1F), followed by the first torsional mode (1T) and higher flexural/torsional vibration shapes (2F, 2T, 3F, and 3T in Figure 8).

In Table 3, the detected vibration modes were compared to past experimental frequencies and damping ratios. Generally, a rather close correlation was observed in terms of vibration frequencies, with average scatter in the order of $\approx$0.6%, hence suggesting the potential of the proposed solution. The exception is represented by the second flexural mode, where the MEMS estimations underestimate the past experiment up to 2% of the reference value.

In terms of damping ratios for the same detected modes, the MEMS measurements led to a higher uncertainty with respect to the past EMA predictions, see Table 3. In general, however, the predicted damping ratios were found to lie in the range of 0.5–1% and to suggest a certain reliability of MEMS measurements, given the actual sensitivity of damping estimations to several parameters ([44,54–58], etc.).

Table 3. EMA vibration frequencies and damping ratios for the first six fundamental modes, as obtained from MEMS measurements and past experiments (TEST0, see [44]). Key: F = flexural; T = torsional; $\Delta = 100 \times (f_{MEMS} - f_{TEST0})/f_{TEST0}$.

Vibration Mode		***f* (Hz)**		**Δ (%)**	**ξ (%)**	
n°	**Order/Type**	**MEMS**	**TEST0**		**MEMS**	**TEST0**
1	1/F	1.678	1.665	0.78	0.28	1.2 ± 0.5
2	1/T	2.659	2.669	−0.37	1.91	0.6 ± 0.5
3	2/F	3.340	3.411	−2.08	0.29	0.7 ± 0.2
4	2/T	4.777	4.750	0.57	0.47	0.4 ± 0.0
5	3 / F	5.307	5.261	0.87	0.39	0.7 ± 0.2
6	3 / T	7.353	7.336	0.23	0.78	0.9 ± 0.2

Careful consideration, based on the available test measurements, was indeed spent for the correlation of the flexural and torsional vibration shapes of the bridge with past literature measurements. Given the limited number of control points, modal correlation was carried out by considering the FEA vibration shapes reported in [44], where the accuracy of such an advanced numerical model was emphasised.

For dynamic identification purposes, the MAC (modal assurance criterion) coefficients were calculated for the MEMS experimental data to the past FEA predictions (in Table 4, the graphical representation of the so calculated MAC values is proposed as a function of the *i*-th mode number). Given the *i*-th vibration shape, in particular, the MAC value is conventionally determined as:

$$MAC_i = \frac{\left[\sum_{j=1}^{n} \phi_{ij}\phi_{ij}^{*}\right]^2}{\sum_{j=1}^{n} \phi_{ij}^2 \sum_{j=1}^{n} \phi_{ij}^{*2}} \quad (5)$$

where ϕ_{ij} and ϕ^*_{ij} are the vibration modal shapes, n the grid point numbers.

According to Equation (5), the MAC values vary from 0 to 1, meaning that there is no similarity between the compared modes, or that the examined modal shapes are consistent.

Table 4. EMA (MEMS) vibration frequencies, and correlation with past FEA results [44]. Key: F = flexural; T = torsional; $\Delta = 100 \times (f_{MEMS} - f_{FEA})/f_{FEA}$.

MEMS			**EMA (MEMS)-to-FEA Modal Correlation**		
Vibration Mode			**MAC (Equation (5))**	***f*$_{FEA}$**	**Δ**
n°	**Order/Type**	***f*$_{MEMS}$ (Hz)**	**(%)**	**(Hz)**	**(%)**
1	1/F	1.678	99.7	1.619	3.64
2	1/T	2.659	99.1	2.691	−1.19
3	2/F	3.340	96.8	3.238	3.15
4	2/T	4.777	84.4	4.718	1.25
5	3/F	5.307	76.1	5.296	0.21
6	3/T	7.353	82.9	7.372	−0.26

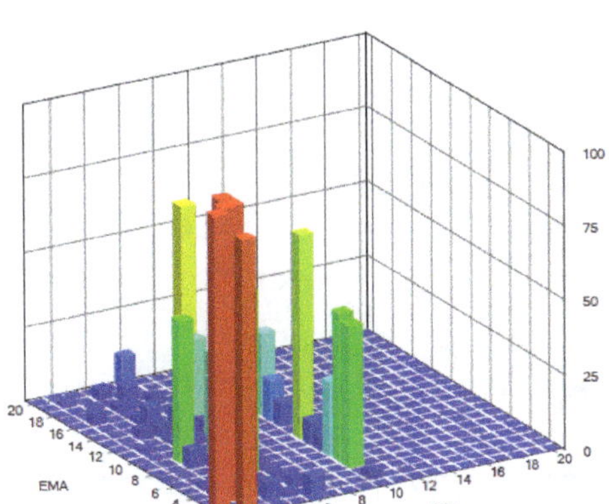

As shown in Table 4, a rather close correlation was generally observed for the experimentally-detected vibration modes, for both flexural and torsional shape types, and especially for the lowest ones. MAC values proved the reliability of test measurements, even with major scatter for higher and complex vibration shapes, with MAC > 96.8 for the first fundamental modes. Given the actual goals and limitations of MAC estimations (see for example [59,60]), the collected results can be considered as well-representative of the potential of MEMS sensors.

Frequency results were also found to have close correlation with FEA calculations, being experimentally estimated with mostly a limited scatter (in the order of 1–3%) and with major discrepancies (3.5%) in the case of the first and third modes only. The scatter for these vibration modes (corresponding to the 1F and 2F flexural shapes) could be affected by local effects of the pier, since resulting in a flexible end support for the FEA deck. Based on the limited number of control points, however, a reasonable accuracy of the prototyped instrumentation can, again, be deducted, even if additional testing and assessment are required.

The good qualitative correlation between experimental and FEA modal shapes is further emphasised in Table 5, in the form of 3D axonometric views for the detected vibration modes, as obtained from further post-processing of modal shape amplitudes [61]. There, in particular, the normalized modal displacements (vertical component only) are assigned at each grid control point, hence, the input takes the form of a table with nodal coordinates and normalized deformations (EMA and FEA estimations, corresponding to red and blue deformed shapes of Table 5).

Table 5. EMA (MEMS)-to-FEA modal correlation and vibration shapes (3D axonometric view).

Mode #	EMA (MEMS, in Red)-to-FEA (Blue) Modal Correlation	FEA Modal Shape
1st (1F)	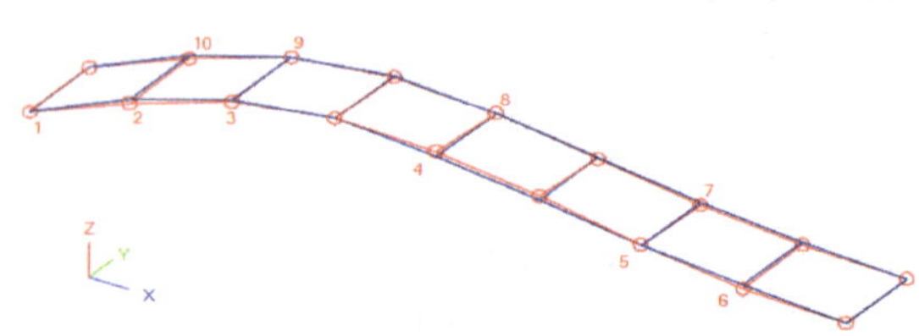	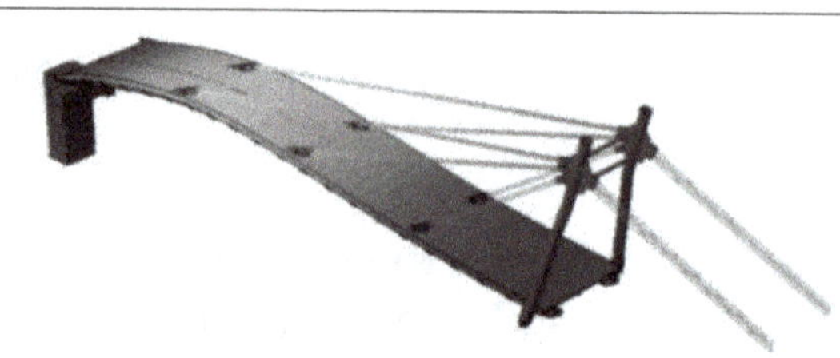
2nd (1T)	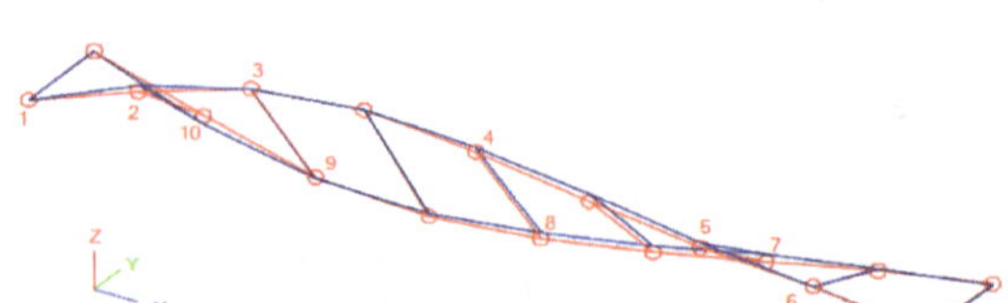	
3rd (2F)	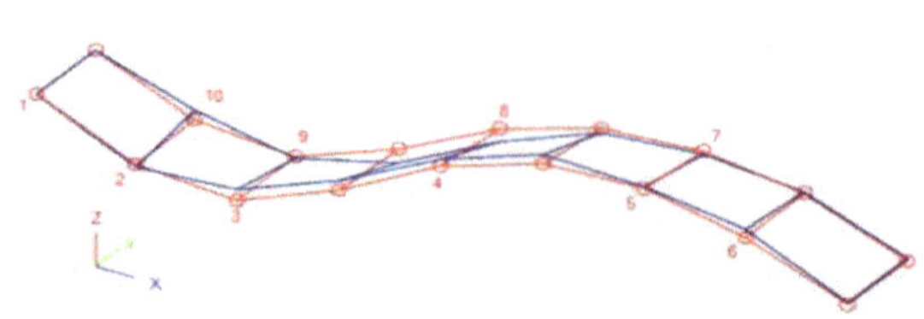	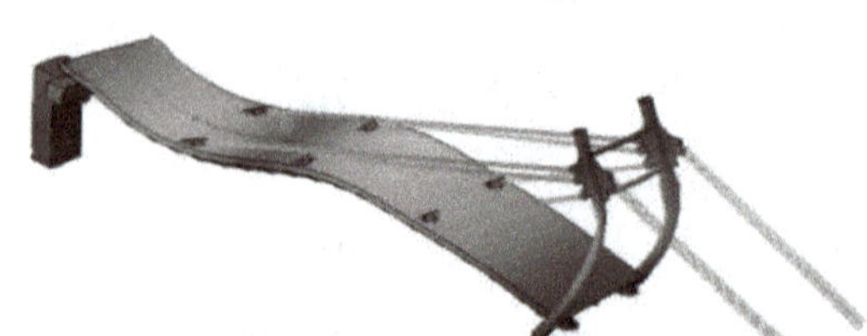
4th (2T)	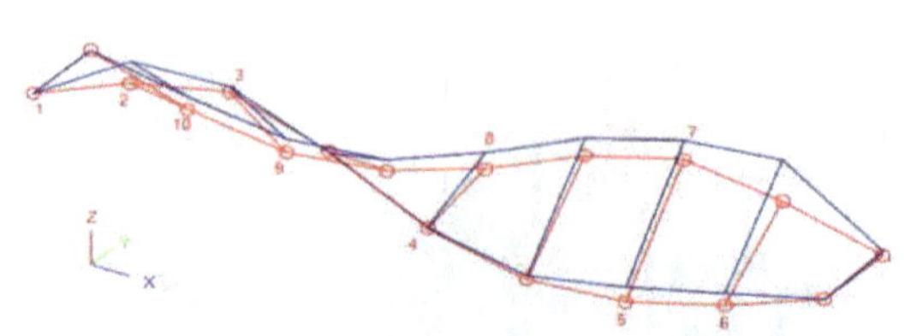	

Table 5. *Cont.*

Mode #	EMA (MEMS, in Red)-to-FEA (Blue) Modal Correlation	FEA Modal Shape
5th (3F)	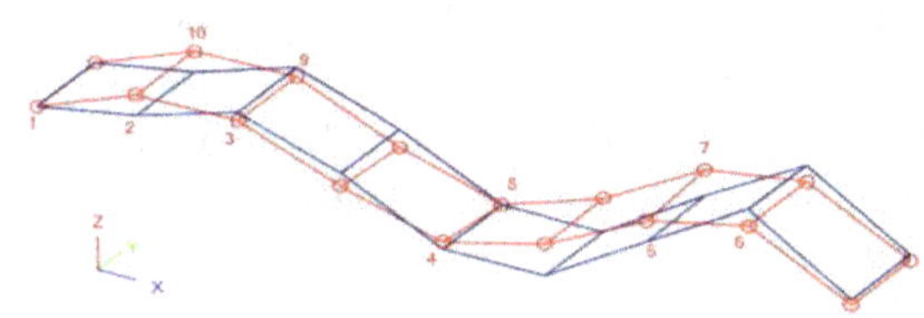	
6th (3T)	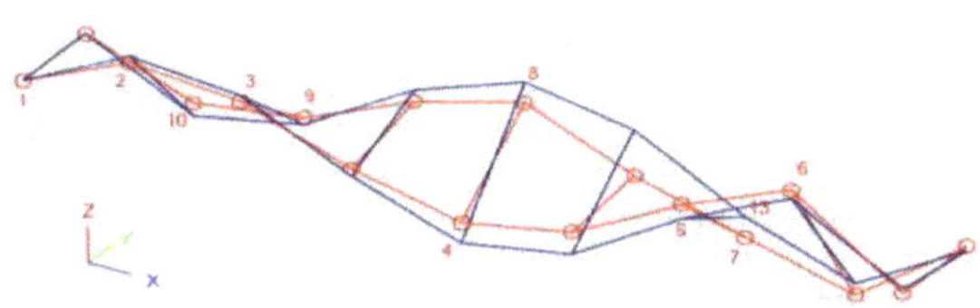	

5. Conclusions

In this paper, original self-made Micro Electro-Mechanical System-based (MEMS) accelerometers have been prototyped and validated via laboratory and on-site experimental tests. To this aim, laboratory experimental comparisons have been first reported, so to assess the expected accuracy of MEMS-based measurements towards traditional accelerometers commercially available on the market, including noise level assessment. Based on the observed close correlation between the tested instruments, a full-scale application has been then reported. As a case study, the cable-stayed bridge in Pietratagliata (Italy) has been taken into account. The capability and potential of MEMS accelerometers has been assessed on the basis of Experimental Modal Analysis (EMA) testing and Finite Element Analytical (EMA) estimations derived from past literature efforts. As shown, the prototyped MEMS accelerometers proved to offer reliable estimations for the dynamic features of the bridge, hence confirming their potential use for structural monitoring in the form of low-cost, but practical, instruments. In this regard, further investigations are also expected to verify the reliability of MEMS estimations for different structural typologies.

Author Contributions: All the authors wrote and revised the paper collaboratively. The presented self-made MEMS accelerometers are the extended result of E.B. Master Thesis project. For the paper drafting, C.B. and M.I. conceived of and pursued the literature survey on MEMS-based sensors. E.B. managed the laboratory and on-site MEMS measurements. C.B. carried out the past FEA studies, while S.N. supervised the overall project.

Acknowledgments: The research results discussed in this paper have been orally presented at the GNGTS National Conference (November 2017, Trieste, Italy). In this regard, Dr. Luigi Bregant (Assistant Professor in Applied Mechanics for Machinery at University of Trieste, Department of Engineering and Architecture, Italy), is acknowledged for his technical support during the laboratory tests.

References

1. EN 1991. Eurocode 1: Actions on Structures. European Standard. European Committee for Standardization (CEN): Brussels, Belgium, 2010. Available online: http://www.phd.eng.br/wp-content/uploads/2015/12/en.1991.1.4.2005.pdf (accessed on 27 July 2018).
2. Nakano, Y.; Maeda, M.; Kuramoto, H.; Murakami, M. Guideline for post-earthquake damage evaluation and rehabilitation of RC buildings in Japan. In Proceedings of the 13th WCEE World Conference on Earthquake Engineering, Vancouver, BC, Canada, 1–6 August 2004.

3. Xue, Q.; Chen, C.C.; Chen, K.C. Damage loss assessment for the basic earthquake insurance claim of residential RC buildings in Taiwan. *J. Build. Apprais.* **2011**, *6*, 213–226. [CrossRef]
4. Pozzi, M.; Zonta, D.; Trapani, D.; Athanasopoulos, N.; Amditis, A.J.; Bimpas, M.; Garetsos, A.; Stratakos, Y.E.; Ulieru, D. MEMS-based sensors for post-earthquake damage assessment. *J. Phys. Conf. Ser.* **2011**, *305*, 012100. [CrossRef]
5. Menderes, A.; Erener, A.; Sarp, G. Automatic Detection of Damaged Buildings after Earthquake Hazard by Using Remote Sensing and Information Technologies. *Procedia Earth Planet. Sci.* **2015**, *15*, 257–262. [CrossRef]
6. Ubertini, F.; Comanducci, G.; Cavalagli, N. Vibration-based structural health monitoring of a historic bell-tower using output-only measurements and multivariate statistical analysis. *Struct. Health Environ* **2016**, *15*, 438–457. [CrossRef]
7. Di Tommaso, A.; Gentilini, C.; Castellazzi, G. Structural Interpretation of Data from Static and Dynamic Structural Health Monitoring of Monumental Buildings. *Key Eng. Mater.* **2017**, *747*, 431–439.
8. Ramos, L.; Marques, L.; Lourenço, P.; De Roeck, G.; Campos-Costa, A.; Roque, J. Monitoring historical masonry structures with operational modal analysis: Two case studies. *Mech. Syst. Sig. Process.* **2010**, *24*, 1291–1305. [CrossRef]
9. Jouan, B.; Rudolph, J.; Bergholz, S. Structural Health Monitoring Solutions for Power Plants. In Proceedings of the EWSHM—7th European Workshop on Structural Health Monitoring, Nantes, France, 8–11 July 2014; pp. 623–630.
10. Lemu, H.G. Assessment and analysis of structural health monitoring techniques for rotating machines. In Proceedings of the 27th International Ocean and Polar Engineering Conference, San Francisco, CA, USA, 25–30 June 2017.
11. Schubert, F.; Frankenstein, B.; Fröhlich, K.J.; Kuttner, M.; Lamek, B.; Schwenkkros, J.; Kerkhof, K.; Petricevic, R. Structural Health Monitoring of Industrial Piping Systems Based on Guided Elastic Waves. *DGZfP* **2007**, *32*, 14–16. Available online: https://pdfs.semanticscholar.org/05e1/ed16583d17c76793c3a65bec592b91ffad15.pdf (accessed on 26 July 2018).
12. Giglio, M.; Manes, A.; Sbarufatti, C. MEMS for structural health monitoring in aircraft. In *MEMS for Automotive and Aerospace Applications*; Woodhead Publishing: Sawston, UK; Cambridge, UK, 2013; pp. 220–244. ISBN 978-0-85709-118-5.
13. Giurgiutiu, V. Structural Health Monitoring (SHM) of aerospace components. In *Polymer Composites in the Aerospace Industry*; Woodhead Publishing: Sawston, UK; Cambridge, UK, 2015; pp. 449–507, ISBN 978-0-85709-523-7.
14. Mecocci, A.; Peruzzi, G.; Pozzebon, A.; Vaccarella, P. Architecture of a hydroelectrically powered wireless sensor node for underground environmental monitoring. *IET Wirel. Sens. Syst.* **2017**, *7*, 123–129. [CrossRef]
15. Bennet, P.; Soga, K.; Wassel, I.; Fidler, P.; Abe, K.; Kobayashi, Y.; Vanicek, M. Wireless sensor network for underground railway applications: Case studies in Prague and London. *Smart Struct. Syst.* **2010**, *6*, 619–639. [CrossRef]
16. Cochran, E.S.; Lawrence, J.F.; Kaiser, A.; Fry, B.; Chung, A.; Christensen, C. Comparison between low-cost and traditional MEMS accelerometers: A case study from the M7.1 Darfield, New Zealand, aftershock deployment. *Ann. Geophys.* **2011**, *54*, 728–737.
17. Evans, J.R.; Allen, R.M.; Chung, A.I.; Cochran, E.S.; Guy, R.; Hellweg, M.; Lawrence, J.F. Performance of Several Low-Cost Accelerometers. *Seismol. Res. Lett.* **2014**, *85*, 147–158. [CrossRef]
18. Cigada, A.; Lurati, M.; Redaelli, M.; Vanali, M. Mechanical performance and metrological characterization of MEMS accelerometers and application in modal analysis. In Proceedings of the IMAC XXV International Modal Analysis Conference, Orlando, FL, USA, 19–22 February 2007; pp. 236–244.
19. Sun, Z.; Chen, D.; Chen, J.; Deng, T.; Li, G.; Xu, C.; Wang, J. A MEMS based electrochemical seismometer with low cost and wide working bandwidth. *Procedia Eng.* **2016**, *168*, 806–809. [CrossRef]
20. Benevicius, V.; Ostasevicius, V.; Gaidys, R. Identification of capacitive MEMS accelerometer structure parameters for human body dynamics measurements. *Sensors* **2013**, *13*, 11184–11195. [CrossRef] [PubMed]
21. Ciuti, G.; Ricotti, L.; Menciassi, A.; Dario, P. MEMS Sensor Technologies for Human Centred Applications in Healthcare Physical Activities, Safety and Environmental Sensing: A Review on Research Activities in Italy. *Sensors* **2015**, *153*, 6441–6468. [CrossRef] [PubMed]
22. Chaudhury, S.B.; Sengupta, M.; Mukherjee, K. Vibration Monitoring of Rotating Machines Using MEMS Accelerometer. *Int. J. Sci. Eng. Res.* **2014**, *2*, J2013358.

23. Jimenez, S.; Cole, M.O.T.; Keogh, P.S. Vibration sensing in smart machine rotors using internal MEMS accelerometers. *J. Sound Vib.* **2016**, *377*, 68–75. [CrossRef]
24. Pedotti, L.A.S.; Zago, R.M.; Fruett, F. Instrument based on MEMS accelerometer for vibration and unbalance analysis in rotating machines. In Proceedings of the 1st INSCIT—International Symposium on Instrumentation Systems, Circuits and Transducers, Belo Horizonte, Brazil, 29 August–3 September 2016; pp. 25–30. [CrossRef]
25. Son, J.-D.; Ahn, B.-H.; Ha, J.-M.; Choi, B.-K. An availability of MEMS-based accelerometers and current sensors in machinery fault diagnosis. *Measurement* **2016**, *94*, 680–691. [CrossRef]
26. Bassoli, E.; Vincenzi, L.; Bovo, M.; Mazzotti, C. Dynamic identification of an ancient masonry bell tower using MEMS-based acquisition system. In Proceedings of the 2015 EESMS Workshop on Environmental, Energy and Structural Monitoring Systems, Trento, Italy, 9–10 July 2015; Paper number 15347414. [CrossRef]
27. Clementi, F.; Pierdicca, A.; Milani, G.; Gazzani, V.; Poiani, M.; Lenci, S. Numerical model upgrading of ancient bell towers monitored with a wired sensor network. In Proceedings of the 10th International Masonry Conference (IMC), Milan, Italy, 9–11 July 2018.
28. Feng, M.; Fukuda, Y.; Mizuta, M.; Ozer, E. Citizen Sensors for SHM: Use of Accelerometer Data from Smartphones. *Sensors* **2015**, *15*, 2980–2998. [CrossRef] [PubMed]
29. Wargantiwar, N.K.; Barbade, A.S.; Shingade, A.P.; Shire, A.N. Wireless Earthquake Alarm Design based on MEMS Accelerometer. *Int. Adv. Res. J. Sci. Eng. Technol.* **2017**, *4*, 128–132.
30. Kok, R.; Furlong, C.; Putniewicz, R.J. Development of a Wireless MEMS Inertial System for Health Monitoring of Structures. In *Materials Research Society Symposium Proceedings*; Cambridge University Press: Cambridge, UK, 2003; Volume 785. [CrossRef]
31. Jung, J.W.; Moon, D.J.; Jung, J.W.; Lee, B.L.; Lee, S.J. A performance test of a 3-axis accelerometer and modal analysis. In Proceedings of the FIG Congress 2014—Engaging the Challenges, Enhancing the Relevance, Kuala Lumpur, Malaysia, 16–21 June 2014.
32. Rajashri, P.P.; Chaudhari, V.D.; Rane, K.P. ARM based 3-axis seismic data acquisition system using Accelerometer sensor and Graphical User Interface. *Int. J. Eng. Res. Gen. Sci.* **2015**, *3*, 833–838.
33. Beskhyroun, S.; Ma, Q. Low-Cost Accelerometers for Experimental Modal Analysis. In Proceedings of the 15th WCEE—World Conference in Earthquake Engineering, Lisbon, Portugal, 24–28 September 2012; Available online: http://www.iitk.ac.in/nicee/wcee/article/WCEE2012_0771.pdf (accessed on 26 July 2018).
34. Spencer, B.F.; Ruiz-Sandoval, M.; Kurata, N. Smart sensing technology for structural health monitoring. In Proceedings of the 13th WCEE—World Conference on Earthquake Engineering, Vancouver, BC, Canada, 1–6 August 2004.
35. Torfs, T.; Sterken, T.; Brebels, S.; Santana, J.; van den Hoven, R.; Spiering, V.; Bertsch, N.; Trapani, D.; Zonta, D. Low Power Wireless Sensor Network for Building Monitoring. *IEEE Sensors J.* **2013**, *13*, 909–915. [CrossRef]
36. Renjan Raj, V.C. Wireless Sensor Network for Building Monitoring. *Int. J. Eng. Sci.* **2014**, *3*, 13–18.
37. Pradeepkumar, N.J.; Ramesh, R.M.; Shalini, K.S.; Sujatha, H.R.; Hemanth Kumar, C.S. Smart System Sensor Network for Building Monitoring. *SSRG IJECE* **2015**, *2*, 116–121.
38. Picozzi, M.C.; Milkereit, C.; Zulfikar, C.; Fleming, K.; Ditommaso, R.; Erdik, M.; Zschau, J.; Fischer, J.; Safak, E.; Ozel, O.; et al. Wireless technologies for the monitoring of strategic infrastructures: An ambient vibration test on the Fatih Sultan Mehmet suspension bridge in Istanbul, Turkey. *Bull. Earthq. Eng.* **2009**, *8*, 671–691. [CrossRef]
39. Domaneschi, M.; Limongelli, M.P.; Martinelli, L. Structural damage localization in a suspension bridge under seismic excitation. In Proceedings of the 15th WCEE Conference—World Conference on Earthquake Engineering, Lisbon, Portugal, 24–28 September 2012.
40. Domaneschi, M.; Limongelli, M.P.; Martinelli, L. Interpolation damage detection method on a suspension bridge model: influence of sensors disturbances. *Key Eng. Mater.* **2013**, *569–570*, 734–741. [CrossRef]
41. Dashti, S.; Bray, J.D.; Reilly, J.; Glaser, S.; Bayen, A. iShake: The Reliability of Phones as Seismic Sensors. In Proceedings of the World Conference Earthquake Engineering, Lisbon, Portugal, 24–28 September 2012.
42. D'Alessandro, A.; D'Anna, G. Suitability of Low-Cost Three-Axis MEMS Accelerometers in Strong-Motion Seismology: Tests on the LIS331DLH (iPhone) Accelerometer. *Bull. Seismol. Soc. Am.* **2013**, *103*, 2906–2913. [CrossRef]
43. Kong, Q.; Allen, R.M.; Schreier, L.; Kwon, Y.-W. MyShake: A smartphone seismic network for earthquake early warning and beyond. *Sci. Adv.* **2016**, *2*. [CrossRef] [PubMed]

44. Bedon, C.; Dilena, M.; Morassi, A. Ambient vibration testing and structural identification of a cable-stayed bridge. *Meccanica* **2016**, *51*, 2777–2796. [CrossRef]
45. Rice, J.A.; Spencer, B.F. Flexible Smart Sensor Framework for Autonomous Full-Scale Structural Health Monitoring. NSEL Report Series, Report n. NSEL-018, August 2009. Available online: https://core.ac.uk/download/pdf/4822684.pdf (accessed on 1 July 2018).
46. Federici, F.; Alesii, R.; Colarieti, A.; Faccio, M.; Graziosi, F.; Gattulli, V.; Potenza, F. Design of Wireless Sensor Nodes for Structural Health Monitoring Applications. *Procedia Eng.* **2014**, *87*, 1298–1301. [CrossRef]
47. Gattulli, V.; Graziosi, F.; Federici, F.; Potenza, F.; Colarieti, A. Structural Health Monitoring of the Basilica S. Maria di Collemaggio. In Proceedings of the 5th International Conference on Structural Engineering, Cape Town, South Africa, 2–4 September 2013.
48. Antonacci, E.; Ceci, A.; Colarieti, V.; Gattulli, V.; Graziosi, F.; Lepidi, M.; Potenza, F. Dynamic testing and health monitoring via wireless sensor networks in the post-earthquake assessment of structural condition at L'Aquila. In Proceedings of the Eurodyn 2011—8th European Conference on Structural Dynamics, Leuven, Belgium, 4–6 July 2011; pp. 2440–2447.
49. ISM400—Multimetric Imote2 Sensor Board—Datasheet and User's Guide. Available online: http://shm.cs.uiuc.edu (accessed on 1 July 2018).
50. *ABAQUS Computer Software*; Simulia: Johnston, RI, USA, 2018.
51. SMIT. Available online: http://smit.atlss.lehigh.edu/?page_id=23 (accessed on 1 March 2018).
52. Chang, M.; Leonard, R.L.; Pakzad, S.N. SMIT User's Guide. Release 1.0, 2012. Available online: http://smit.atlss.lehigh.edu/wp-content/uploads/2012/07/SMIT-Users-Guide.pdf (accessed on 26 July 2018).
53. Chang, M.; Pakzad, S.N. Observer Kalman Filter Identification for Output-Only Systems Using Interactive Structural Modal Identification Toolsuite. *J. Bridge Eng.* **2014**, *19*, 04014002. [CrossRef]
54. Frizzarin, M.; Feng, M.Q.; Franchetti, P.; Soyoz, S.; Modena, C. Damage detection based on damping analysis of ambient vibration data. *Struct. Control Health Monit.* **2010**, *17*, 368–385. [CrossRef]
55. Turek, M.E.; Bentura, C.E.; Shawwaf, K. Vibration testing of bridge stay cables to obtain damping values. *Struct. Dyn.* **2011**, *3*, 331–340.
56. Gonzalez, A.; Obrien, E.J.; McGetrick, P.J. Identification of damping in a bridge using a moving instrumented vehicle. *J. Sound Vib.* **2012**, *331*, 4115–4131. [CrossRef]
57. Liu, Y.; Ge, Y.; Cao, F.; Zhou, Y.; Wang, S. Statistics and identification of mode-dependent structural damping of cable-supported bridges. In Proceedings of the APCWE-VIII—The 8th Asia-Pacific Conference on Wind Engineering, Chennai, India, 10–13 December 2013.
58. Bedon, C.; Morassi, A. Dynamic testing and parameter identification of a base-isolated bridge. *Eng. Struct.* **2014**, *60*, 85–99. [CrossRef]
59. Allemang, R.J.; Brown, D.L. A correlation Coefficient for Modal Vector Analysis. In Proceedings of the 1st International Modal Analysis (IMAC) Conference, Orlando, FL, USA, 8–10 November 1982; pp. 110–116.
60. Allemang, R.J. The Modal Assurance Criterion—Twenty Years of Use and Abuse. *Sound Vib.* **2003**, *37*, 14–23.
61. DDS. FEMtools Computer Software. Customer Support Documentations. Available online: www.femtools.com (accessed on 26 July 2018).

5

Software Defined Networks in Industrial Automation

Khandakar Ahmed [1,*], Jan O. Blech [2], Mark A. Gregory [3] and Heinz W. Schmidt [2]

1 Discipline of IT, College of Engineering & Science, Victoria University, Footscray Park Campus, Melbourne, VIC 3011, Australia
2 School of Science (Computer Science), RMIT University, Melbourne, VIC 3001, Australia; joblech@gmail.com (J.O.B.); heinz.schmidt@rmit.edu.au (H.W.S.)
3 School of Engineering (Electrical and Computer Systems), RMIT University, Melbourne, VIC 3001, Australia; mark.gregory@rmit.edu.au
* Correspondence: khandakar.e.ahmed@ieee.org

Abstract: Trends such as the Industrial Internet of Things and Industry 4.0 have increased the need to use new and innovative network technologies in industrial automation. The growth of industrial automation communications is an outcome of the shift to harness the productivity and efficiency of manufacturing and process automation with a minimum of human intervention. Due to the ongoing evolution of industrial networks from Fieldbus technologies to Ethernet, a new opportunity has emerged to harness the benefits of Software Defined Networking (SDN). In this paper, we provide a brief overview of SDN in the industrial automation domain and propose a network architecture called the Software Defined Industrial Automation Network (SDIAN), with the objective of improving network scalability and efficiency. To match the specific considerations and requirements of having a deterministic system in an industrial network, we propose two solutions for flow creation: the Pro-active Flow Installation Scheme and the Hybrid Flow Installation Scheme. We analytically quantify the proposed solutions that alleviate the overhead incurred from the flow setup. The analytical model is verified using Monte Carlo simulations. We also evaluate the SDIAN architecture and analyze the network performance of the modified topology using the Mininet emulator. We further list and motivate SDIAN features and report on an experimental food processing plant demonstration featuring Raspberry Pi as a software-defined controller instead of traditional proprietary Programmable Logic Controllers. Our demonstration exemplifies the characteristics of SDIAN.

Keywords: controller; industry network; Open Flow; Software Defined Networking; Programmable Logic Controller

---

1. Introduction

Networking large automated machines is a recent focus for industrial automation and one challenge is the connectivity with traditional automation machinery that is not designed to support more than local computer connectivity. Industrial networks can be highly decentralized, rigid and complex to manage due to the tight coupling of the automation data and control plane that is often embedded within the equipment. The computing and communication nodes are often configured individually when the plant is setup and interconnections remain static thereafter. The traditional industrial communications hierarchical structure consists of three network levels with various networking technologies and protocols that limits what can be achieved and adds complexity due to localized configuration. The traditional structure requires offline manual network control and management, which is time-consuming, error-prone and introduces complexity. It hinders the ability to make live changes to the configuration and feature set as the production line is shifted from one task to another. The resolution of medium access control (MAC) address and virtual routing address during data forwarding in an industrial network can lead to challenges, including integrating software and devices from different vendors.

Legacy industrial communications is a challenge to be overcome as part of the fourth-generation industry revolution (FGIR). FGIR is underpinned by the principle of intelligent manufacturing (IM) enabling customized production. To ensure that a smooth transition occurs between production tasks, IM aims to reconstruct the industrial plant by decoupling the manufacturing entities. To attain optimal production, a goal of FGIR is to utilize live monitoring of machine status, environmental values and manufacturing parameters to carry out advanced management, control and fault detection. The outcomes of FGIR will assist with maintenance scheduling to reduce downtime. The future industrial network will connect a varying range of industrial machinery within one or more locations that could change over time. To facilitate FGIR, the current heterogeneous hierarchical localized network structure should be replaced with IP-based networking to provide flexible real-time communications and simplified data mapping. There is also a requirement to change the configuration of the industrial machines and production systems as the production tasks change. It is in this context that future industrial facility networks should embrace Software-Defined Networks (SDNs) to provide flexible programmatic capabilities. The research gap that this paper addresses is the introduction of SDN and IP-based networking into an industrial automation setting to provide flexibility and programmability while maintaining the features and capabilities expected for a real-time communications environment.

1.1. *Software Defined Network*

SDNs [1–3] separate the networks control logic (the control plane) from the underlying routers and switches that forward the traffic (the data plane). With the separation of the control and data planes, network switches become simple forwarding devices, and the control logic is implemented in a logically centralized controller, simplifying policy enforcement, and network (re)configuration and evolution [4]. Therefore, the most promising and possibly profitable benefit of SDNs is their potential in making the network directly programmable. SDNs become a hot topic at within cloud and enterprise networks in about 2010. To our knowledge, SDN solutions are new to the industrial automation domain. SDNs permit reusable configurations and designs that improve system performance. SDNs complement and build on technologies such as industrial Ethernet [5–7], wireless technologies [8,9], and network technologies with guaranteed timing behavior for real-time (e.g., [10]) communication. SDNs can be characterized by: (1) decoupling the control plane from the data plane within network devices; (2) providing programmability for network services; (3) taking forwarding decisions based on flow instead of destination; (4) hosting control logic in an external network component called controller or Network Operating System (NOS); and (5) running software applications on top of the NOS to interact with the underlying data plane devices. With the realization of the aforementioned characteristics of SDNs, the current "touch many, configure many" model is being evolved into "touch one, configure many" [11].

1.2. *Brief History of Industrial Networks*

Dedicated industry networks, e.g., Fieldbus System, dominated the early days of industrial automation. The reduction of the communication gap in the lower level of the automation pyramid was the essence of the dedicated communication infrastructure. However, the complexity of coupling different communication technologies and protocols used across different communication layers was one of the fundamental motives to adopt a solution. Table 1 presents the timeline of the progression of industrial automation networks with unsustainable disruptions that come from the evolution of computer networks. In the 2000s, the Internet technologies evolved and became commercially successful raising the possibility of plausible disruption with the inclusion of Ethernet-based networks and IP. However, due to the lack of guaranteed real-time capabilities, the phenomena of having

Ethernet-based industry networks did not occur and the emergence of dedicated industry networks continued. Later some of the Ethernet-based approaches including Powerlink, PROFINET, EtherCAT, to name a few, emerged to meet the low-latency requirements, in particular, for motion control applications. In the early 2000s, network evolution occurred with the integration of wireless networking. The IEEE 802 protocol family was aggressively adopted to realize the flexibility afforded by connecting machines and devices wirelessly. The typical use of wireless networks in the automation industry was limited due to the need for wired networks to provide reliable real-time communications. We have yet to see the full use of Wireless Sensor Networks (WSN) in industry automation though it is now a mature technology.

Until recently, industrial communication was a mixture of Fieldbus, Ethernet and wireless solutions that has become complex, difficult to upgrade or change and remains a challenge to be overcome before industrial automation can take a significant step forward. New networking approaches that have evolved include the Internet of Things (IoT) and Cyber-Physical Systems (CPS), both of which should find a place within future industrial automation solutions. The idea behind CPS being used in industrial automation is to create an industrial ecosystem allowing more comprehensive and more fine-grained interconnections between machines and systems. Moving business logic into the cloud is a promising trend in the application layer of the information processing pyramid. There are two well-known reference architectures for industrial IoT including the Reference Architecture Model for Industry 4.0 (RAMI4.0) [12] and the Industrial Internet Reference Architecture (IIRA) [13]. RAMI 4.0 uses three dimensions including the lifecycle, physical world and the mapping of IT-based business models in describing the space of the fourth industrial revolution. Some of the leading industry sector companies based in Germany initiated and are driving RAMI 4.0. On the other hand, the Industrial Internet Consortium developed IIRA in the U.S. IIRA focuses on four different viewpoints including functional, usage, business, and implementation.

1.3. SDNs in Industrial Automation

In transitioning to a software-defined network, the key challenges involve changing the traditional practices in industrial automation on the factory floor [14,15]. That means providing relevant employees with the tools and knowledge to support new, more intelligent infrastructure and systems.

Cronberger [16] and Kalman et al. [17] first considered and discussed the use of SDN in industrial automation networks. Cronberger investigated the potential of SDN through a conceptual framework whereas Kalman et al. saw SDN as a possible evolution for future industrial Ethernet planning and extensions towards using Layer 3 networks and wireless solutions. In 2015, we first proposed the integration of SDN in industrial automation by reforming the current industry communication pyramid to become a single Ethernet-based solution in a conceptual paper [14]. In [18], the authors proposed an application-aware industrial Ethernet by exploiting the capabilities of SDN in collecting topology information and application requirements. A newly developed routing and scheduling algorithm uses the received information to generate network configuration autonomously. This configuration is later installed in the network through north and southbound communication, and an enhanced TDMA approach is used to facilitate real-time communication. D. Li et al. [15] proposed a single IP-based solution that can respond to the dynamic change of product orders by adaptively reconfiguring the networks. The architecture promised to guarantee real-time data transmission, enable plug-and-play, and support wireless access with seamless handover capability. In [19] the authors reviewed SDN to draw a correlation between the requirement of the industrial network and existing work. In [20] we continue the evaluation of SDN for future industrial automation networks. This work extended the Ryu controller for direct multicast routing of industrial traffic in a cyclic switched Ethernet network setup. The experiment was conducted in an IEC 61499 compliant development environment. The experiment result shows that there is a promising opportunity to have a flexible and reliable network that is also suitable for real-time traffic. Table 2 summarizes the current state of the art of SDN in industrial automation.

Table 1. Industrial communication protocols timeline.

Industrial Communication Protocols							Computer Networks	
Protocol Name		Published by	Place	Com. Tech	Year		Protocol Name	Year
Modbus	Modular Bus	Modicon (now Schneider Electric)	United States	Master/Slave	1979	1970–1980	ARPANET	1970
							Ethernet	1973
							ISO/OSI	1978
PROWAY	Process Data Highway	Working Group 6			1981	1981–1990		
FIP	French Initiatice	factory instrumentation protocol	France	Producer/Consumer	1982		MAP	1980
Bitbus	BIT Fieldbus	Intel Corporation	USA	Master/Slave	1983			
HART	Highway Addressable Remote Transducer	FieldComm Group	USA	Master/Slave	1985			
CAN	Controller Area Network	Robert Bosch GmbH	Detroit, Michigan	Producer/Consumer, Peer to Peer	1985		Internet	1981
P-NET	Process Network	Process-Data Silkeborg ApS	Denmark	Master/Slave	1987			
INTERBUS	INTERBUS	Phoenix Contact	Germany	Master/Slave	1987		MMS	1985
PROFIBUS	The Federal Ministry of Education and Research (BMBF)	process field bus	Germany	Master/Slave, Peer to Peer	1989		Ubiq. Comp	1988
EIB	European Installation Bus (EIB)	EIB Association	Europe	Master/Slave	1991	1991–2000		
Asi	Actuator Sensor Interface	AS-International	Germany	Master/Slave	1992		WWW	1992
SDS	Smart Distributed System	Honeywell	USA	Master/Slave	1993			
DeviceNet	Connecting Devices	Allen-Bradley	USA	Producer/Consumer	1993		2G GSM	1996
FF								
ControlNet	Real-Time Control Network	Rockwell Automation	USA	Producer/Consumer	1995		WLAN	1997
TTP	Time-Triggered-Protocol	Vienna University of Technology	Vienna, Austria	Master/Slave	1998		IoT	1997
Powerlink	Ethernet Powerlink	B&R Industrial Automation GmbH	Austria	Producer/Consumer	2001	2001–2010	Bluetooth	2003
Modbus/TCP	Modbus RTU protocol with a TCP interface that runs on Ethernet	Modicon (now Schneider Electric)	United States	Master/Slave	2001		SOAP	2003
PROFINET	Process Field Net	Profibus & PROFINET International	Germany	Real-Time Ethernet	2001		3G: UMTS	2001
EtherCAT	Ethernet for control automation technology	Beckhoff Automation	Germany	Master/Slave	2003		ZigBee	2003
ISA 100.11a	Wireless Systems for Industrial Automation	International Society of Automation	Worldwide	NIL	2009		3G: HSPA	2005
							UWB	2008
Wire. HART	Wireless HART	HART Communication Foundation	USA	Master/Slave	2007		6loWPAN	2009
							4G: LTE	2010

Table 2. Software-Defined Network Timeline.

Framework/Concept	Brief Description	Year Published
Software Defined Industrial Network [16]	Reflects the possibility of bringing programming capability in industrial network through the use of SDN. A theoretical framework is provided	2014
Outlook on Future Possibilities [17]	Possible evolution of industrial Ethernet using SDN	2014
SDNPROFINET [14]	Proposed to transform the typical communication architecture of PROFINET integrating SDN	2015
SDN-based TDMA in IE [18]	SDN approach is used to formulate an application-aware Industrial Ethernet Based on TDMA	2016
SDIN [15]	Propose a new Software Defined Industry Network (SDIN) architecture to achieve high reliability, low latency, and low energy consumption in Industrial Networks	2016
Challenge and Opportunities [19]	Prospect of future industrial network by means of SDN	2016
Direct Multicast Routing [20]	Evaluates SDN for deterministic communication in distributed industrial automation systems	2017
SDIAN [21]	Software-defined industry automation networks	2017

1.4. Contributions

The contribution of this paper can be summarized as

1. We investigate the research gap that exists for IP-based networking in industrial automation and introduce a novel industrial network framework based on an SDN communication architecture.
2. We propose two solutions for flow creation in relieving the incurred overhead due to the flow setup cost in SDN.
3. We render an optimal latency model based on a meticulous flow analysis using L_1-Norm Optimization to calculate the shortest path. It verifies the quantified model using a Monte Carlo simulation.
4. We validate the proposed scheme by running an experiment in an emulated environment using Mininet [22].
5. We exploit the merits of the proposed framework by presenting an ongoing test bed implementation. The investigation is conducted on a food processing demonstrator.

1.5. Paper Organization

The remainder of this paper is organized as follows. Section 2 presents the architecture, communication framework and flow creation of SDIAN. In Section 3, we examine the flow analysis and present an optimal latency model of the proposed solution. Section 4 exhibits the stochastic analysis of the model formulated in Section 3. In Section 5, the network performance of the target mesh topology is shown using a modelled emulation scenario and a report on the experimental setup in a food processing plant demonstrator is presented. Finally, Section 6 concludes the paper. This manuscript is the extended version of the paper presented in [21].

2. Architecture and Framework

In this section, we first introduce the architecture and three-layer SDIAN framework. Then we describe our proposed flow installation scheme.

2.1. System Model

In this section, we present the conceptual architecture of the proposed SDIAN and the packet dissemination model with the plant components. Figure 1 shows the remolded version of a standard plant hierarchy that incorporates SDN features and builds an intelligent industrial automation network.

In this transformed architecture, within the three hierarchical levels (Control Plane, Plant Level and Field Level), traditional proprietary Programmable Logic Controllers (PLCs) are replaced with the open Raspberry Pi (RPi) systems running the Rasbian operating system, a Linux flavor, and using open-source language for software-defined automation control. Sensors and actuators are interfaced with field level RPis, except the direct I/Os, which are interfaced directly within the plant level hierarchy. A script running on the RPi-based PLCs can receive and send interrupts from the sensors and to the actuators through I/O pins. The scripts written for the RPis replicate the behavior of traditional PLCs. The data layer communication is illustrated using group-1 messages (1A–1E) shown in Figure 1. In this scenario, when an object is detected on the conveyor belt, RPI-PL-1 receives an interrupt and invokes the robotic arm via a reply interrupt. This interrupt is sent through the output pin. In this case, the response of the arm is to deliver the object to another conveyor belt within a limited time constraint. Likewise, group 2 messages (2A–2C) are used to present the control layer communication. In this scenario, a remote SDIAN administrator updates control applications deployed on the controller. After receiving updates, the controller adaptively pushes the information to the associated RPis. Based on the updated instructions received, RPis update the data plane behavior accordingly.

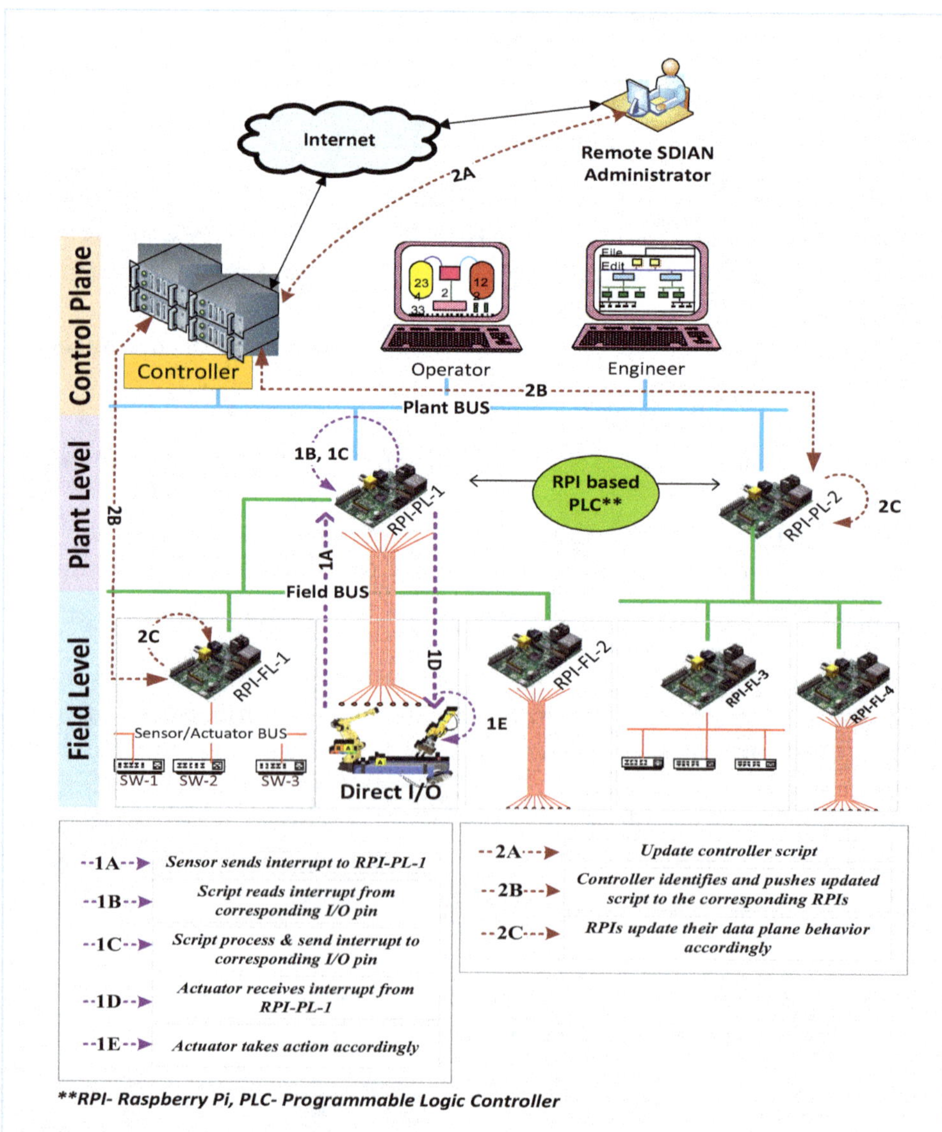

Figure 1. SDIAN (Software Defined Industrial Automation Network) conceptual architecture.

2.2. SDIAN Communication Framework

Figure 2 shows the three-layer SDIAN communication framework. Sensors, actuators, and RPis reside in the data plane, while the logically centralized but physically distributed controllers reside in the control plane. RPis are responsible for receiving packets from sensors and instruct the corresponding actuators to take actions based on the respective flow retrieved from the flow table or corresponding controller. In this framework, RPis are connected through a mesh topology. We deliberately use a mesh topology to map the requirements of the food processing plant, which is presented in Section 5. In the case of a flow table miss [23], an RPi sends a Packet-In message to the controller sitting in the control plane. After getting the Packet-In message, the controller instructs the RPi by sending a Packet-Out/Flow-MOD message. This communication between data plane and control plane happens through the southbound interface (SBI) of the control plane. A task or application is created in the application (also called service management-control) plane that explicitly uses northbound interface (NBI) to translate the business use case, network requirements and, behavior programmatically and logically to the controller. The users are responsible for defining the attributes of a task. Table 3 presents the summary of the different components of the SDIAN architecture.

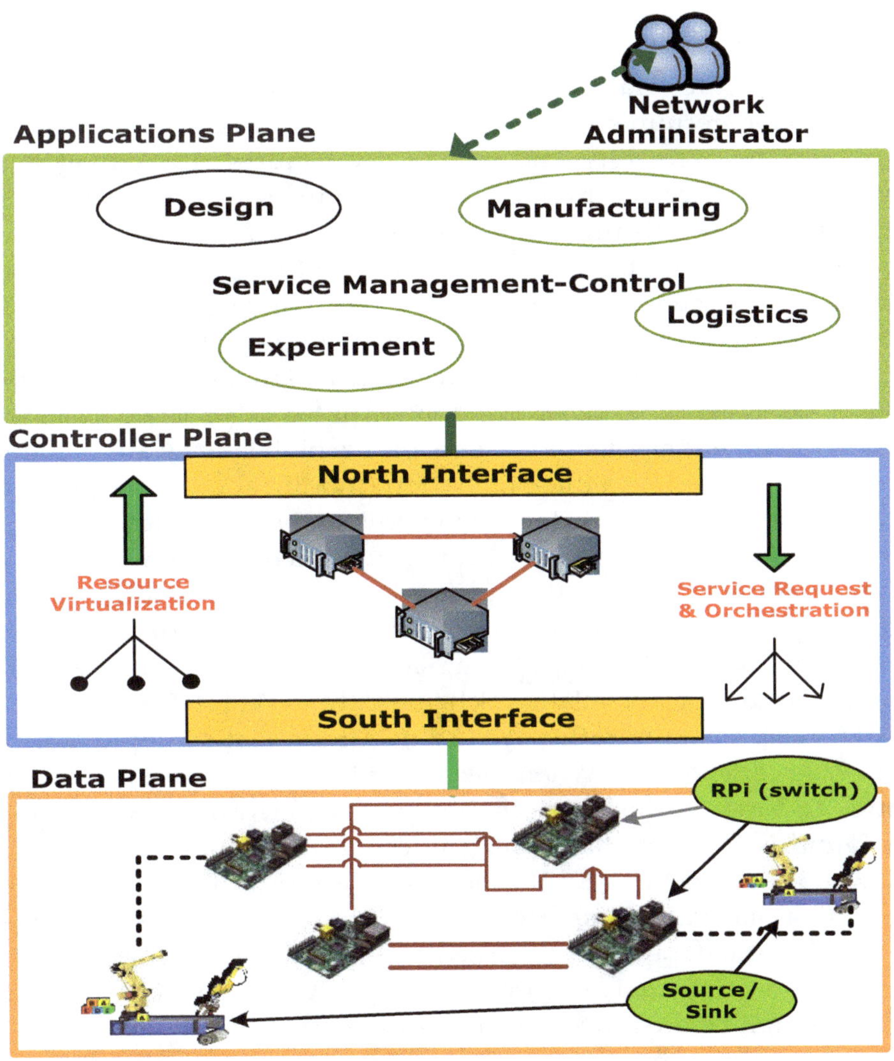

Figure 2. SDIAN communication framework.

Table 3. SDIAN (Software Defined Industrial Automation Network) architectural components.

Component	Task	Layer
RPi	Receive and send interrupt to sensors and actuators	Data Plane
Sensors	Sends an interrupt to an associated RPi immediately after sensing an object	Data Plane
Actuators	Executes the explicitly specified action immediately after receiving an interrupt from RPi	Data Plane
Southbound Interface (SBI)	Interface between data and controller plane. The functions realized through this interface include, but not limited to: (i) programmatic control of all forwarding operations (ii) monitoring (iii) network statistics (iv) advertisement and (v) event notification	Between Control and Data Plane
Controller	*Manage/control* network services. It consists of NBI and SBI agents and control logic. A logically *centralized* but physically *distributed*	Control Pane
Northbound Interface (NBI)	Interface between application and controller plane. It typically provides an abstract view of the network and enables direct expression of network requirements and behavior	Between Application and Control Plane
Applications	Programs in execution that explicitly translate the *business use case*, network *requirements* and, *behavior* programmatically and logically to the controller	Application Plane

2.3. Creating Flows

Unlike other networks, industrial networking environments have specific considerations and requirements to fabricate a deterministic system. These include—real-time network performance, remote access, onsite security, reliability, and ease of use features and manageability. The unique features, when compared to other communication environments, represent significant disparities and pose both challenges and opportunities when implementing SDN-based industrial Ethernet infrastructure. By the inclusion of SDN, there is an inherent opportunity to resolve the reliability, manageability and ease of use issues that are a challenge to achieving real-time performance. Due to the fundamental hardware attributes of switch and software implementation inefficiencies, the latency of flow installations is higher than in traditional network installations. In the case of a flow table miss, there is a higher latency to resolve what should be done with the first packet. From the empirical study provided in [22], it was identified that the root causes of this high latency are as follows: (a) outbound latency, i.e., the latency incurred due to the installation/modification/deletion of forwarding rules, (b) inbound latency, i.e., the latency to send packet events to the controller can be high, in particular, when the switch simultaneously processes forwarding rules received from the controller.

We provide two solutions for flow creation, from which the network administrators can determine the appropriate flow mapping based on their predilection and the application requirements. In the first solution, we use the innovative idea of mixing reactive and pro-active flow installation methods. This is referred to as a Hybrid Flow Installation Scheme (HFIS). With HFIS we cater for non-real-time traffic, in other words, delay tolerant traffic. We use two immediately deployable techniques: Flow Engineering (FE) and Rule Offload (RO). When a switch in the control-level network of a plant receives a packet from control and monitoring devices, it starts by performing a table lookup in the flow table. If a match is found with a flow table entry, it applies the action set associated with the flow as per the Open Flow 1.3 specification [22]. In the case of a table miss, when the controller receives a Packet-In message, it first calculates the shortest route (FE) to reach the destination and then sends the respective Packet-Out/Flow-Mod messages to all switches across this route (RO). Therefore, the packet transmission latency is increased by only one inbound and outbound event irrespective of the number of relay nodes it goes through before it reaches the destination.

The precise synchronization of processes underpins today's manufacturing industry, and therefore, the network must be enhanced to ensure consistent real-time performance in transporting deterministic delay-sensitive traffic. Data must be prioritized based on QoS parameters to ensure that critical information is received first. To tackle this problem, in the second solution, we propose to use a *Pro-active Flow Installation Scheme (PFIS)* catering for delay-sensitive traffic by providing precise synchronization. In this case, we adopted the direct RO method. The controller sends the flow installation packet for all pre-determined critical delay-sensitive traffic to the switches immediately after switch discovery. This pre-installation happens during the convergence of the network. For further clarification, we present the SDIAN packet dissemination model in Figures 3 and 4. In Figure 3, the packet exchange is classified into two categories—Non-Real-Time (NRT) communication and real-time (RT) communication. We apply HFIS for NRT and PFIS for RT. Figure 4 illustrates the working mechanism of PFIS. Please note that in the test bed implementation the data channel and control channel are separate, but for drawing simplification this is not portrayed in Figure 4. As shown in Figure 4a, the switch S1 receives a data packet from a field level device. For this packet, there is a table miss, therefore, the switch sends a control packet (Packet-In) request to the controller. Based on the header information, the controller determines the shortest path for this packet and responds with Packet-Out to all the intermediate switches along this path (Figure 4b). Therefore, as shown in Figure 4c, there is no further table miss as all the intermediate switches along the path pre-install the flow into the flow table before the packet arrives.

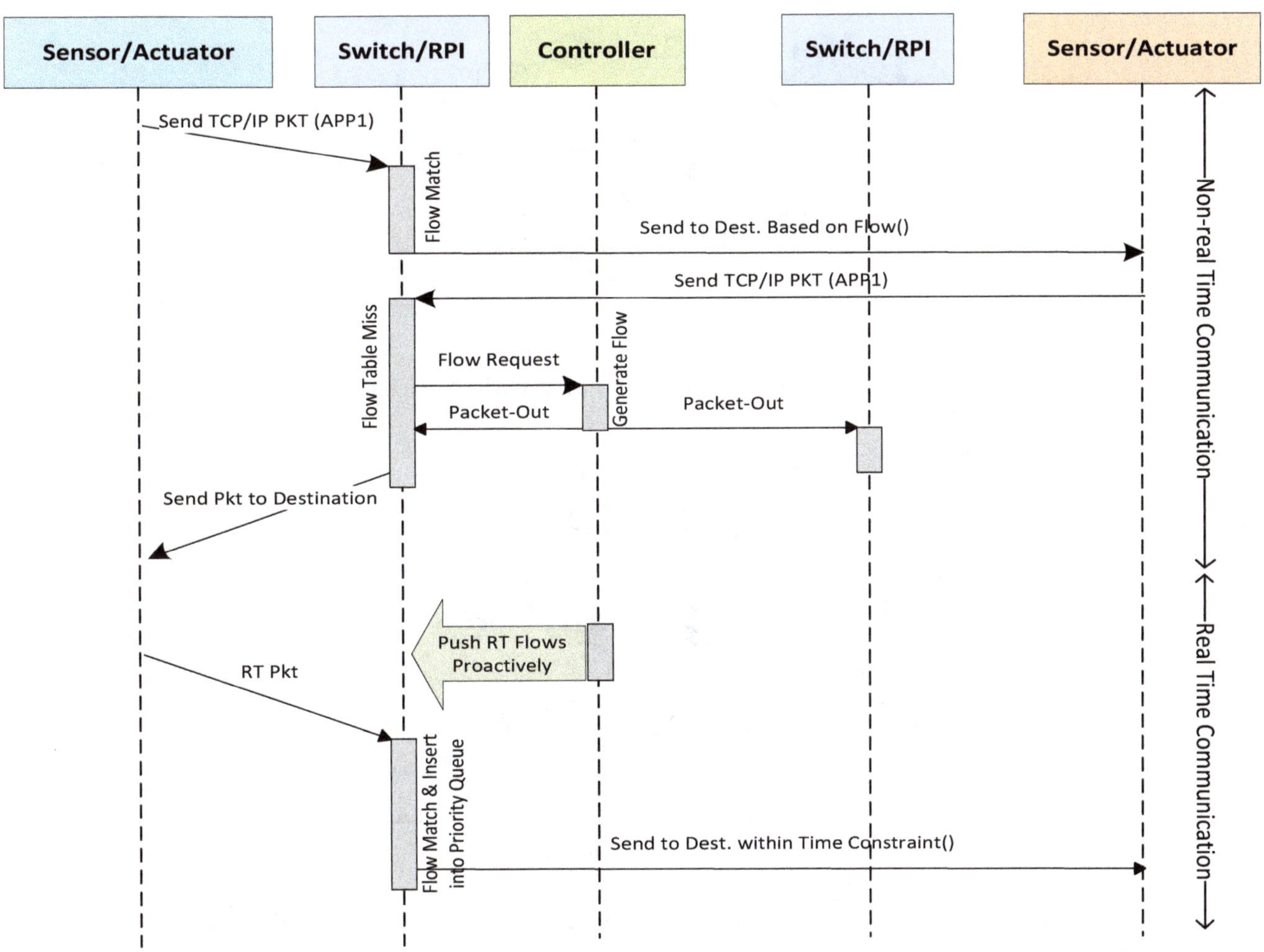

Figure 3. SDIAN packet dissemination model.

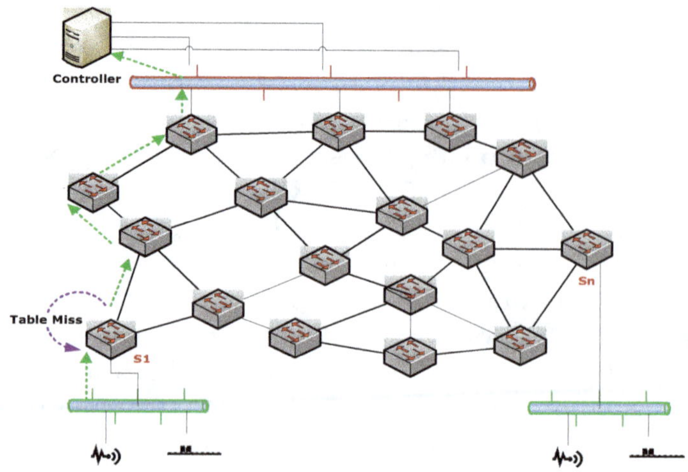

(**a**) Table Look Up

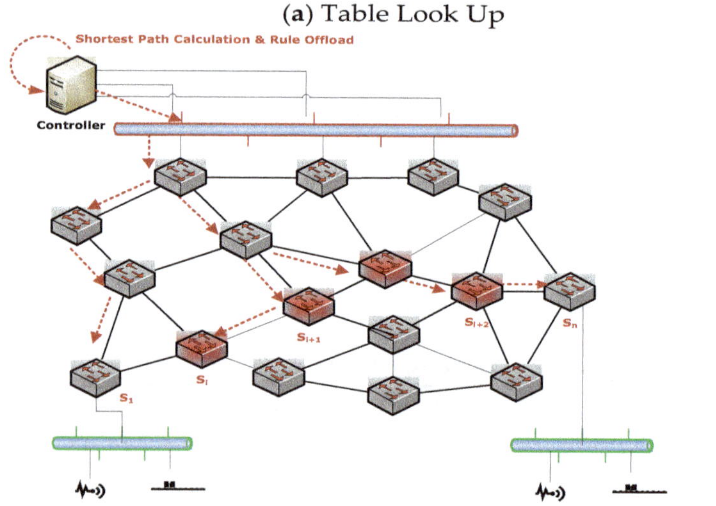

(**b**) Flow Engineering (FE) and Rule Offload (RO)

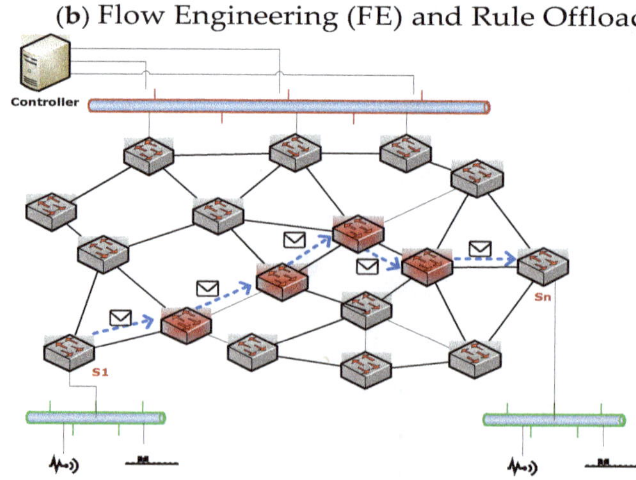

(**c**) Reaches Destination

Figure 4. Working mechanism of HFIS (Hybrid Flow Installation Scheme). (**a**) Table Look Up; (**b**) Flow Engineering (FE) and Rule Offload (RO); (**c**) Reaches Destination.

3. Flow Analysis

In this section, we first illustrate the basic notation used to represent the data layer of the control network of a plant. Since the control channel is separated from the data channel, we kept the graph representation of the control channel out of the scope of this paper and assumed that each switch could reach the controller in single hop fashion using a secured and fast directly connected control channel. Now, we formulate the shortest path routing as the flow optimization problems in a network that is realized by the controller based on the discovered topology. Finally, we compute the model for determining optimal latency to reach the destination.

3.1. Data Layer: Basic Notations

We represent our n-node data plane of the control network of a plant by an undirected graph, $G = (S, L, X)$, where $S = \{s_1, s_{2,\cdots}, s_n\}$ is the set of switches, L is the set of links, and X is an $n \times n$ matrix defined by $\{x_{ij} | (i,j) \in L\}$, where each (i, j)-th entry, denoted by x_{ij}, represents the positive weight of a link $(i, j) \in L$. Due to the undirected nature of the graph, (i, j) and (j, i) designate the same link, i.e., $x_{ij} = x_{ji}$. When $(i, j) \notin L$, delineate $x_{ij} = 0$ fabricating the weight matrix $X == [x_{ij}]$ into symmetric. We also define that X is a 0-1 matrix, i.e., all links have a unit weight, therefore, G refers to a simple graph and, X is the respective adjacency matrix.

Consider $d = [s_1, s_n]$, $s_1, s_n \in S$ denotes the source-destination switch pair in the network G and $F^d : S \times S \rightarrow \mathbf{R}^+$ function defines the amount of traffic ($f^{(d)}$- unit) that traverse from s_1 (source) to s_n (destination) subject to the following constraints:

(1) along network links:

$$\text{if } (i,\ j) \notin L \text{ then } F_{ij}^d = 0 \tag{1}$$

(2) along one direction:

$$\text{if } F_{ij}^d > 0 \text{ then } F_{ji}^d = 0 \tag{2}$$

(3) at source s_1:

$$f^{(d)} + \sum_{k=1}^{n} F_{ks_1}^d = \sum_{j=1}^{n} F_{s_1 j}^d \tag{3}$$

relay node $i \neq s_1,\ s_n$:

$$\sum_{j=1}^{n} F_{ij}^d = \sum_{k=1}^{n} F_{ki}^d \tag{4}$$

(4) at destination s_n:

$$\sum_{k=1}^{n} F_{ks_n}^d = \sum_{j=1}^{n} F_{s_n j}^d + f^{(d)} \tag{5}$$

The constraint in Equation (1) ensures that for each link $(i, j) \notin L$, $F_{ij}^d = 0$ and in particular, for each undirected link $(i, j) \in L$, the constraint in Equation (2) says if $F_{ij}^d > 0$ then $F_{ji}^d = 0$ or if $F_{ji}^d > 0$ then $F_{ij}^d = 0$. The traffic constraints defined in Equations (3)–(5) state that the amount of $f^{(d)}$ unit traffic sent by source s_1 is received by destination s_n at the exact number. The amount of traffic entering and leaving a relay switch is same.

Considering a set of intermediate or relay switches $S_{F(d)} \subset S$ and a corresponding subset of links $L_{F(d)} \subset L$ to carry the given $f^{(d)}$ unit traffic from source s_1 to destination s_n, we induce a directed (or oriented) sub-graph of G, $G_{F(d)} = (S_{F(d)}, L_{F(d)})$. $G_{F(d)}$ is a directed acyclic graph (DAG, we refer to it as a routing graph) that routes the traffic from source s_1 to destination s_n. The traffic could split or merge across the nodes of $G_{F(d)}$ to travel across multiple paths. We define $F^{d'}$ to refer the collection of flows, in other words, all functions that satisfy the constraints in Equations (1)–(5).

In the following subsection, we derive the shortest path routing strategy by minimizing L_1-norm of traffic between a given source-destination pair. We build this model based on the fabrication of two well-known results [24,25] presented in [26].

Shortest Path Routing (L_1-Norm Optimization)

For simplicity and clarity of notation, we assume that $f^{(d)} = 1$, F_{ij} equivalently specifies the traffic function $F^{(d)}$, $s_1 = 1$, and $s_n = n$. Therefore, we define the following L_1-norm (L_1 Primal) flow optimization problem that can be solved using linear programming (LP).

$$\min_{F^d} \sum_{i=1}^{n} \sum_{j=1}^{n} x_{ij} F_{ij} \quad \text{s.t. (1)–(5)} \tag{6}$$

To comply with the constraints specified in Equations (1)–(5), (6) can more specifically be stated as

$$\sum_{j:(i,j)\in L} F_{ij} - \sum_{k:(k,i)\in L} F_{ki} = \begin{cases} 1 & \text{if } i = s_1 \\ 0 & \text{if } i, j \neq s_1, s_n \\ -1 & \text{if } i = s_n \end{cases} \tag{7}$$

where, $F_{ij} \geq 0$ and $1 \leq i, j \leq n$.

Hence, the optimization problem presented in (6) minimized the weighted L_1-norm. Based on the flow conservation constraints presented in Equation (7), we consider the dual (L_1 Dual) of (6) in terms of Lagrange multipliers (U_i's) to find the shortest path routing

$$\max_{U} U_1 \tag{8}$$

Subject to,

$$U_n = 0 \text{ and } U_i - U_j \leq x_{ij}, \forall_{ij} \in L \tag{9}$$

Assuming F^* and U^* refer to the optimal traffic solution for the primal and dual problem respectively, we derive the following relations between $F_{ij}^{*\prime}s$ and $U_i^{*\prime}s$

$$\text{if } F_{ij}^* > 0, \text{ then } U_i^* - U_j^* = x_{ij} \tag{10}$$

and

$$\text{if } F_{ij}^* = 0, \text{ then } U_i^* - U_j^* < x_{ij} \tag{11}$$

Based on these relations, we can define the following properties of the optimal solution ($U_i^{*\prime}s$) of the dual problem.

Lemma 1. *Let P_1 and P_2 (alternative to P_1) are two different paths from source (s_1) to destination (s_n) to carry the traffic. If for each link $(i, j) \in P_1$, $U_i^* - U_j^* < x_{ij}$ then P_1 is not the shortest path and $U_{s_1}^* < \sum_{(i,j)\in P_1} x_{ij}$. On the other hand, the alternative path P_2 is a shortest path if for each link $(i, j) \in P_2$, $U_i^* - U_j^* = x_{ij}$ and $U_{s_1}^* = \sum_{(i,j)\in P_2} x_{ij}$.*

It is evident from the above Lemma that for any switch S_i on a shortest path, $U_{s_i}^*$ is the shortest path distance from the switch S_i to the destination S_n. All intermediate switches including S_i and S_n are the elements of $S_{F(d)}^*$ that form the shortest routing graph $G_{F(d)}^*$.

3.2. Optimal Latency Model: Hybrid

In this subsection, we derive the optimal latency model for HFIS. In HFIS, a packet traverses across the shortest path to reach the destination switch.

Let α denotes the total latency for a packet to reach from source S_1 to destination S_n, α_{in} refers to the inbound latency, α_{ou} is the outbound latency, $\alpha_p^{S_k}$ is the single hop propagation delay of a packet travelling from S_k to S_{k+1}. We consider γ is the average time taken by a controller to process a Packet-In message and β is the control channel latency i.e., time taken by a Packet-In/Packet-Out message to travel between a switch and a controller. To this end, our target is to minimize the value of α, and therefore, the optimization model of latency can be stated as:

$$\min_{\alpha}\left\{\alpha_{in}+\alpha_{ou}+\sum_{k=1}^{m}\left\{\alpha_{P}^{S_k}\right\}+2\times\beta+\gamma\right\} \tag{12}$$

where, m is the number of hops i.e., the total number of switches in the shortest routing graph $G^*_{F(d)}$ and $S_k \in S^*_{F(d)}$, where $S^*_{F(d)} = \{S_i, S_{i+1}, \ldots\ldots, S_{i+m}\}$.

According to HFIS, only the first switch generates the packet event to the controller; then all switches, including the first switch, along the path receive and install the flow instruction. Therefore, there is only one inbound, one outbound, two control channels and one Packet-In resolution latency. Considering a consistent and deterministic link state and performance among all switches, we assume $\alpha_P^{S_k} \cong \alpha_P^{S_{k+1}} \cong \alpha_P^{S_{k+2}} \cdots \cong \alpha_P^{S_{k+m}} \cong \alpha_p$, where α_p is the average propagation delay, therefore, we can rewrite Equation (12) as follows

$$\min_{\alpha}\{\alpha_{in}+\alpha_{ou}+m\times\alpha_p+2\times\beta+\gamma\} \tag{13}$$

Lemma 2. *During the lifetime of a packet, if it traverses across the shortest path, then latency $\alpha \propto \alpha_p$, i.e., $\alpha = m\times\alpha_p+K$, where $K = \alpha_{in}+\alpha_{ou}+2\times\beta+\gamma$.*

Lemma 2 asserts that in the entire journey of a packet, there is no more than one table miss regardless of the number of hops across the path. Therefore, one table miss generates only one Packet-In event incurring single inbound (α_{in}) and outbound (α_{ou}) latency with the associated control channel (β) and Packet-In processing time by controller (γ).

3.3. Optimal Latency Model: Pro-Active

According to the second solution, the controller will pro-actively offload the rule to all switches immediately after the deployment of an application. A network administrator deploys an application through the application plane. The application plane creates a particular flow and sends it to the controller in the control plane through the northbound interface. The controller then floods the flow across all the switches within the respective domain. The value of the associated *Idle timeout* and *Hard timeout* [23], in this case, are set to zero i.e., Flow entry is considered permanent, and it does not timeout unless it is removed with a flow table modification message of type OFPFC_DELETE [23]. When a switch receives a packet of this kind, the switch gets an obvious *table match* and therefore, apply the action accordingly. This pre-offloading of flows eventually eradicates the control channel communication entirely during the lifetime of a packet in the data plane.

Lemma 3. *With the PFIS, if a packet travels across the shortest path, then the latency is calculated as $\alpha \propto \alpha_p$ i.e., $\alpha = m\times\alpha_p+K$, where $K \cong \alpha_{in}+\alpha_{ou}+2\times\beta+\gamma \cong 0$.*

Lemma 3 asserts that in the entire journey of a packet, there is no *table miss* regardless of the number of hops across the path. Therefore, there is no Packet-In event, i.e., inbound (α_{in}) and outbound (α_{ou}) latency with the associated control channel (β) and Packet-In processing time by the controller (γ) are equivalent to zero.

4. Stochastic Analysis of SDIAN

To validate the analytical approach presented in Section 3, we perform an extensive Monte Carlo simulation with 10,000 runs. In each run, we use a randomized distribution of inbound (α_{in}) and outbound (α_{ou}) flow latency, control channel latency (β), data channel latency (α_p) and packet processing time (γ) by a controller. The distribution of α_{in} and α_{ou} is fabricated from the outcome of a comprehensive measurement study [27,28] conducted using four types of production SDN switches. The distribution of inbound latency is a Chi-squared distribution attributed by a mean of 1.853 ms, a median of 0.71 ms and a standard deviation of 6.71 ms. The outbound delay is less variable and skewed with the same mean and median of 1.09 ms and a standard deviation of 0.18 ms. Assuming the simulation is running with a ten (10) switch control network for a single small-scale plant, the number of hops (m) in the shortest path calculation is varied between 1 and 10 and the distribution is a normal distribution with a mean of six (6) and standard deviation of three (3). The Round Trip Time (RTT) between two switches (α_p) and between controller and switch (β) is negligible (≈ 0.1 ms). The influx rate of *Packet-In* messages from switches to the controller determines the time (β) taken by a controller to process a packet and therefore the distribution of (β) is a normal distribution with a mean of 5.49 µs and standard deviation of 2.86 µs.

Figure 5a–c respectively shows the histogram of the Monte Carlo simulation results of three flow installation schemes: hybrid, reactive and pro-active. The bin size in Figure 5a,b is 5 ms whereas in it is 0.035 ms. The ascendancy acquired by using HFIS and PFIS over the Reactive Flow Installation Scheme (RFIS) is discernible. 95% of packets are resolved within 3.28 ms using the HFIS and within 0.19 ms using the PFIS. Table 4 presents the summary simulation result statistics as shown in Figure 5.

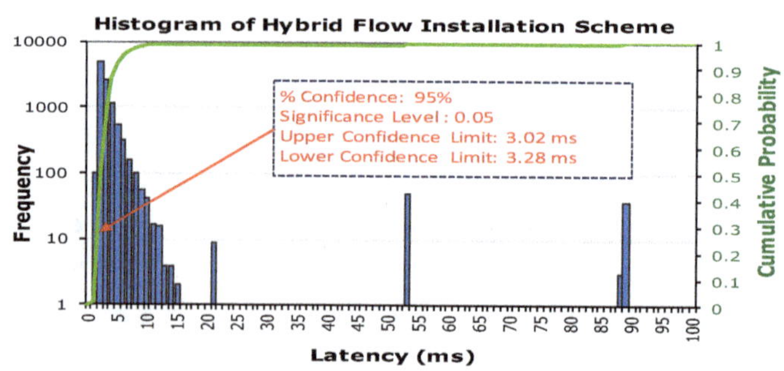

(a)

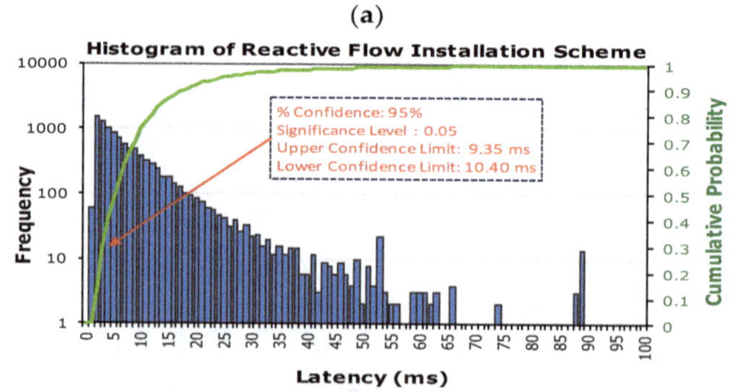

(b)

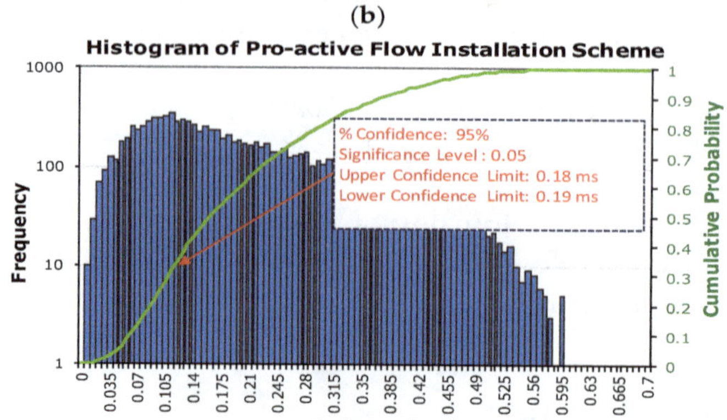

(c)

Figure 5. Histogram of monte carlo simulation results of (**a**) Hybrid Flow Installation Scheme (HFIS) (**b**) Reactive Flow Installation Scheme (RFIS) and (**c**) Pro-active Flow Installation Scheme (PFIS).

Table 4. Summary statistics of stochastic analysis.

Sample Size		**HFIS**	**RFIS**	**PFIS**
		10,000	**10,000**	**10,000**
Central Tendency	**Mean**	3.15533	9.84786	0.19091
	Median	2.03766	5.40382	0.163
	StErr	0.06706	0.27245	0.00119
Spread	**StDev**	6.7095	26.7239	0.1189
	Max	88.7062	883.67201	0.598
	Min	0.6773	0.6773	0.003
	Range	88.0288	882.9946	0.595
	Q(0.75)	3.0157	10.2242	0.264
	Q(0.25)	1.5196	2.7774	0.098
	Q Range	1.4960	7.4467	0.166
Shape	Skewness	10.3932	15.9369	0.8368
	Kurtosis	118.0304	330.8579	0.0093
Quantiles, Percentiles, Intervals	90% Interval	Q(0.05) = 1.17 Q(0.95) = 6.01	Q(0.05) = 1.34 Q(0.95) = 24.9	Q(0.05) = 0.04 Q(0.95) = 0.42
	95% Interval	Q(0.025) = 1.08 Q(0.975) = 7.84	Q(0.025) = 1.22 Q(0.975) = 35.61	Q(0.025) = 0.03 Q(0.975) = 0.47
95% CI for the Mean	Upper Limit	3.0210	9.5795	0.1883
	Lower Limit	3.2839	10.7175	0.1930

To see the implication of Lemma 2, we repeat the simulation with the number of switches varying between ($5 \leq S \leq 100$). After performing all the Monte Carlo simulation runs, we average the results obtained for each value of S as shown in Figure 6. Figure 6a shows that with HFIS, the total flow installation latency ($K = \alpha_{in} + \alpha_{ou} + 2 \times \beta + \gamma$) is constant irrespective of the network size; therefore, the total latency (α) is directly proportional to $m \times \alpha_p$. Figure 6b presents the latency for RFIS. In the worst-case scenario for RFIS, each switch in a route could have packet flow table miss with the associated flow setup cost. Therefore, the total latency is dominated by the flow installation overhead ($\alpha \cong K, m\alpha_p \ll K$). The PFIS latency results are shown in Figure 6c. Since the respective flow is installed across all switches before the arrival of any data packet, there is no table miss. As in the HFIS case α is directly proportional to $m\alpha_p$ with $\alpha \cong m\alpha_p$ and $K \cong 0$.

Discussion

The results presented in Figures 5 and 6 and Table 4, highlight that the PFIS confers the lowest latency as the overhead from flow establishment is, in fact, close to zero. Regarding HFIS, the cost for flow setup is constant regardless of the network size. The upper and lower limits of the 95% Confidence Interval (CI) for HFIS in a network of ten (10) switches are 3.02 ms and 3.28 ms respectively, indicating a stable deterministic condition. For consistent RT performance in transporting deterministic delay-sensitive traffic, we can apply PFIS, while we can use HFIS to provision the rest of the traffic sustaining the dynamic behavior of the SDN network. In a nutshell, the latency bound for RFIS, HFIS and PFIS are 0.025–0.975, 1.08–7.84 and 1.22–35.61 ms respectively with 95% confidence.

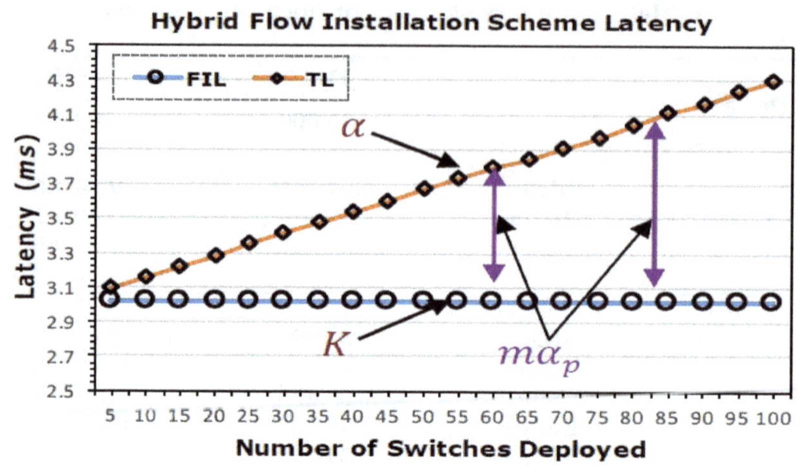

(a)

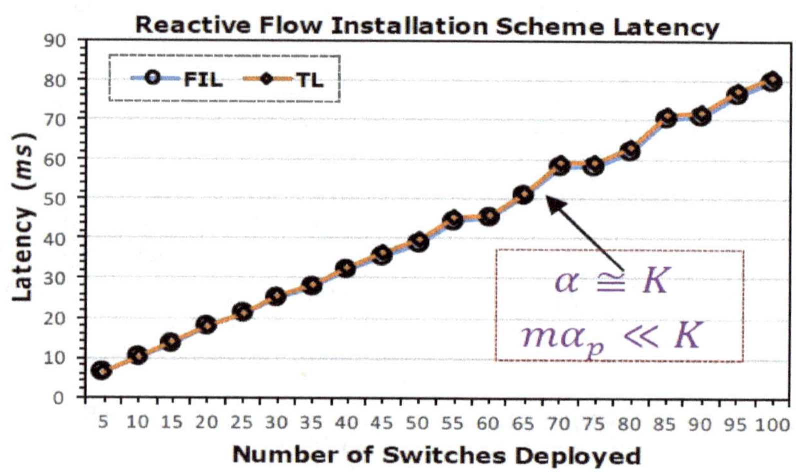

(b)

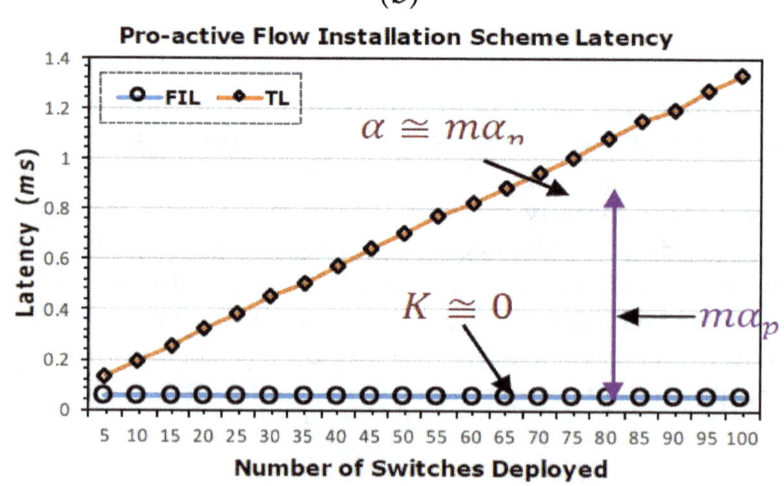

(c)

Figure 6. Mean Latency for varied number of switches (**a**) Hybrid (**b**) Reactive (**c**) Pro-active. FIL—Flow Installation Latency, TL—Total Latency.

5. Experiments

In this section, we first present the network performance of the target mesh topology using a modelled emulation scenario and then report on an experimental setup with the adaptive configuration in a food processing plant demonstrator.

5.1. Emulation Environment

For further validation of our proposed scheme, we run another experiment in an emulated environment using Mininet. Although the accuracy of Mininet cannot be taken for granted particularly for large scale topologies, the SDN community adopts it widely. In our case, we are essentially interested in looking at the expediency of our proposed solution before we investigate it with limited functionality in a real testbed. Therefore, we deploy a small mesh network of five (5) switches and a Ryu controller [29] as shown in Figure 7. The Ryu controller is tailored to incorporate the three flow installation schemes, and Spanning Tree Protocol (STP) is implemented to discard any possibility of creating a loop. We generate the plant level network packets from openflow switch#2 (source) to openflow switch#5 (sink) and vice versa. We varied the rate of packets generated from source to sink and measure the latency and success rate for the three flow installation schemes. We present the results in Figures 8 and 9. In Figure 8, we present the latency for each flow installation scheme against a varying number of packets generated per second. The latency of RFIS increases linearly with the increase in the number of packets while PFIS and HFIS show a similar pattern. The latency bound of PFIS and HFIS are 1–3 ms and 3–7 ms respectively, therefore for this setup, the guaranteed delay is <7 ms. Figure 9 shows the success rate of the three flow installation schemes against a varying packet rate. The success rate for PFIS and HFIS varies from 98–99% and 97–99%.

From the results it was found that the HFIS retains a consistent low latency and high success rate as well as maintaining the flexibility and dynamic behavior of SDN.

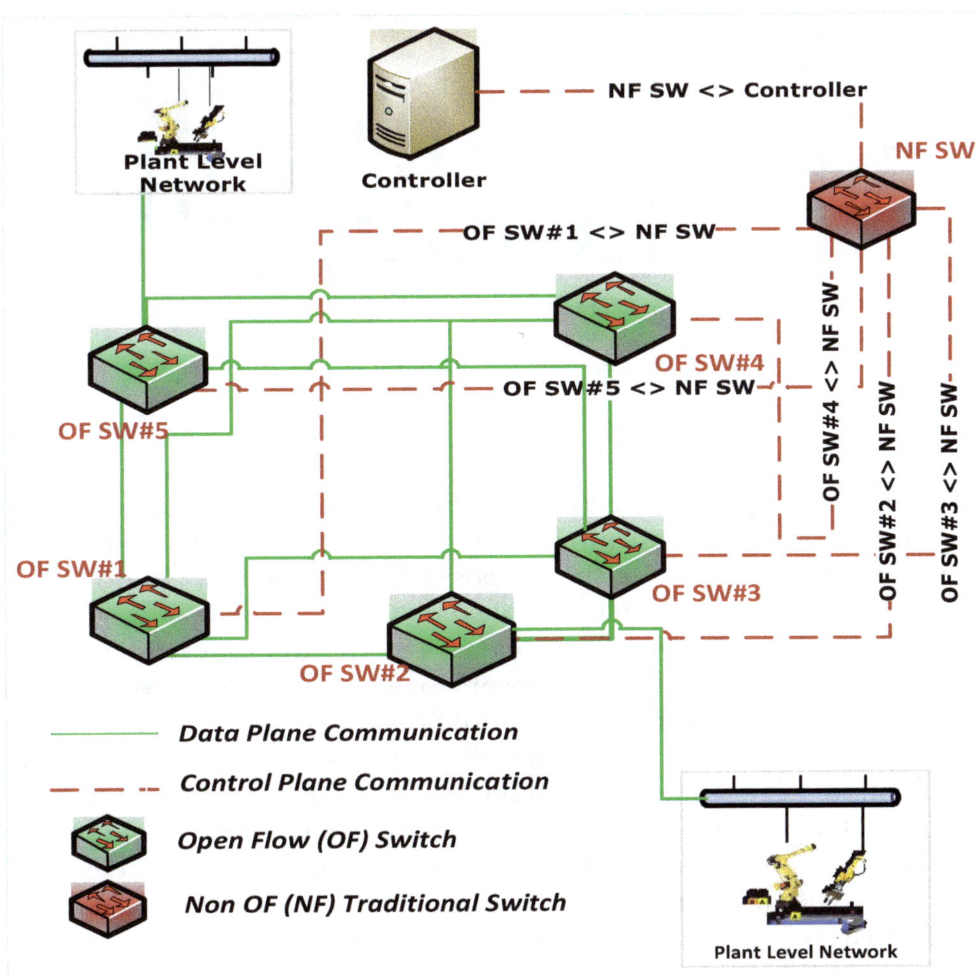

Figure 7. Network setup in Mininet.

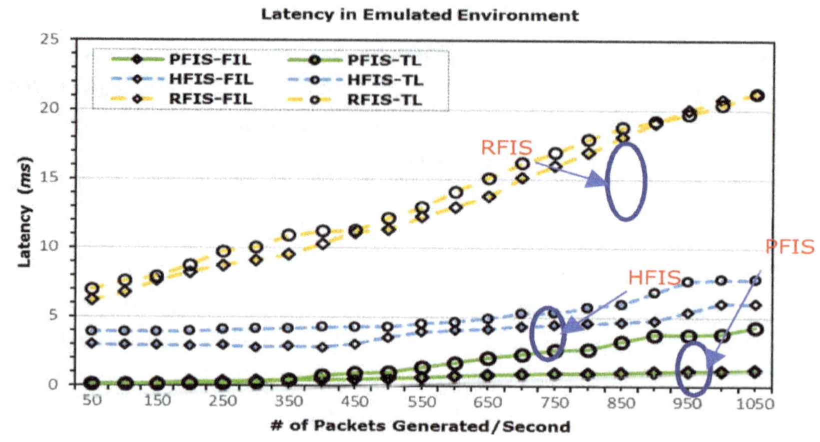

Figure 8. Latency in emulated environment.

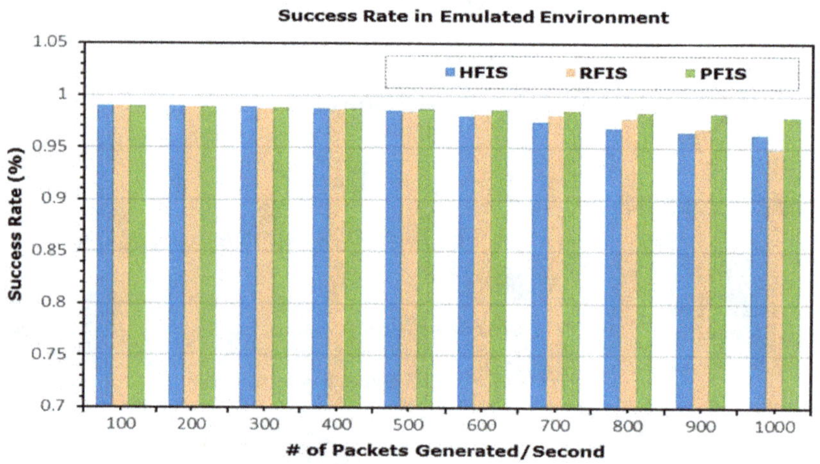

Figure 9. Success rate in emulated environment.

5.2. Test Bed Implementation

The demonstrator bottling plant comprises sensors and actuators such as conveyor belts, physical and vacuum grippers, robots and a turning table. We designed and implemented the test bed experiment to study the performance of the proposed SDIAN model. To do so, we transformed some parts of the demonstrator plant to be controlled by the RPis, while other parts rely on classical PLC solutions and vendor specific robot controllers. The portion that was controlled by the RPis includes a conveyor belt carrying bottle caps, sensors to detect when a cap arrives and a robotic arm as an actuator that will pick the cap and restore it into the designated location. The behavior of the sensors and actuators are determined by the controller and accordingly the script is pushed into the RPi. In this experiment, we have replaced two of the traditional PLCs with RPI-based PLCs to control a small set of sensors/actuators mounted on the Festo plant demonstrator. As shown in Figure 10, we interface two RPi-based PLCs (RPI-1 and RPI-2) with one of the gear boxes from the food demonstrator plant to get connectivity with a set of sensors and actuators. The two RPi-based PLCs are connected to a controller through a control channel. We use a python script to read, write and process signals from/to the I/O pin of the RPi. The python script replicates the standard behavior of traditional PLCs. We deploy a controller application in the controller to facilitate flow control communication between controller and RPi-based PLCs.

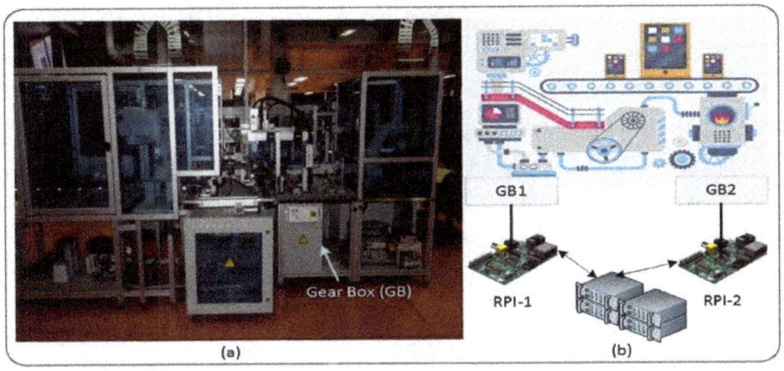

Figure 10. (**a**) Festo-based food processing plant demonstrator; (**b**) deployment diagram of the test bed.

Figure 11 presents the collage of a few snapshots of our test bed setup. It briefly demonstrates the different stages of the experiment. Clockwise from top left: a python script running on an RPi-based PLC replicates a traditional PLC, interfacing of RPis with sensors/actuators through the gear box, a robotic arm picking the desired object based on the instruction received from the corresponding RPi, and placing the object into a designated conveyor belt.

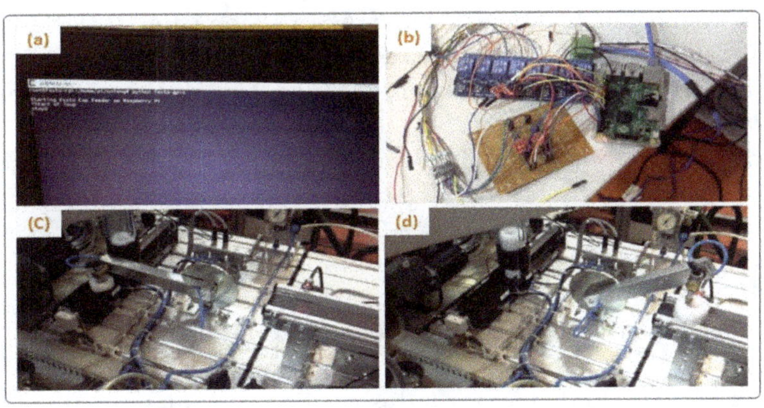

Figure 11. Clockwise from top left: (**a**) python script (**b**) interfacing with gear box (**c**,**d**) task execution by robotic arm.

The supplementary video clip demonstrates that the transformed architecture is working in a small-scale testbed experiment.

6. Conclusions

In this paper, we have explained the characteristics of SDN in the context of industrial automation. We highlighted the design of two flow installation schemes to precisely synchronize the industrial automation processes as well as presenting the potential benefits and opportunities of SDN. Furthermore, we have presented our architectural model that utilizes SDN and brought this into the context of an ongoing demonstrator project. Future work comprises the use of our demonstrator in current industry and academic projects. We are addressing both the challenges of the industrial automation hardware as well as integrating SDN into the communications utilizing software configurable devices.

Limitations of the demonstrator constrain evaluation of the proposed framework at this stage; however, the results obtained provide support for the approach and future work. For simplification, we limit our work to wired network technologies although it is evident that the approach could be extended to the integration of wireless (e.g., sensor network) with the wired network to achieve a unified architecture. The inclusion of wireless networks will introduce challenges including the seamless integration between the controllers across the wired and wireless domains. We also have limited our scope to one plant; therefore, the validation of using multiple controllers across multiple plants and the east-west communication are left unexplored and identified as future work. In the proposed framework, each RPi-based PLC is also used as an SDN switch, in future we may consider the use of lightweight SDN switches such as Zodiac FX, which could reduce the chance of bottlenecks across the RPis and clearly separate the forwarding devices from underlying field level sensors and actuators.

Author Contributions: K.A. was responsible for conceptualization, data curation, investigation, methodology, validation, visualization, and writing the original draft. M.G. reviewed and edited the manuscript. J.O.B. and H.S. participated in conceptualization and reviewed the final manuscript.

Acknowledgments: The authors wish to thank the Virtual Experiences Laboratory (VXLab), RMIT University. VxLab ensures smooth access to the Festo-based food processing plant demonstrator in conducting test bed implementation. This work is also supported in part by the Australia-India Research Centre for Automation Software Engineering (AICAUSE), RMIT University.

References

1. Xia, W.; Wen, Y.; Foh, C.H.; Niyato, D.; Xie, H. A Survey on Software-Defined Networking. *IEEE Commun. Surv. Tutor.* **2015**, *17*, 27–51. [CrossRef]
2. Huang, T.; Yu, F.R.; Zhang, C.; Liu, J.; Zhang, J.; Liu, J. A Survey on Large-scale Software Defined Networking (SDN) Testbeds: Approaches and Challenges. *IEEE Commun. Surv. Tutor.* **2016**, *19*. [CrossRef]
3. Hu, F.; Hao, Q.; Bao, K. A Survey on Software-Defined Network and OpenFlow: From Concept to Implementation. *IEEE Commun. Surv. Tutor.* **2014**, *16*, 2181–2206. [CrossRef]
4. Raghavan, B.; Casado, M.; Koponen, T.; Ratnasamy, S.; Ghodsi, A.; Shenker, S. Software-defined internet architecture: Decoupling architecture from infrastructure. In Proceedings of the 11th ACM Workshop on Hot Topics in Networks, Redmond, WA, USA, 29–30 October 2012.
5. Skeie, T.; Johannessen, S.; Holmeide, O. Timeliness of real-time IP communication in switched industrial Ethernet networks. *IEEE Trans. Ind. Inform.* **2006**, *2*, 25–39. [CrossRef]
6. Decotignie, J.D. Ethernet-Based Real-Time and Industrial Communications. *Proc. IEEE* **2005**, *93*, 1102–1117. [CrossRef]
7. Rojas, C.; Morell, P.; Sales, D.E. Guidelines for Industrial Ethernet infrastructure implementation: A control engineer's guide. In Proceedings of the 2010 IEEE-IAS/PCA 52nd Cement Industry Technical Conference, Colorado Springs, CO, USA, 28 March–1 April 2010; pp. 1–18.
8. Gungor, V.C.; Hancke, G.P. Industrial Wireless Sensor Networks: Challenges, Design Principles, and Technical Approaches. *IEEE Trans. Ind. Electron.* **2009**, *56*, 4258–4265. [CrossRef]
9. Hou, L.; Bergmann, N.W. Novel Industrial Wireless Sensor Networks for Machine Condition Monitoring and Fault Diagnosis. *IEEE Trans. Instrum. Meas.* **2012**, *61*, 2787–2798. [CrossRef]
10. Kopetz, H.; Ademaj, A.; Grillinger, P.; Steinhammer, K. The time-triggered Ethernet (TTE) design. In Proceedings of the Eighth IEEE International Symposium on Object-Oriented Real-Time Distributed Computing (ISORC'05), Seattle, WA, USA, 18–20 May 2005; pp. 22–33.
11. Cronberger, D. Software Defined Networks. 2015. Available online: http://www.industrial-ip.org/en/industrial-ip/convergence/software-defined-networks (accessed on 20 July 2018).
12. RAMI 4.0. Retrieved in August 2017. Available online: https://www.zvei.org/en/subjects/industry-4-0/thereference-architectural-model-rami-40-and-the-industrie-40-component/ (accessed on 20 July 2018).
13. IIRA. Retrieved in August 2017. Available online: http://www.iiconsortium.org/ (accessed on 20 July 2018).
14. Ahmed, K.; Blech, J.O.; Gregory, M.A.; Schmidt, H. Software Defined Networking for Communication and Control of Cyber-Physical Systems. In Proceedings of the 2015 IEEE 21st International Conference on Parallel and Distributed Systems (ICPADS), Melbourne, VIC, Australia, 14–17 December 2015; pp. 803–808.
15. Li, D.; Zhou, M.T.; Zeng, P.; Yang, M.; Zhang, Y.; Yu, H. Green and reliable software-defined industrial networks. *IEEE Commun. Mag.* **2016**, *54*, 30–37. [CrossRef]
16. Cronberger, D. The software defined industrial network. *Ind. Ethernet Book* **2014**, *84*, 8–13.

17. Kalman, G.; Orfanus, D.; Hussain, R. Overview and future of switching solutions for industrial Ethernet. *Int. J. Adv. Netw. Serv.* **2014**, *7*, 206–215.
18. Schweissguth, E.; Danielis, P.; Niemann, C.; Timmermann, D. Application-aware Industrial Ethernet Based on an SDN-supported TDMA Approach. In Proceedings of the 2016 IEEE World Conference on Factory Communication Systems (WFCS), Aveiro, Portugal, 3–6 May 2016.
19. Henneke, D.; Wisniewski, L.; Jasperneite, J. Analysis of realizing a future industrial network by means of Software-Defined Networking (SDN). In Proceedings of the 2016 IEEE World Conference on Factory Communication Systems (WFCS), Aveiro, Portugal, 3–6 May 2016.
20. Schneider, B.; Zoitl, A.; Wenger, M.; Blech, J.O. Evaluating Software-Defined Networking for Deterministic Communication in Distributed Industrial Automation Systems. In Proceedings of the 2017 22nd IEEE International Conference on Emerging Technologies and Factory Automation (ETFA), Limassol, Cyprus, 12–15 September 2017.
21. Ahmed, K.; Nafi, N.S.; Blech, J.O.; Gregory, M.A.; Schmidt, H. Software defined industry automation networks. In Proceedings of the 2017 27th International Telecommunication Networks and Applications Conference (ITNAC), Melbourne, VIC, Australia, 22–24 November 2017; pp. 1–3.
22. Open Networking Fundation. Software-Defined Networking: The New Norm for Networks. Available online: https://www.opennetworking.org/images/stories/downloads/sdn-resources/white-papers/wp-sdn-newnorm.pdf (accessed on 5 March 2018).
23. Opennetworking.org. OpenFlow—Open Networking Foundation. 2016. Available online: https://www.opennetworking.org/sdn-resources/openflow (accessed on 10 April 2018).
24. Kelly, F.P. Network routing. *Philos. Trans. R. Soc. Lond. A Math. Phys. Eng. Sci.* **1991**, *337*, 343–367. [CrossRef]
25. Yufei, W.; Zheng, W.; Leah, Z. Internet traffic engineering without full mesh overlaying. In Proceedings of the INFOCOM 2001. Twentieth Annual Joint Conference of the IEEE Computer and Communications Societies, Anchorage, AK, USA, 22–26 April 2001; Volume 1, pp. 565–571.
26. Li, Y.; Zhang, Z.L.; Boley, D. From Shortest-Path to All-Path: The Routing Continuum Theory and Its Applications. *IEEE Trans. Parallel Distrib. Syst.* **2014**, *25*, 1745–1755. [CrossRef]
27. He, K.; Khalid, J.; Gember-Jacobson, A.; Das, S.; Prakash, C.; Akella, A.; Li, L.E.; Thottan, M. Measuring control plane latency in SDN-enabled switches. In Proceedings of the 1st ACM SIGCOMM Symposium on Software Defined Networking Research, Santa Clara, CA, USA, 17–18 June 2015.
28. Blenk, A.; Basta, A.; Zerwas, J.; Reisslein, M.; Kellerer, W. Control Plane Latency With SDN Network Hypervisors: The Cost of Virtualization. *IEEE Trans. Netw. Serv. Manag.* **2016**, *13*, 366–380. [CrossRef]
29. Community, R.S.F. Ryu SDN Framework. 2014. Available online: https://osrg.github.io/ryu/ (accessed on 10 April 2018).

6

Big Sensed Data Meets Deep Learning for Smarter Health Care in Smart Cities

Alex Adim Obinikpo † and Burak Kantarci *,†

School of Electrical Engineering and Computer Science, University Ottawa, Ottawa, ON K1N 6N5, Canada; aobin064@uottawa.ca

* Correspondence: burak.kantarci@uottawa.ca

† These authors contributed equally to this work.

Abstract: With the advent of the Internet of Things (IoT) concept and its integration with the smart city sensing, smart connected health systems have appeared as integral components of the smart city services. Hard sensing-based data acquisition through wearables or invasive probes, coupled with soft sensing-based acquisition such as crowd-sensing results in hidden patterns in the aggregated sensor data. Recent research aims to address this challenge through many hidden perceptron layers in the conventional artificial neural networks, namely by deep learning. In this article, we review deep learning techniques that can be applied to sensed data to improve prediction and decision making in smart health services. Furthermore, we present a comparison and taxonomy of these methodologies based on types of sensors and sensed data. We further provide thorough discussions on the open issues and research challenges in each category.

Keywords: wearable sensors; biosensors; smart health; deep learning; machine learning; analytics

1. Introduction

Smart cities are built on the foundation of information and communication technologies with the sole purpose of connecting citizens and technology for the overall improvement of the quality of lives. While quality of life includes ease of mobility and access to quality healthcare, amongst others, effective management and sustainability of resources, economic development and growth complement the fundamental requirements of smart cities. These goals are achieved by proper management and processing of the data acquired from dedicated or non-dedicated sensor networks. In most cases, the information gathered is refined continuously to produce other information for efficiency and effectiveness within the smart city.

In the Internet of Things (IoT) era, the interplay between mobile networks, wireless communications and artificial intelligence is transforming the way that humans live and survive via various forms of improvements in technological advancements, more specifically improved computing power, high performance processing and huge memory capacities. With the advent of cyber-physical systems, which comprise the seamless integration of physical systems with computing and communication resources, a paradigm shift from the conventional city concept towards a smart city design has been coined as the term smart city. Basically, a smart city is envisioned to be ICT driven and capable of offering various services such as smart driving, smart homes, smart living, smart governance and smart health, just to mention a few [1]. ICT-driven management and control, as well as the overwhelming use of sensors in smart devices for the good and well-being of citizens yields the desired level of intelligence to these services. Besides continuously informing the citizens, being liveable and ensuring the well-being of the citizens are reported among the requirements of smart cities [2]. Therefore, the smart city concept needs transformation of health services through sensor and IoT-enablement of medical devices, communications and decision support. This ensures availability, ubiquity and personal customization of services, as well as ease of access to these services. As stated in [3], the criteria for smartness of an environment are various. As a corollary to this statement, the authors investigate a smart city in the following dimensions: (1) the technology angle (i.e., digital, intelligent, virtual, ubiquitous and information city), (2) the people angle (creative city, humane city, learning city and knowledge city) and (3) the community angle (smart community).

Furthermore, in the same vein, Anthopoulus (see [4]) divides the smart city into the following eight components: (1) smart infrastructures where facilities utilize sensors and chips; (2) smart transportation where vehicular networks along with the communication infrastructure are deployed for monitoring purposes; (3) smart environments where ICTs are used in the monitoring of the environment to acquire useful information regarding environmental sustainability; (4) smart services where ICTs are used for the the provision of community health, tourism, education and safety; (5) smart governance, which aims at proper delivery of government services; (6) smart people that use ICTs to access and increase humans' creativity; (7) smart living where technology is used for the improvement of the quality of life; and (8) smart economy, where businesses and organizations develop and grow through the use of technology. Given these components, a smart health system within a smart city appears to be one of the leading gateways to a more productive and liveable structure that ensures the well-being of the community.

Assurance of quality healthcare is a social sustainability concept in a smart city. Social sustainability denotes the liveability and wellbeing of communities in an urban setting [5]. The large population in cities is actually a basis for improved healthcare services because one negligence or improper health service might lead to an outbreak of diseases and infections, which might become epidemic, thereby costing much more in curtailing them; however, for these health services to be somewhat beneficial, the methods and channels of delivery need to be top-notch. There have been various research studies into the adequate method required for effective delivery of healthcare. One of these methods is smart health. Smart health basically is the provision of health services using the sensing capabilities and infrastructures of smart cities. In recent years, smart health has gained wide recognition due to the increase of technological devices and the ability to process the data gathered from these devices with minimum error.

As recent research states, proper management and development of smart health is the key to success of the smart city ecosystem [6]. Smart health involves the use of sensors in smart devices and specifically manufactured/prototyped wearable sensors/bio-patches for proper monitoring of the health status of individuals living within a smart city, as shown in Figure 1. This example depicts a scenario for air quality monitoring to ensure healthier communities. Smart city infrastructure builds upon networked sensors that can be either dedicated or non-dedicated. As an integral component of the smart city, smart health systems utilize devices with embedded sensors (as non-dedicated sensing components) for environmental and ambient data collection such as temperature, air quality index (AQI) and humidity. Besides, wearables and carry-on sensors (as the dedicated sensing components) are also utilized to acquire medical data from individuals. Both dedicated and non-dedicated sensory data are transmitted to the data centres as the inputs of processing and further decision making processes. To achieve this goal, the already existing smart city framework coupled with IoT networking needs to be leveraged.

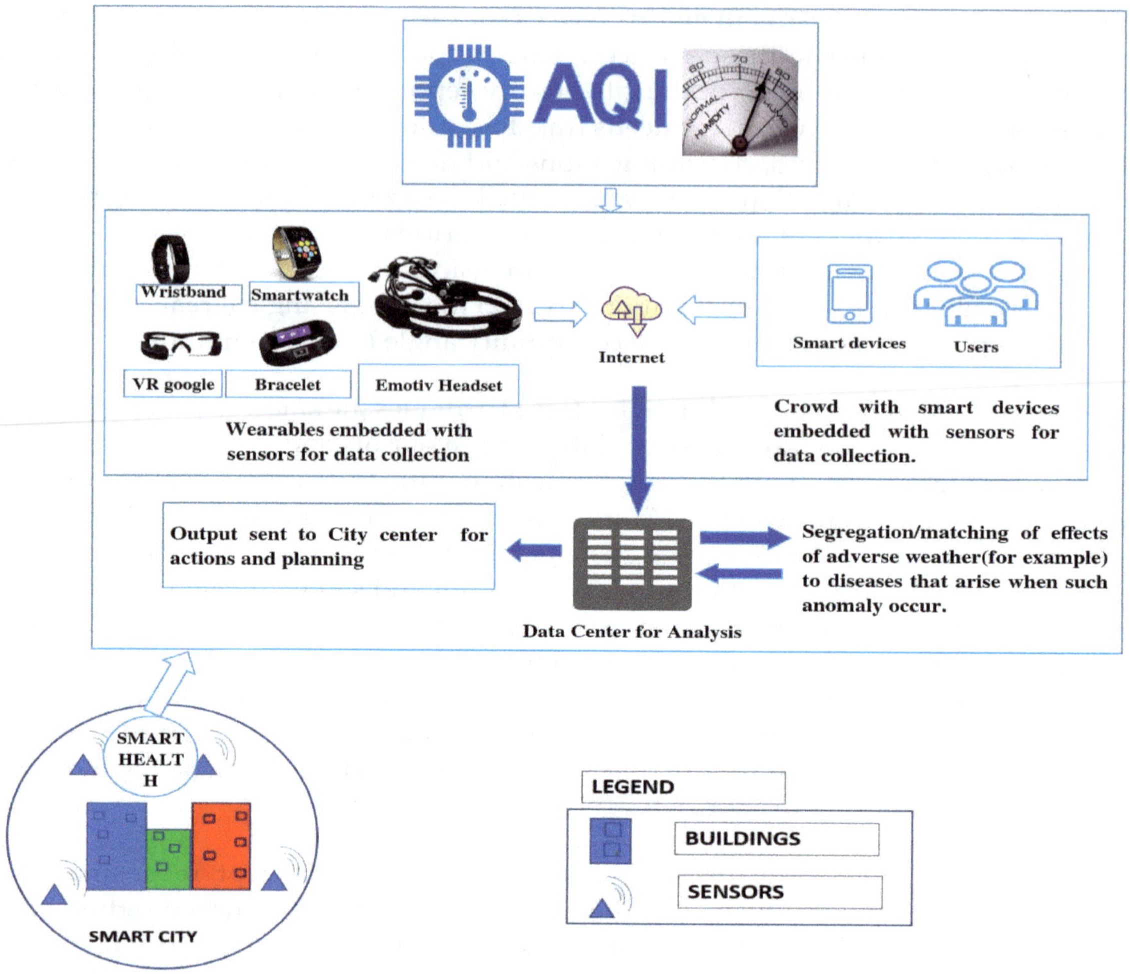

Figure 1. Smart health embedded within a smart city. An example scenario is illustrated to detect the air quality indicator to ensure healthier communities (figure produced by Creately Online Diagram, Cinergix Pvt. Ltd., Mentone, Australia).

In other words, a smart city needs to provide the required framework for smart health to grow rapidly and achieve its aim. Ensuring the quality of big sensed data acquisition is one key aspect of smart health challenges, and the ability to leverage these data is an important aspect of smart health development, as well as building a sustainable smart city structure [7–9]. Applications of smart health within smart cities are various. For example, Zulfiqar et al. [10] proposed an intelligent system for detecting and monitoring patients that might have voice complication issues. This is necessary since a number of services within the smart cities are voice enabled. As such, any disorder with the voice might translate into everyday service problems within the smart city. The same problem has been studied by the researchers in [11], where voice data and ECG signals were used as the inputs to the voice pathology detection system. Furthermore, with the aim of ensuring air quality within a smart city, the researchers in [12] developed a cloud-based monitoring system for air quality to ensure environmental health monitoring. The motivation for environmental health is ensuring the wellness of communities in a smart city for sustainability. The body area network within a smart city can be used for ECG monitoring with the aim of warning an individual of any heart-related problem, especially cardiac arrest [13], and also helps in determining the nature of human kinematic actions with the aim of ensuring improved quality of healthcare whenever needed [14]. Data management of patients is also one of the applications of smart health in smart cities that is of paramount importance. As discussed in [15], proper management of patient records both at the data entry and application levels ensures that patients get the required treatment when due, and this also helps in the development of personalized medicine applications [16]. Furthermore, the scope of smart health in smart cities is not limited to the physiologic phenomena in human bodies; it also extends to the environment and physical building blocks of the smart infrastructure. Indeed, the consequences of mismanaged environment and/or physical structures are potentially unhealthy users and communities, which is not a sustainable case for a smart city. To address this problem, the researchers in [17] created a system to monitor the structural health within a city using wireless sensor networks (WSN). It is worth mentioning that structural health monitoring also leads to inferential decision making services.

With these in mind, calls for new techniques that will ensure proper health service delivery have emerged. As an evolving concept, machine intelligence has attracted the healthcare sector by introducing effective decision support mechanisms applied to sensory data acquired through various media such as wearables or body area networks [18]. Machine learning techniques have undergone substantial improvements during the evolution of artificial intelligence and its integration with sensor and actuator networks. Despite many incremental improvements, deep learning has arisen as the most powerful tool thanks to its high level abstraction of complex big data, particularly big multimedia data [19].

Deep learning (DL) derives from conventional artificial neural networks (ANNs) with many hidden perceptron layers that can help in identifying hidden patterns [20]. Although having many hidden perceptron layers in a deep neural network is promising, when the concept of deep learning was initially coined, it was limited mostly by computational power of the available computing systems. However, with the advent of the improved computational capability of computing systems, as well as the rise of cloudified distributed models, deep learning has become a strong tool for analysing sensory data (particularly multimedia sensory data) and assisting in long-term decisions. The basic idea of deep learning is trying to replicate what the human brain does in most cases. Thus, in a sensor and actuator network setting, the deep learning network receives sensory input and iteratively passes it to subsequent layers until a desirable output is met. With the iterative process, the weights of the network links are adapted so as to match the input with the desirable output during the training process. With the widespread use of heterogeneous sensors such as wearables, medical imaging sensors, invasive sensors or embedded sensors in smart devices to acquire medical data, the emergence and applicability of deep learning is quite visible in modern day healthcare, from diagnosis to prognosis to health management.

While shallow learning algorithms enforce shallow methods on sensor data for feature representation, deep learning seeks to extract hierarchical representations [21,22] from large-scale data. Thus, the idea is using deep architectural models with multiple layers of non-linear transformations [23]. For instance, the authors in [24] use a shallow network with a covariance representation on the 3D sensor data in order to recognize human action from skeletal data. On the other hand, in the study in [25], the AlexNet model and the histogram of oriented gradients features are used to obtain deep features from the data acquired through 3D depth sensors.

In this article, we provide a thorough review of the deep learning approaches that can be applied to sensor data in smart health applications in smart cities. The motivation behind this study is that deep learning techniques are among the key enablers of the digital health technology within a smart city framework. This is due to the performance and accuracy issues experienced by conventional machine learning techniques under high dimensional data. Thus, it is worth noting that deep learning is not a total replacement of machine learning, but an effective tool to cope with dimensionality issues in several applications such as smart health [26]. To this end, this article aims to highlight the emergence of deep learning techniques in smart health within a smart city ecosystem and at the same time to give future directions by discussing the challenges and open issues that are still pertinent. In accordance with these, we provide a taxonomy of sensor data acquisition and processing techniques in smart health applications. Our taxonomy and review of deep learning approaches pave the way for providing insights for deep learning algorithms in particular smart health applications.

This work is organized as follows. In Section 2, we briefly discuss the transition from conventional machine learning methodologies to the deep learning methods. Section 3 gives a brief overview of the use of deep learning techniques on sensor network applications and major deep learning techniques that are applied on sensory data, while Section 4 provides insights for the smart health applications where deep learning can be used to process and interpret sensed data. Section 5 presents outstanding challenges and opportunities for the researchers to address in big sensed health data by deep learning. Finally, Section 6 concludes the article by summarizing the reviewed methodologies and future directions.

2. Analysis of Sensory Data in E-Health

2.1. Conventional Machine Learning on Sensed Health Data

With the advent of the WSN concept, machine learning has been identified as a viable solution to reduce the capital and operational expenditures on the design of the network, as well as to improve the lifetime of the network [27]. Presently, the majority of the machine learning techniques use a combination of feature extraction and modality-specific algorithms that are used to identify/recognize handwriting and/or speech [28]. This normally requires a dataset that is big in volume and powerful computing resources to support tremendous amount of background tasks. Furthermore, despite tedious efforts, there are always bound to be certain issues, and these perform poorly in the presence of inconsistencies and diversity in the dataset. One of the major advantages of machine learning in most cases is feature learning where a machine is trained on some datasets and the output provides valuable representation of the initial feature.

Applications of machine learning algorithms on sensory data are various such as telemedicine [18,29,30], air quality monitoring [31], indoor localization [32] and smart transportation [33]. However, conventional machine learning still has certain limitations such as inability to optimize non-differentiable discontinuous loss functions or not being able to obtain results following a feasible training duration at all times. These and many other issues encountered by machine learning techniques paved the way for deep learning as a more robust learning tool.

2.2. Deep Learning on Sensed Health Data

In [34], deep learning is defined as a collection of algorithmic procedures that 'mimic the brain'. More specifically, deep learning involves learning of layers via algorithmic steps. These layers enable the definition of hierarchical knowledge that derives from simpler knowledge [35]. There have been several attempts to build and design computers that are equipped with the ability to think. Until recently, this effort has been translated into rule-based learning, which is a 'top down' approach that involves creating rules for all possible circumstances [36]. However, this approach suffers from scalability since the number of rules is limited while its rule base is finite.

These issues can be remedied by adopting learning from experience instead of rule-based learning via a bottom-up approach. Labelled data form the experience. Labelled data are used as training input to the system where the training procedure is built upon past experiences. The learning from experience approach is well suited for applications such as spam filtering. On the other hand, the majority of the data collected by multimedia sensors (e.g., pictures, video feeds, sounds, etc.) are not properly labelled [37].

Real-world problems that involve processing of multimedia sensor data such as speech or face recognition are challenging to represent digitally due to the possibly infinite the problem domain. Thus, describing the problem adequately suffers especially in the presence of multi-dimensional features, which in turn leads to an increase in the volume of the space in such a way that the available data become sparse, and training on sparse data would not lead to meaningful results. Nevertheless, such 'infinite choice' problems are common in the processing of sensory data that are mostly acquired from multimedia sensors [38]. These issues pave the way for deep learning as deep learning algorithms have to work with hard and/or intuitive problems, which are defined with no or very few rules on high dimensional features. The absence of a rule set enforces the system to learn to cope with unforeseen circumstances [34].

Another characteristic of deep learning is the discovery of intricate structure in large datasets. To achieve this, deep learning utilizes a back propagation algorithm to adjust the internal parameters in each layer based on the representation of the parameters in the previous layer [34]. As such, it can be stated that representation learning is possible on partially labelled or unlabelled sensory data.

Acharya et al. [39] used deep learning techniques (specifically CNN) in the diagnosis and detection of coronary artery disease from the signals acquired from the electrocardiogram (ECG) and achieved an accuracy of 94.95%. The authors in [38] proposed the use of deep neural networks (DNN) for the active and automatic classification of ECG signals. Furthermore, in order to detect epileptic conditions early enough, deep learning with edge computing for the localization of epileptogenicity using electroencephalography (EEG) data has been proposed in [40]. Emotional well-being is a key state in the life of humans. With this in mind, the authors in [41] classified positive and negative emotions using deep belief networks (DBN) and data from EEG. The aim is to accurately capture the moment an emotional swing occurs. Their work yielded an 87.62% classification accuracy. The authors in [42] designed a BGMonitor for detecting blood glucose concentration and used a multi-task deep learning approach to analyse and process the data, and to make further inferences. The research yielded an accuracy of 82.14% when compared to the conventional methods.

3. Deep Learning Methods and Big Sensed Data

3.1. *Deep Learning on Sensor Network Applications*

Sensors are key enablers of the objects (things) of the emerging IoT networks [43]. The aggregation of sensors forms a network whose purpose amongst others is to generate and aggregate data for inferential purposes. The data generated from sensors need to be fine-tuned prior to undergoing any analytics procedure. This has led to various methods to formulate proper and adequate processing of sensed data from sensor and actuator networks. These methods are dependent on the type and applications of the sensed data. Deep learning (one of such methods) can be applied on sensor and actuator network applications to process data generated from sensors effectively and efficiently [44]. The output of a deep learning network can be used for decision making. Costilla-Reyes et al. used a convolutional neural network to learn spatio-temporal features that were derived from tomography sensors, and this yielded an effective and efficient way of performance classification of gait patterns using a limited number of experimental samples [45]. Transportation is another important service in a smart city architecture. The ability to acquire quality (i.e., high value) data from users is key to developing a smart transportation architecture. As an example study, the authors in [46] developed a mechanism using a deep neural network to learn the transportation modes of mobile users. In the same study, the integration of a deep learning-driven decision making system with a smart transportation architecture has been shown to result in 95% classification accuracy. Besides these, deep learning helped in the power usage pattern analysis of vehicles using the sensors embedded in smart phones [47]. The goal of the study in [47] is the timely prediction of the power consumption behaviour of vehicles using smart phones. The use of sensors in healthcare has been leading to significant achievements, and deep learning is being used to leverage the use of sensors and actuators for proper healthcare delivery. For instance, in assessing the level of Parkinson's disease, Eskofier et al. [48] used convolutional neural networks (CNN) for the classification and detection of the key features in Parkinson's disease based on data generated from wearable sensors. The results of the research proved that deep learning techniques work well with sensors when compared to other methods. Moreover, deep learning techniques, particularly CNNs, were used in estimating energy consumption while using wearable sensors for health condition monitoring [49]. Furthermore, by using sensory data generated from an infrared distance sensor, a deep learning classifier was developed for fall detection especially amongst the elderly population [50]. Besides these, with respect to security, combining biometric sensors and CNNs has resulted in a more robust approach for spoofing and security breach detection in digital systems [51]. Yin et al. [52] used a deep convolution network for proper visual object recognition as another application area of deep learning in the analysis of big sensed data, whereas for early detection of deforestation, Barreto et al. [53] proposed using a multilayer perceptron technique. In both studies, the input data are acquired via means of remote sensing.

3.2. Major Deep Learning Methods in Medical Sensory Data

In this subsection, we discuss the major deep learning methods that are used in e-health applications on medical sensory data. The following are the major deep learning methods in e-health, and a table of all notation used is given as Table 1.

Table 1. Basic notations used in the article. Notations are grouped into three categories: stand-alone symbols, vectors between units of different layers and symbols for functions.

Notations	Definition
x	Samples
y	Outputs
v	Visible vector
h	Hidden vector
q	State vector
W	Matrix of weight vectors
M	Total number of units for the hidden layer
w_{ij}	Weights vector between hidden unit h_j and visible unit v_i
S_j	Binary state of a vector
s_i^q	Binary state assigned to unit i by state vector q
Z	Partition factor
d_j	Biased weights for the j-th hidden units
c_i	Biased weights for the i-th visible units
z_i	Total i-th inputs
v_i	Visible unit i
w_{kj}^2	Weight vector from the k-th unit in the hidden Layer 2 to the j-th output unit
w_{ji}^1	Weight vector from the j-th unit in the hidden Layer 1 to the i-th output unit
W_{ji}^1	Matrix of weights from the j-th unit in the hidden Layer 1 to the i-th output unit
$E(q)$	Energy of a state vector q
σ	activation function
$P_r(q)$	Probability of a state vector q
$E(v,h)$	Energy function with respect to visible and hidden units
$pdf(v,h)$	Probability distribution with respect to visible and hidden units

3.2.1. Deep Feedforward Networks

Deep feed-forward networks can be counted among the first generation deep learning models and are based on multilayer perceptrons [34,54]. Basically, a feed-forward network aims at approximating a function f^*. A mapping $y = f(x;\Theta)$ is defined by a feed-forward network to learn the value of the Θ parameters by approximating with respect to the best function. Information flow in these networks is usually from the variables x being evaluated with respect to the outputs y.

During training, the aim is to ensure matching of $f(x)$ to $f^*(x)$, where each example of x is accompanied by a label $y \approx f^*(x)$. In most cases, the learning algorithm decides how to use these layers in order to get the best approximation of f^*. Since these layers do not obtain the desired output from the training data, they are referred to as hidden layers such as Layer 2 in the illustration in Figure 2 below.

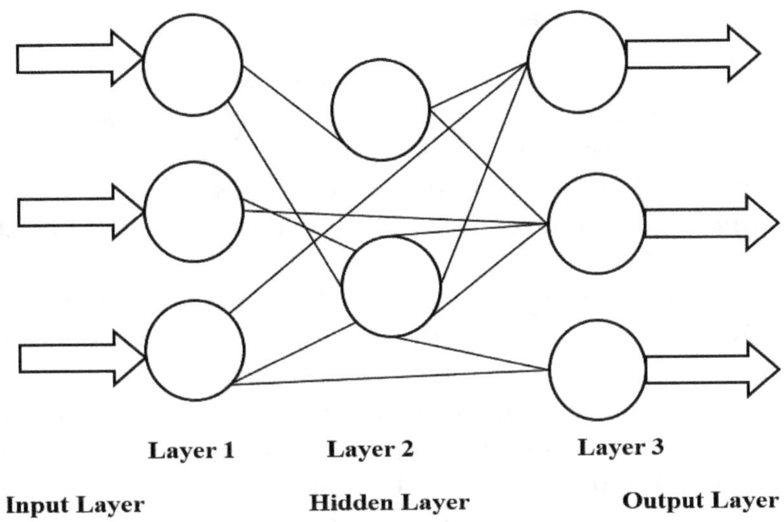

Figure 2. A basic deep feed-forward network.

A typical model for the deep forward network is described as follows. Given K outputs $y_1, \ldots, y_k$ for a given input x and the hidden layer, which consists of M units, then the output is formulated as shown in Equation (1) where σ is the activation function, $W_{ji}^{(1)}$ denotes the matrix of weights from unit j of hidden Layer 1 to the output unit i. In the equation, $w_{ji}^{(1)}$ and $w_{ki}^{(2)}$ stand for the weight vector from unit j of hidden Layer 1 to output unit i, and the weight vector from unit k of hidden Layer 2 to output unit i, respectively.

$$y_k(x, w) = \sigma\left(\sum_{j=1}^{M} w_{kj}^{(2)} h\left(\sum_{j=1}^{M} w_{ji}^{(1)} x_i + W_{ji}^{(1)}\right) + w_{ki}^{(2)}\right) \tag{1}$$

3.2.2. Autoencoder

An autoencoder is a neural network that is trained to copy its input to its output [22,55]. In most cases, an autoencoder is implemented as a three-layer neural network (see Figure 3) by directly connecting output units back to input units. In the figure, every output i is linked back to input i. The hidden layer h in the autoencoder represents the input by a code. Thus, a minimalist description of the network can be made by two main components as follows: (1) an encoder function $h = f(x)$; (2) a decoder function that is used to reconstruct the input, $r = g(h)$. Previously, autoencoders were used for dimensionality reduction or feature learning, but currently, the main purpose of autoencoder use is generative modelling because of the connection between autoencoders and latent variables.

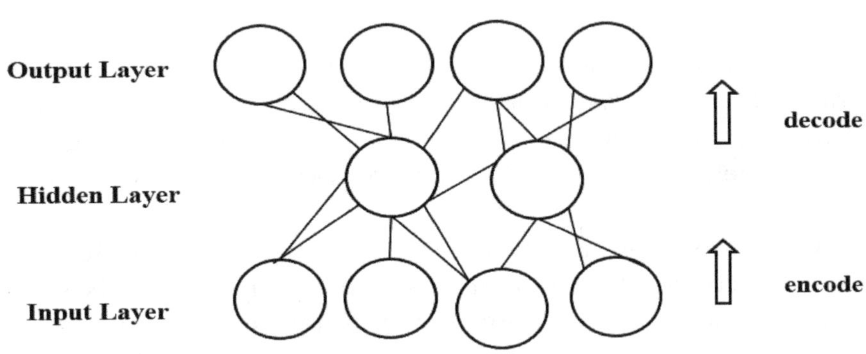

Figure 3. Autoencoder network.

Below are the various types of autoencoders.

- Undercomplete autoencoders [54] are suitable for the situation where the dimension of the code is less than the dimension of the input. This phenomenon usually leads to the inclusion of important features during training and learning.
- Regularized autoencoders [56] enable training any architecture of autoencoder successfully by choosing the code dimension and the capacity of the encoder/decoder based on the complexity of the distribution to be modelled.
- Sparse autoencoders [54] have a training criterion with a sparsity penalty, which usually occurs in the code layer with the purpose of copying the input to the output. Sparse autoencoders are used to learn features for another task such as classification.
- Denoising autoencoders [22] change the reconstruction error term of the cost function instead of adding a penalty to the cost function. Thus, a denoising autoencoder minimizes $L(x, g(f(\tilde{x})))$, where $\tilde{x}$ is a copy of x that has been distorted by noise.
- Contractive autoencoders [57] introduce an explicit regularizer on $h = f(x)$ making the derivatives of f as small as possible. The contractive autoencoders are trained to resist any perturbation of the input; as such, they map a neighbourhood of input points to a smaller neighbourhood of output points.

3.2.3. Convolutional Neural Networks

Convolutional neural networks (CNNs) replace matrix multiplication with convolutions in at least one of their layers [54,58,59]. CNNs have multiple layers of fields with small sets of neurons where an input image is partially processed [60]. When the outputs of these sets of neurons are tiled, their input regions overlap, leading to a new representation of the original image with higher resolution. This sequence is repeated in each sublayer. It is also worth mentioning that the dimensions of a CNN are mostly dependent on the size of the data.

The CNN architecture consists of three distinct layers: (1) the convolutional layer, (2) pooling layer and (3) fully-connected layer. Although it is not a requirement for the CNNs, as illustrated in Figure 4, fully-connected layers can follow a number of convolutional and subsampling layers.

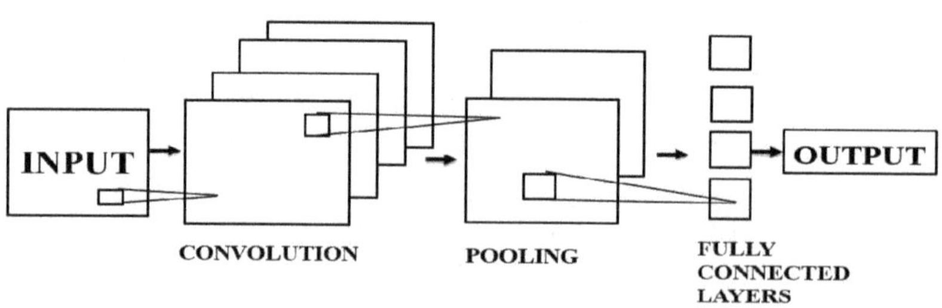

Figure 4. CNN architecture.

- Convolutional layer: The convolutional layer takes an $m \times m \times r$ image as the input, where m and r denote the height/width of the image and the number of channels, respectively. The convolutional layer contains k filters (or kernels) of size $n \times n \times q$, where $n < m$ and q can be less than or equal to the number of channels r (i.e., $q \leq r$). Here, q may vary for each kernel, and the feature map in this case has a size of $m - n + 1$.
- Pooling layers: These are listed as a key aspect of CNNs. The pooling layers are in general applied following the convolutional layers. A pooling layer in a CNN subsamples its input. Applying a max operation to the output of each filter is the most common way of pooling. Pooling over the complete matrix is not necessary. With respect to classification, pooling gives an output matrix with a fixed size thereby reducing the dimensionality of the output while keeping important information.

- Fully-connected layers: The layers here are all connected, i.e., both units of preceding and subsequent layers are connected

3.2.4. Deep Belief Network

The deep belief network (DBN) is a directed acyclic graph that builds on stochastic variables. It is a type of neural network composed of latent variables connected between multiple layers [36,61]. Despite the connections between layers, there are no connections between units within each layer. It can learn to reconstruct its inputs, then is trained to perform classification. In fact, the learning principle of DBNs is "one layer at a time via a greedy learning algorithm".

The properties of DBN are:

- Learning generative weights is through a layer-by-layer process with the purpose of determining the dependability of the variables in layer ℓ on the variables in layer ℓ' where ℓ denotes the index of any upper layer.
- Upon observing data in the bottom layer, inferring the values of the latent variables can be done in a single attempt.

It is worth noting that a DBN with one hidden layer implies a restricted Boltzmann machine (RBM). To train a DBN, first an RBM is trained using constructive divergence or stochastic maximum likelihood. The second RBM is then trained to model the defined distribution by sampling the hidden units in the first RBM. This process can be iterated as many times as possible to add further layers to the DBN.

3.2.5. Boltzmann Machine

As a special type of neural network, the Boltzmann machine (BM) consists of nodes that are connected symmetrically as shown in Figure 5, where neurons help a BM make on/off decisions [62]. In order to identify features that exhibit complex data regularities, a BM utilizes learning algorithms that are well-suited for search and learning problems.

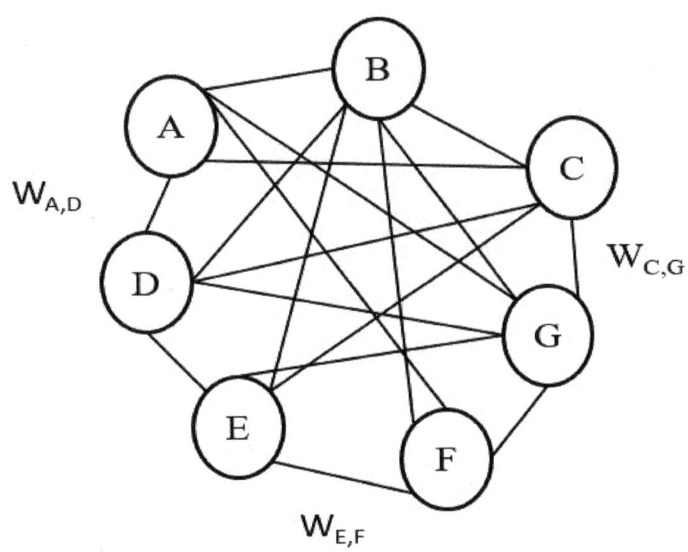

Figure 5. Boltzmann network.

To visualize the operation of a BM better, suppose unit i is able to continuously change its state. First, the node calculates the total input z_i, which is formulated as the sum of b_i (bias) and all the weights on the connections from other units as formulated in Equation (2). In the equation, w_{ij} denotes the weights on connections between i and j, and s_j is defined as in Equation (3). The probability for unit i is formulated as shown in Equation (4).

$$z_i = b_i + \sum_j s_j w_{ij} \tag{2}$$

$$s_j = \begin{cases} 1, & if \quad j \quad is \quad on \\ 0, & if \quad j \quad is \quad off \end{cases}. \tag{3}$$

$$P_r(s_i = 1) = \frac{1}{1 + e^{-z_i}} \tag{4}$$

If all neurons are updated sequentially, the network is expected to reach a BM distribution with state vector probability q and energy $E(q)$ as shown in Equations (5) and (6), respectively

$$P_r(q) = \frac{e^{-E(q)}}{\sum_u e^{-E(u)}} \tag{5}$$

$$E(q) = -\sum_i s_i^q b_i - \sum_{i<j} s_i^q s_j^q w_{ij} \tag{6}$$

With a view toward discarding the local optima, the weights on the connections could be chosen in such a way that each energy of the individual vectors represents the cost of these vectors. Learning in BM takes place in two different manners; either with hidden units or without hidden units.

There are various types of BM, and some of them are listed below:

Conditional Boltzmann machines model the data vectors and their distribution in such a way that any extension, no matter how simple it is, leads to conditional distributions.

Mean field Boltzmann machines compute the state of a unit based on the present state of other units in the network by using the real values of mean fields.

Higher-order Boltzmann machines have structures and learning patterns that can accept complex energy functions.

Restricted Boltzmann machines (RBMs): Two types of layers (i.e., visible vs. hidden) are included in the RBMs with no two similar connections [21,55].

In order to obtain unbiased elements from the set $\langle s_i s_j \rangle_{data}$, the hidden units h need to be conditionally independent of the visible unit v. However, heavy computation is required to get unbiased samples from $\langle s_i s_j \rangle_{data}$ [63].

Mathematically, the energy function of an RBM is given as formulated in Equation (7) and has a probability distribution as shown in Equation (8). In the equations, d_j and c_i stand for the biased weights for the hidden and visible units, respectively, whereas Z in the probability distribution function denotes the partition factor.

$$E(v, h) = -\sum_i c_i v_i - \sum_j d_j h_j - \sum_i \sum_j v_i w_{i,j} h_j \tag{7}$$

$$pdf(v, h) = \frac{1}{Z} e^{-E(v,h)} \tag{8}$$

Upon learning one hidden layer, the outcome can be treated as the input data required for training another RBM. This in turn leads to cascaded learning of multiple hidden layers, thus making the entire network be viewed as one model with improvements done on the lower bound [64].

4. Sensory Data Acquisition and Processing Using Deep Learning in Smart Health

Accurate data acquisition and processing is key to effective healthcare delivery. However, ensuring the accuracy of the acquired data has been one of the typical challenges in smart healthcare systems. This is due to the nature of data needed for quality assurance of healthcare delivery and the methods used for data acquisition. This phenomenon in the acquisition of sensory data in healthcare applications has led to the development of innovative techniques for data acquisition with the aim of complementing the "already in use", but upgraded methods. Due to these improvements in data acquisition methods, processing and interpreting sensory data have experienced an upward improvement, as well. These improvements have recently translated into improved quality of healthcare delivery. In this section, we briefly discuss the methods of sensory health data acquisition and processing as they relate to deep learning. Furthermore, we discuss how the generated sensory data types are processed using deep learning techniques. Figure 6a presents a brief taxonomy of data acquisition and processing. Data acquisition is performed mainly via wearables and probes as dedicated sensors and via built-in sensors of mobile devices as non-dedicated sensors.

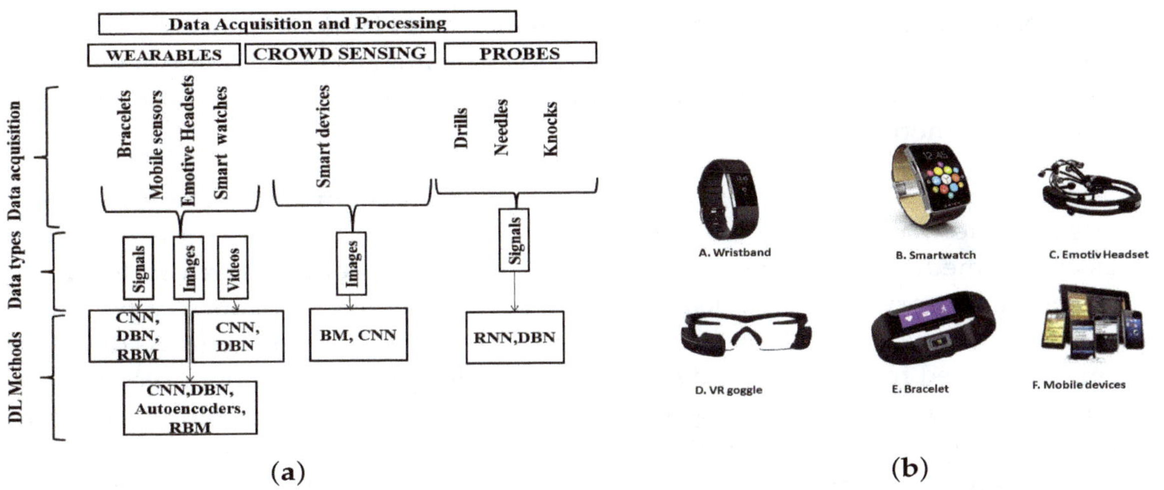

Figure 6. Data acquisition methods and processing techniques. (**a**) Taxonomy of sensory data acquisition and processing techniques; (**b**) types of wearables/carry-ons.

4.1. Sensory Data Acquisition and Processing via Wearables and Carry-Ons

Wearables and carry-ons have appeared as crucial components of personalized medicine aiming at performance and quality improvement in healthcare delivery. All wearables are equipped with built-in sensors that are used in data acquisition, and these smart devices with built-in sensors are in various forms as shown in the Figure 6b. The sensors in smart watches acquire data for the heart rate, movement, blood oxygen level and skin temperature. The virtual reality (VR) goggle captures video/image data, whereas the emotive headset senses mostly the brain signals. The wrist band and bracelets sense heart rate, body mass index (BMI), movement data (i.e., accelerometer and gyroscope) and temperature. Mobile devices provide non-dedicated sensing by acquiring sensory data regarding location, movement and BMI. While they all have similar functions, sensing functions mainly depend on the situations or needs for which they are required. The sensors embedded in these devices are the main resources for data acquisition and generation. The data output can be in the form of signals, images or videos; all with various importance and usefulness.

Processing of data generated from wearables and carry-ons is done based on the data types. Deep learning techniques used in this regards are also dependent on the data types and intended applications. This section aims to discuss the various deep learning methods used to process various data types generated from wearables and carry-ons.

- Image processing: Deep learning techniques play a major role in image processing for health advancements. Prominent amongst these methods are CNN, DBN, autoencoders and RBM. The authors in [65] use CNNs to help create a new network architecture with the aim of multi-channel data acquisition and also for supervised feature learning. Extracting features from brain images (e.g., magnetic resonance imaging (MRI), functional Magnetic resonance imaging (fMRI)) can help in early diagnosis and prognosis of severe diseases such as glioma. Moreover, the authors in [66] use DBN for the classification of mammography images in a bid to detect calcifications that may be the indicators of breast cancer. With high accuracy achieved in the detection, proper diagnosis of breast cancer becomes possible in radiology. Kuang and He in [67] modified and used DBN for the classification of attention deficit hyperactivity disorder (ADHD) using images from fMRI data. In a similar fashion, Li et al. [68] used the RBM for training and processing the dataset generated from MRI and positron emission tomography (PET) scans with aim of accurately diagnosing Alzheimer's disease. Using deep CNN and clinical images, Esteva et al. [69] were able to detect and classify melanoma, which is a type of skin cancer. According to their research, this method outperforms the already available skin cancer classification techniques. In the same context, Peyman and Hamid [70] showed that CNN performs better in the preprocessing of clinical and dermoscopy images in the lesion segmentation part of the skin. The study argues that CNN requires less preprocessing procedure when compared to other known methods.

- Signal processing: Signal processing is an area of applied computing that has been evolving since its inception. Signal processing is an utmost important tool in diverse fields including the processing of medical sensory data. As new methods are being improved for accurate signal processing on sensory data, deep learning, as a robust method, appears as a potential technique used in signal processing. For instance, Ha and Choi use improved versions of CNN to process the signals derived from embedded sensors in mobile devices for proper recognition of human activities [71]. Human activity recognition is an important aspect of ubiquitous computing and one of the examples of its application is the diagnosis and provision of support and care for those with limited movement ability and capabilities. The authors in [72] propose applying a CNN-based methodology on sensed data for the prediction of sleep quality. In the corresponding study, the CNN model is used with the objective of classifying the factors that contribute to efficient and poor sleeping habits with wearable sensors [72]. Furthermore, deep CNN and deep feed-forward networks on the data acquired via wearable sensors are used for the classification and processing of human activity recognition by the researchers in this field [73].

- Video processing: Deep learning techniques are also used for processing of videos generated from wearable devices and carry-ons. Prominent amongst these applications is the human activity recognition via CNNs to process video data generated by wearables and/or multimedia sensors [74–76].

4.2. *Data Acquisition via Probes*

Data acquisition using probes was an early stage data gathering technique. Their development has been made possible using technological enhancements attached to these probing tools. With these enhancements, acquiring sensory readings of medical data has become possible. Probes can be in the form of needles, drills and sometimes knocks, with feedbacks generated via the technological enhancements attached to the probing tools. Probes have seen a revolution since modern science and traditional medicine are being harmonized, with both playing vital roles in healthcare delivery.

Data generated via probes are usually in the form of signals. Processing probe data requires certain deep learning techniques that are augmented in most cases for this purpose. As an example, Cheron et al. [77] used the invasive electrode injection method to acquire signals needed to formulate the kinematic relation between electromyography (EMG) and the trajectory of the arm during movements. To this end, the authors used a dynamic recurrent neural network to process these signals and showed the correlation between EMG and arm trajectory.

4.3. Data Acquisition via Crowd-Sensing

Mobile crowd-sensing is a non-dedicated sensing concept where data are acquired using built-in sensors of smart devices. As the capabilities of smart devices such as smartphones and tablets have tremendously improved during the last decade, any smartphone today is equipped with tens of built-in sensors. Thus, ubiquity, improved sensing, computing and communication capabilities in mobile smart devices have enabled these devices to be used as data acquisition tools in critical applications including smart health and emergency preparedness [78,79]. This type of data acquisition involves users moving towards a particular location and being implicitly recruited to capture required data by the built-in sensors in their smart devices [80]. Crowdsensing envisions a robust data collection approach where users have the leverage and the ability to choose and report more data for experimental purposes in real time [81]. Consequently, this type of data acquisition increases the amount of data required for any purpose especially for the applications under smart health and smart cities [82]. Moreover, it is worth noting that crowd-sensed data are big especially in volume and velocity; hence, application-specific and effective processing methods are required to analyse crowd-sensed datasets. Application-specific data analytics techniques are required for the following reason in mobile crowd-sensing: Besides volume and velocity, the variety and heterogeneity of sensors in mobile crowd-sensing are also phenomenal, which results in producing a gigantic amount of data, which might be partially labelled in some cases.

As an example smart health application where mobile crowd-sensed data are used, Pan et al. [83] have introduced AirTick, which utilizes crowd-sensed image data to obtain air quality information. To this end, AirTick applies Boltzmann machines as the deep learning method on the crowd-sensed image to process the data for eventual results. Furthermore, the authors in [84] have introduced a proposal for cleaner and healthier neighbourhoods and have developed a mobile crowd-sensing application called SpotGarbage, which allows users to capture images of garbage in their locality and send them to the data hub where the data are analysed. In the SpotGarbage application, CNNs are used as the deep learning method to analyse the crowd-sensed images.

Based on the review of different data acquisition techniques and the corresponding deep learning methodologies applied to sensed data under those acquisition techniques, a brief review is presented in Table 2. The table provides a summary of data acquisition techniques, their corresponding data types and some examples of the deep learning techniques used.

Table 2. Summary of data acquisition methods, data types and examples of deep learning technique used. Three types of data acquisition categories are defined, which acquire images, one-dimensional signals and videos. CNN, DBN, restricted Boltzmann machine (RBM) and BM are the deep learning methods that are used to analyse big sensed data.

Data Acquisition Technique	Data Type	Deep Learning Technique
Wearables	Image	CNN [65,69,70], DBN [66,67], RBM [68],
	Signal	CNN [71–73]
	Video	CNN [74–76]
Probes	Signal	RNN [77]
Crowd-sensing	Image	BM [83], CNN [84]

5. Deep Learning Challenges in Big Sensed Data: Opportunities in Smart Health Applications

Deep learning can assist in the processing of sensory data by classification, as well as prediction via learning. Training a dataset by a deep learning algorithm involves prediction, detection of errors and improving prediction quality with time. Based on the review of the state of the art in the previous sections, it can be stated that integration of sensed data in smart health applications with deep learning yields promising outcomes. On the other hand, there are several challenges and open issues that need to be addressed prior to realization of such integration. As those challenges arise from the nature of deep learning, sensor deployment and sensory data acquisition, addressing those challenges paves the way towards robust smart health applications. As a corollary, in this section, we introduce challenges and open issues in the integration of deep learning with smart health sensory data; and pursuant to these, we present opportunities to cope with these challenges in various applications with the integration of deep learning and the sensory data provided.

5.1. Challenges and Open Issues

Deep learning techniques have attracted researchers from many fields recently for sustainable and efficient smart health delivery in smart environments. However, it is worth noting that the application of deep learning techniques on sensory data still experiences challenges. Indeed, in most smart health applications, CNNs have been introduced as revolutionized methodologies to cope with the challenges that deep learning networks suffer. Thanks to the improvements in CNNs, to date, CNN has been identified as the most useful tool in most cases when smart health is involved.

To be able to fully exploit deep learning techniques on medical sensory data, certain challenges need to be addressed by the researchers in this field. The challenges faced by deep learning techniques in smart health are mostly related to the acquisition, quality and dimensionality of data. This is due to the fact that inferences or decisions are made based on the output/outcome of processed data. As seen in the previous section, data acquisition takes place on heterogeneous settings, i.e., various devices with their own sampling frequency, operating system and data formats. The heterogeneity phenomenon generally results in a data plane with huge dimensions. The higher the dimension gets, the more difficult the training of the data, which ultimately leads to a longer time frame for result generation. Moreover, determining the depth of the network architecture in order to get a favourable result is another challenge since the depth of the network impacts the training time, which is an outstanding challenge to be addressed by the researchers in this field.

Value and trustworthiness of the data comprise another challenge that impacts the success of deep learning algorithms in a smart health setting. As any deep learning technique is supposed to be applied to big sensed health data, novel data acquisition techniques that ensure the highest level of trustworthiness for the acquired data are emergent.

Uncertainty in the acquired sensor data remains a grand challenge. A significant amount of the acquired data is partially labelled or unlabelled. Therefore, novel mechanisms to quantify and cope with the uncertainty phenomenon in the sensed data are emergent to improve the accuracy, as well as the efficiency of deep learning techniques in smart health applications.

Furthermore, data acquisition via non-dedicated sensors is also possible in smart cities sensing [85]. In the presence of non-dedicated sensors for data acquisition in smart health, it is a big challenge to know how much data should be acquired prior to processing. Since dedicated and non-dedicated sensors are mostly coupled in smart health applications, determining the amount of data required from different wearables becomes a grand challenge, as well. Recent research proposes the use of compressing sensing methods in participatory or opportunistic sensing via mobile devices [86]; hence, data explosion in non-dedicated acquisition can be prevented. However, in the presence of non-dedicated sensing system, dynamic determination of the number of wearables/sensors that can ensure the desired amount of data is another open issue to be addressed prior to analysing the big sensed data via deep learning networks.

In addition to all this, ensuring trustworthiness of the acquired sensory data prior to deep learning analysis remains an open issue, while auction and game theoretic solutions have been proposed to increase user involvement in the trustworthiness assurance stage of the data that are acquired via non-dedicated sensors [87,88]. In the presence of a collaboration between dedicated and non-dedicated sensors in the data acquisition, coping with the reliability of the non-dedicated end still requires efficient solutions. It is worth noting that deep learning can also be used for behaviour analysis of non-dedicated sensors in such an environment with the aim of eliminating unreliable sensing sources in the data acquisition.

Last but not least, recent research points out the emergence of IoT-driven data acquisition systems [78,89].

5.2. Opportunities in Smart Health Applications for Deep Learning

In this section, we discuss some of the applications of deep learning. To this end, we categorize these applications into three main groups for easy reference. Table 3 shows a summary of these applications together with the deep learning methods used under the three categories, namely medical imaging, bioinformatics and predictive analysis. The table is a useful reference to select the appropriate deep learning technique(s) while aiming to address the challenges and open issues in the previous subsection.

Table 3. Smart health applications with their respective deep learning techniques on medical sensory data. Applications are grouped into three categories: Medical imaging, bioinformatics and predictive analysis. Each application addresses multiple problems on sensed data through various deep learning techniques. DNN, deep neural network.

Application	Problem	Deep Learning Techniques	References
Medical Imaging	Neural Cells Classification	CNN	[65]
	3D brain reconstruction	Deep CNN	[90]
	Brain Tissue Classification	DBN	[67,68]
	Tumour Detection	DNN	[65,66]
	Alzheimer's Diagnosis	DNN	[91]
Bioinformatics	Cancer Diagnosis	Deep Autoencoder	[92]
	Gene Classification	DBN	[93]
	Protein Slicing	DBN	[94,95]
Predictive Analysis	Disease prediction and analysis	Autoencoder	[96]
		RNN	[97]
		CNN	[97,98]

- Medical imaging: Deep learning techniques have actually helped the improvement of healthcare through accurate disease detection and recognition. An example is the detection of melanoma. To do this, deep learning algorithms learn important features related to melanoma from a group of medical images and run their learning-based prediction algorithm to detect the presence or likelihood of the disease.

  Furthermore, using images from MRI, fMRI and other sources, deep learning has been able to help 3D brain construction using autoencoders and deep CNN [90], neural cell classification using CNN [65], brain tissue classifications using DBN [67,68], tumour detection using DNN [65,66] and Alzheimer's diagnosis using DNN [91].
- Bioinformatics: The applications of deep learning in bioinformatics have seen a resurgence in the diagnosis and treatment of most terminal diseases. Examples of these could be seen in cancer diagnosis where deep autoencoders are used using gene expression as the input data [92]; gene selection/classification and gene variants using micro-array data sequencing with the aid of deep belief networks [93]. Moreover, deep belief networks play a key role in protein slicing/sequencing [94,95].
- Predictive analysis: Disease predictions have gained momentum with the advent of learning-based systems. Therefore, with the capability of deep learning to predict the occurrence of diseases accurately, predictive analysis of the future likelihood of diseases has experienced significant progress. Particular techniques that are used for predictive analysis of diseases are autoencoders [96], recurrent neural networks [97] and CNNs [97,98]. On the other hand, it is worth mentioning that in order to improve the accuracy of prediction, sensory data monitoring medical phenomena have to be coupled with sensory data monitoring human behaviour. Coupling of data

acquired from medical and behavioural sensors helps in conducting effective analysis of human behaviour in order to find patterns that could help in disease predictions and preventions.

6. Conclusions

With the growing need and widespread use of sensor and actuator networks in smart cities, there is a growing demand for top-notch methods for the acquisition and processing of big sensed data. Among smart city services, smart healthcare applications are becoming a part of daily life to prolong the lifetime of members of society and improve quality of life. With the heterogeneous and various types of data that are being generated on a daily basis, the existence of sensor and actuator networks (i.e., wearables, carry-ons and other medical sensors) calls for effective acquisition of sensed data, as well as accurate and efficient processing to deduce conclusions, predictions and recommendations for the healthiness state of individuals. Deep learning is an effective tool that is used in the processing of big sensed data especially under these settings. Although deep learning has evolved from the traditional artificial neural networks concept, it has become an evolving field with the advent of improved computational power, as well as the convergence of wired/wireless communication systems. In this article, we have briefly discussed the growing concept of smart health within the smart city framework by highlighting its major benefits for the social sustainability of the smart city infrastructure. We have provided a comprehensive survey of the use of deep learning techniques to analyse sensory data in e-health and presented the major deep learning techniques, namely deep feed-forward networks, autoencoders, convolutional neural networks, deep belief networks, Boltzmann machine and restricted Boltzmann machine. Furthermore, we have introduced various data acquisition mechanisms, namely wearables, probes and crowd-sensing. Following these, we have also linked the surveyed deep learning techniques to existing use cases in the analysis of medical sensory data. In order to provide a thorough understanding of these linkages, we have categorized the sensory data acquisition techniques based on the available technology for data generation. In the last part of this review article, we have studied the smart health applications that involve sensors and actuators and visited specific use cases in those applications along with the existing deep learning solutions to effectively analyse sensory data. To facilitate a thorough understanding of these applications and their requirements, we have classified these applications under the following three categories: medical imaging, bioinformatics and predictive analysis. In the last part of this review article, we have studied smart health applications that involve sensors and actuators and visited specific problems in those applications along with the existing deep learning solutions to effectively address those problems. Furthermore, we have provided a thorough discussion of the open issues and challenges in big sensed data in smart health, mainly focusing on the data acquisition and processing aspects from the standpoint of deep learning techniques.

Acknowledgments: This material is based upon work supported by the Natural Sciences and Engineering Research Council of Canada (NSERC) under Grant RGPIN/2017-04032.

Author Contributions: Alex Adim Obinikpo and Burak Kantarci conceived and pursued the literature survey on deep learning techniques on big sensed data for smart health applications, reviewed the state of the art, challenges and opportunities, and made conclusions. They both wrote the paper. Alex Adim Obinikpo created the illustrative images.

References

1. Guelzim, T.; Obaidat, M.; Sadoun, B. Chapter 1—Introduction and overview of key enabling technologies for smart cities and homes. In *Smart Cities and Homes*; Obaidat, M.S., Nicopolitidis, P., Eds.; Morgan Kaufmann: Boston, MA, USA, 2016; pp. 1–16.
2. Liu, D.; Huang, R.; Wosinski, M. Development of Smart Cities: Educational Perspective. In *Smart Learning in Smart Cities*; Springer: Singapore, 2017; pp. 3–14.

3. Nam, T.; Pardo, T.A. Conceptualizing smart city with dimensions of technology, people, and institutions. In Proceedings of the 12th Annual International Digital Government Research Conference: Digital Government Innovation in Challenging Times, College Park, MD, USA, 12–15 June 2011; ACM: New York, NY, USA, 2011; pp. 282–291.
4. Anthopoulos, L.G. The Rise of the Smart City. In *Understanding Smart Cities: A Tool for Smart Government or an Industrial Trick?* Springer: Cham, Switzerland, 2017; pp. 5–45.
5. Munzel, A.; Meyer-Waarden, L.; Galan, J.P. The social side of sustainability: Well-being as a driver and an outcome of social relationships and interactions on social networking sites. *Technol. Forecast. Soc. Change* **2017**, in press.
6. Fan, M.; Sun, J.; Zhou, B.; Chen, M. The smart health initiative in China: The case of Wuhan, Hubei province. *J. Med. Syst.* **2016**, *40*, 62.
7. Ndiaye, M.; Hancke, G.P.; Abu-Mahfouz, A.M. Software Defined Networking for Improved Wireless Sensor Network Management: A Survey. *Sensors* **2017**, *17*, 1031.
8. Pramanik, M.I.; Lau, R.Y.; Demirkan, H.; Azad, M.A.K. Smart health: Big data enabled health paradigm within smart cities. *Expert Syst. Appl.* **2017**, *87*, 370–383.
9. Nef, T.; Urwyler, P.; Büchler, M.; Tarnanas, I.; Stucki, R.; Cazzoli, D.; Müri, R.; Mosimann, U. Evaluation of three state-of-the-art classifiers for recognition of activities of daily living from smart home ambient data. *Sensors* **2015**, *15*, 11725–11740.
10. Ali, Z.; Muhammad, G.; Alhamid, M.F. An Automatic Health Monitoring System for Patients Suffering from Voice Complications in Smart Cities. *IEEE Access* **2017**, *5*, 3900–3908.
11. Hossain, M.S.; Muhammad, G.; Alamri, A. Smart healthcare monitoring: A voice pathology detection paradigm for smart cities. *Multimedia Syst.* **2017**, doi:10.1007/s00530-017-0561-x.
12. Mehta, Y.; Pai, M.M.; Mallissery, S.; Singh, S. Cloud enabled air quality detection, analysis and prediction—A smart city application for smart health. In Proceedings of the 2016 3rd MEC International Conference on Big Data and Smart City (ICBDSC), Muscat, Oman, 15–16 March 2016; pp. 1–7.
13. Sahoo, P.K.; Thakkar, H.K.; Lee, M.Y. A Cardiac Early Warning System with Multi Channel SCG and ECG Monitoring for Mobile Health. *Sensors* **2017**, *17*, 711.
14. Kim, T.; Park, J.; Heo, S.; Sung, K.; Park, J. Characterizing dynamic walking patterns and detecting falls with wearable sensors using Gaussian process methods. *Sensors* **2017**, *17*, 1172.
15. Yeh, Y.T.; Hsu, M.H.; Chen, C.Y.; Lo, Y.S.; Liu, C.T. Detection of potential drug-drug interactions for outpatients across hospitals. *Int. J. Environ. Res. Public Health* **2014**, *11*, 1369–1383.
16. Venkatesh, J.; Aksanli, B.; Chan, C.S.; Akyurek, A.S.; Rosing, T.S. Modular and Personalized Smart Health Application Design in a Smart City Environment. *IEEE Internet Things J.* **2017**, *PP*, 1, doi:10.1109/JIOT.2017.2712558.
17. Rajaram, M.L.; Kougianos, E.; Mohanty, S.P.; Sundaravadivel, P. A wireless sensor network simulation framework for structural health monitoring in smart cities. In Proceedings of the 2016 IEEE 6th International Conference on Consumer Electronics-Berlin (ICCE-Berlin), Berlin, Germany, 5–7 September 2016; pp. 78–82.
18. Hijazi, S.; Page, A.; Kantarci, B.; Soyata, T. Machine Learning in Cardiac Health Monitoring and Decision Support. *IEEE Comput.* **2016**, *49*, 38–48.
19. Ota, K.; Dao, M.S.; Mezaris, V.; Natale, F.G.B.D. Deep Learning for Mobile Multimedia: A Survey. *ACM Trans. Multimedia Comput. Commun. Appl.* **2017**, *13*, 34.
20. Yu, D.; Deng, L. Deep Learning and Its Applications to Signal and Information Processing [Exploratory DSP]. *IEEE Signal Process. Mag.* **2011**, *28*, 145–154.
21. Larochelle, H.; Bengio, Y. Classification using discriminative restricted Boltzmann machines. In Proceedings of the 25th International Conference on Machine Learning, Helsinki, Finland, 5–9 July 2008; pp. 536–543.
22. Vincent, P.; Larochelle, H.; Lajoie, I.; Bengio, Y.; Manzagol, P.A. Stacked denoising autoencoders: Learning useful representations in a deep network with a local denoising criterion. *J. Mach. Learn. Res.* **2010**, *11*, 3371–3408.
23. Wang, L.; Sng, D. Deep Learning Algorithms with Applications to Video Analytics for A Smart City: A Survey. *arXiv* **2015**, arXiv:1512.03131.
24. Cavazza, J.; Morerio, P.; Murino, V. When Kernel Methods Meet Feature Learning: Log-Covariance Network for Action Recognition From Skeletal Data. In Proceedings of the 2017 IEEE Conference on Computer Vision and Pattern Recognition Workshops (CVPRW), Honolulu, HI, USA, 21–26 July 2017; pp. 1251–1258.

25. Keceli, A.S.; Kaya, A.; Can, A.B. Action recognition with skeletal volume and deep learning. In Proceedings of the 2017 25th Signal Processing and Communications Applications Conference (SIU), Antalya, Turkey, 15–18 May 2017; pp. 1–4.
26. LeCun, Y.; Bengio, Y.; Hinton, G. Deep learning. *Nature* **2015**, *521*, 436–444.
27. Alsheikh, M.A.; Lin, S.; Niyato, D.; Tan, H.P. Machine Learning in Wireless Sensor Networks: Algorithms, Strategies, and Applications. *IEEE Commun. Surv. Tutor.* **2014**, *16*, 1996–2018.
28. Mohri, M.; Rostamizadeh, A.; Talwalkar, A. *Foundations of Machine Learning*; MIT Press: Cambridge, MA, USA, 2012.
29. Clifton, L.; Clifton, D.A.; Pimentel, M.A.F.; Watkinson, P.J.; Tarassenko, L. Predictive Monitoring of Mobile Patients by Combining Clinical Observations with Data from Wearable Sensors. *IEEE J. Biomed. Health Inform.* **2014**, *18*, 722–730.
30. Tsiouris, K.M.; Gatsios, D.; Rigas, G.; Miljkovic, D.; Seljak, B.K.; Bohanec, M.; Arredondo, M.T.; Antonini, A.; Konitsiotis, S.; Koutsouris, D.D.; et al. PD_Manager: An mHealth platform for Parkinson's disease patient management. *Healthcare Technol. Lett.* **2017**, *4*, 102–108.
31. Hu, K.; Rahman, A.; Bhrugubanda, H.; Sivaraman, V. HazeEst: Machine Learning Based Metropolitan Air Pollution Estimation from Fixed and Mobile Sensors. *IEEE Sens. J.* **2017**, *17*, 3517–3525.
32. Tariq, O.B.; Lazarescu, M.T.; Iqbal, J.; Lavagno, L. Performance of Machine Learning Classifiers for Indoor Person Localization with Capacitive Sensors. *IEEE Access* **2017**, *5*, 12913–12926.
33. Jahangiri, A.; Rakha, H.A. Applying Machine Learning Techniques to Transportation Mode Recognition Using Mobile Phone Sensor Data. *IEEE Trans. Intell. Transp. Syst.* **2015**, *16*, 2406–2417.
34. Schmidhuber, J. Deep Learning in Neural Networks: An Overview. *Neural Netw.* **2014**, *61*, 85–117.
35. Deng, L.; Yu, D. Deep learning: Methods and applications. *Found. Trends Signal Process.* **2014**, *7*, 197–387.
36. Hinton, G.E.; Osindero, S.; Teh, Y.W. A fast learning algorithm for deep belief nets. *Neural Comput.* **2006**, *18*, 1527–1554.
37. Do, T.M.T.; Gatica-Perez, D. The places of our lives: Visiting patterns and automatic labeling from longitudinal smartphone data. *IEEE Trans. Mob. Comput.* **2014**, *13*, 638–648.
38. Al Rahhal, M.M.; Bazi, Y.; AlHichri, H.; Alajlan, N.; Melgani, F.; Yager, R.R. Deep learning approach for active classification of electrocardiogram signals. *Inform. Sci.* **2016**, *345*, 340–354.
39. Acharya, U.R.; Fujita, H.; Lih, O.S.; Adam, M.; Tan, J.H.; Chua, C.K. Automated Detection of Coronary Artery Disease Using Different Durations of ECG Segments with Convolutional Neural Network. *Knowl.-Based Syst.* **2017**, *132*, 62–71.
40. Hosseini, M.P.; Tran, T.X.; Pompili, D.; Elisevich, K.; Soltanian-Zadeh, H. Deep Learning with Edge Computing for Localization of Epileptogenicity Using Multimodal rs-fMRI and EEG Big Data. In Proceedings of the 2017 IEEE International Conference on Autonomic Computing (ICAC), Columbus, OH, USA, 17–21 July 2017; pp. 83–92.
41. Zheng, W.L.; Zhu, J.Y.; Peng, Y.; Lu, B.L. EEG-based emotion classification using deep belief networks. In Proceedings of the 2014 IEEE International Conference on Multimedia and Expo (ICME), Chengdu, China, 14–18 July 2014; pp. 1–6.
42. Gu, W. Non-intrusive blood glucose monitor by multi-task deep learning: PhD forum abstract. In Proceedings of the 16th ACM/IEEE International Conference on Information Processing in Sensor Networks, Pittsburgh, PA, USA, 18–20 April 2017; ACM: New York, NY, USA, 2017; pp. 249–250.
43. Anagnostopoulos, T.; Zaslavsky, A.; Kolomvatsos, K.; Medvedev, A.; Amirian, P.; Morley, J.; Hadjieftymiades, S. Challenges and Opportunities of Waste Management in IoT-Enabled Smart Cities: A Survey. *IEEE Trans. Sustain. Comput.* **2017**, *2*, 275–289.
44. Taleb, S.; Al Sallab, A.; Hajj, H.; Dawy, Z.; Khanna, R.; Keshavamurthy, A. Deep learning with ensemble classification method for sensor sampling decisions. In Proceedings of the 2016 International Wireless Communications and Mobile Computing Conference (IWCMC), Paphos, Cyprus, 5–9 September 2016; pp. 114–119.
45. Costilla-Reyes, O.; Scully, P.; Ozanyan, K.B. Deep Neural Networks for Learning Spatio-Temporal Features from Tomography Sensors. *IEEE Trans. Ind. Electron.* **2018**, *65*, 645–653, doi:10.1109/TIE.2017.2716907.
46. Fang, S.H.; Fei, Y.X.; Xu, Z.; Tsao, Y. Learning Transportation Modes From Smartphone Sensors Based on Deep Neural Network. *IEEE Sens. J.* **2017**, *17*, 6111–6118.

47. Xu, X.; Yin, S.; Ouyang, P. Fast and low-power behavior analysis on vehicles using smartphones. In Proceedings of the 2017 6th International Symposium on Next Generation Electronics (ISNE), Keelung, Taiwan, 23–25 May 2017; pp. 1–4.
48. Eskofier, B.M.; Lee, S.I.; Daneault, J.F.; Golabchi, F.N.; Ferreira-Carvalho, G.; Vergara-Diaz, G.; Sapienza, S.; Costante, G.; Klucken, J.; Kautz, T.; et al. Recent machine learning advancements in sensor-based mobility analysis: Deep learning for Parkinson's disease assessment. In Proceedings of the 2016 IEEE 38th Annual International Conference of the Engineering in Medicine and Biology Society (EMBC), Orlando, FL, USA, 16–20 August 2016; pp. 655–658.
49. Zhu, J.; Pande, A.; Mohapatra, P.; Han, J.J. Using deep learning for energy expenditure estimation with wearable sensors. In Proceedings of the 2015 17th International Conference on E-health Networking, Application & Services (HealthCom), Boston, MA, USA, 14–17 October 2015; pp. 501–506.
50. Jankowski, S.; Szymański, Z.; Dziomin, U.; Mazurek, P.; Wagner, J. Deep learning classifier for fall detection based on IR distance sensor data. In Proceedings of the 2015 IEEE 8th International Conference on Intelligent Data Acquisition and Advanced Computing Systems: Technology and Applications (IDAACS), Warsaw, Poland, 24–26 September 2015; Volume 2, pp. 723–727.
51. Menotti, D.; Chiachia, G.; Pinto, A.; Schwartz, W.R.; Pedrini, H.; Falcao, A.X.; Rocha, A. Deep representations for iris, face, and fingerprint spoofing detection. *IEEE Trans. Inform. Forensics Secur.* **2015**, *10*, 864–879.
52. Yin, Y.; Liu, Z.; Zimmermann, R. Geographic information use in weakly-supervised deep learning for landmark recognition. In Proceedings of the 2017 IEEE International Conference on Multimedia and Expo (ICME), Hong Kong, China, 10–14 July 2017; pp. 1015–1020.
53. Barreto, T.L.; Rosa, R.A.; Wimmer, C.; Moreira, J.R.; Bins, L.S.; Cappabianco, F.A.M.; Almeida, J. Classification of Detected Changes from Multitemporal High-Res Xband SAR Images: Intensity and Texture Descriptors From SuperPixels. *IEEE J. Sel. Top. Appl. Earth Obs. Remote Sens.* **2016**, *9*, 5436–5448.
54. Goodfellow, I.; Bengio, Y.; Courville, A. *Deep Learning*; MIT Press: Cambridge, MA, USA, 2016.
55. Hinton, G.E.; Salakhutdinov, R.R. Reducing the dimensionality of data with neural networks. *Science* **2006**, *313*, 504–507.
56. Alain, G.; Bengio, Y.; Rifai, S. Regularized auto-encoders estimate local statistics. *Proc. CoRR* **2012**, 1–17.
57. Rifai, S.; Bengio, Y.; Dauphin, Y.; Vincent, P. A generative process for sampling contractive auto-encoders. *arXiv* **2012**, arXiv:1206.6434.
58. Abdulnabi, A.H.; Wang, G.; Lu, J.; Jia, K. Multi-Task CNN Model for Attribute Prediction. *IEEE Trans. Multimedia* **2015**, *17*, 1949–1959.
59. Deng, L.; Abdelhamid, O.; Yu, D. A deep convolutional neural network using heterogeneous pooling for trading acoustic invariance with phonetic confusion. In Proceedings of the 2013 IEEE International Conference on Acoustics, Speech and Signal Processing, Vancouver, BC, Canada, 26–31 May 2013; pp. 6669–6673.
60. Aghdam, H.H.; Heravi, E.J. *Guide to Convolutional Neural Networks: A Practical Application to Traffic-Sign Detection and Classification*; Springer: Cham, Switzerland, 2017.
61. Huang, G.; Lee, H.; Learnedmiller, E. Learning hierarchical representations for face verification with convolutional deep belief networks. In Proceedings of the 2012 IEEE Conference on Computer Vision and Pattern Recognition, Providence, RI, USA, 16–21 June 2012; pp. 2518–2525.
62. Ackley, D.H.; Hinton, G.E.; Sejnowski, T.J. A learning algorithm for boltzmann machines. *Cognit. Sci.* **1985**, *9*, 147–169.
63. Salakhutdinov, R.; Mnih, A.; Hinton, G. Restricted Boltzmann machines for collaborative filtering. In Proceedings of the 24th international conference on Machine learning, Corvalis, OR, USA, 20–24 June 2007; pp. 791–798.
64. Ribeiro, B.; Gonçalves, I.; Santos, S.; Kovacec, A. Deep Learning Networks for Off-Line Handwritten Signature Recognition. In Proceedings of the 2011 CIARP 16th Iberoamerican Congress on Pattern Recognition, Pucón, Chile, 15–18 November 2011; pp. 523–532.
65. Nie, D.; Zhang, H.; Adeli, E.; Liu, L.; Shen, D. *3D Deep Learning for Multi-Modal Imaging-Guided Survival Time Prediction of Brain Tumor Patients*; Springer: Cham, Switzerland, 2016.
66. Rose, D.C.; Arel, I.; Karnowski, T.P.; Paquit, V.C. Applying deep-layered clustering to mammography image analytics. In Proceedings of the 2010 Biomedical Sciences and Engineering Conference, Oak Ridge, TN, USA, 25–26 May 2010; pp. 1–4.

67. Kuang, D.; He, L. Classification on ADHD with Deep Learning. In Proceedings of the 2014 International Conference on Cloud Computing and Big Data, Wuhan, China, 12–14 November 2014; pp. 27–32.
68. Li, F.; Tran, L.; Thung, K.H.; Ji, S.; Shen, D.; Li, J. A Robust Deep Model for Improved Classification of AD/MCI Patients. *IEEE J. Biomed. Health Inform.* **2015**, *19*, 1610–1616.
69. Esteva, A.; Kuprel, B.; Novoa, R.A.; Ko, J.; Swetter, S.M.; Blau, H.M.; Thrun, S. Dermatologist-level classification of skin cancer with deep neural networks. *Nature* **2017**, *542*, 115–118.
70. Sabouri, P.; GholamHosseini, H. Lesion border detection using deep learning. In Proceedings of the 2016 IEEE Congress on Evolutionary Computation (CEC), Vancouver, BC, Canada, 24–29 July 2016; pp. 1416–1421.
71. Ha, S.; Choi, S. Convolutional neural networks for human activity recognition using multiple accelerometer and gyroscope sensors. In Proceedings of the 2016 International Joint Conference on Neural Networks (IJCNN), Vancouver, BC, Canada, 24–29 July 2016; pp. 381–388.
72. Sathyanarayana, A.; Joty, S.; Fernandez-Luque, L.; Ofli, F.; Srivastava, J.; Elmagarmid, A.; Arora, T.; Taheri, S. Sleep quality prediction from wearable data using deep learning. *JMIR mHealth uHealth* **2016**, *4*, e125.
73. Hammerla, N.Y.; Halloran, S.; Ploetz, T. Deep, convolutional, and recurrent models for human activity recognition using wearables. *arXiv* **2016**, arXiv:1604.08880.
74. Baccouche, M.; Mamalet, F.; Wolf, C.; Garcia, C.; Baskurt, A. Sequential deep learning for human action recognition. In *International Workshop on Human Behavior Understanding*; Springer: Berlin/Heidelberg, Germany, 2011; pp. 29–39.
75. Ji, S.; Xu, W.; Yang, M.; Yu, K. 3D convolutional neural networks for human action recognition. *IEEE Trans. Pattern Anal. Mach. Intell.* **2013**, *35*, 221–231.
76. Karpathy, A.; Toderici, G.; Shetty, S.; Leung, T.; Sukthankar, R.; Fei-Fei, L. Large-scale video classification with convolutional neural networks. In Proceedings of the 2014 IEEE Conference on Computer Vision and Pattern Recognition, Columbus, OH, USA, 23–28 June 2014; pp. 1725–1732.
77. Cheron, G.; Draye, J.P.; Bourgeios, M.; Libert, G. A dynamic neural network identification of electromyography and arm trajectory relationship during complex movements. *IEEE Trans. Biomed. Eng.* **1996**, *43*, 552–558.
78. Page, A.; Hijazi, S.; Askan, D.; Kantarci, B.; Soyata, T. Research Directions in Cloud-Based Decision Support Systems for Health Monitoring Using Internet-of-Things Driven Data Acquisition. *Int. J. Serv. Comput.* **2016**, *4*, 18–34.
79. Guo, B.; Han, Q.; Chen, H.; Shangguan, L.; Zhou, Z.; Yu, Z. The Emergence of Visual Crowdsensing: Challenges and Opportunities. *IEEE Commun. Surv. Tutor.* **2017**, *PP*, 1, doi:10.1109/COMST.2017.2726686.
80. Ma, H.; Zhao, D.; Yuan, P. Opportunities in mobile crowd sensing. *IEEE Commun. Mag.* **2014**, *52*, 29–35.
81. Haddawy, P.; Frommberger, L.; Kauppinen, T.; De Felice, G.; Charkratpahu, P.; Saengpao, S.; Kanchanakitsakul, P. Situation awareness in crowdsensing for disease surveillance in crisis situations. In Proceedings of the Seventh International Conference on Information and Communication Technologies and Development, Singapore, 15–18 May 2015; p. 38.
82. Cardone, G.; Foschini, L.; Bellavista, P.; Corradi, A.; Borcea, C.; Talasila, M.; Curtmola, R. Fostering participaction in smart cities: a geo-social crowdsensing platform. *IEEE Commun. Mag.* **2013**, *51*, 112–119.
83. Pan, Z.; Yu, H.; Miao, C.; Leung, C. Crowdsensing Air Quality with Camera-Enabled Mobile Devices. In Proceedings of the Twenty-Ninth IAAI Conference, San Francisco, CA, USA, 6–9 February 2017; pp. 4728–4733.
84. Mittal, G.; Yagnik, K.B.; Garg, M.; Krishnan, N.C. SpotGarbage: Smartphone app to detect garbage using deep learning. In Proceedings of the 2016 ACM International Joint Conference on Pervasive and Ubiquitous Computing, Heidelberg, Germany, 12–16 September 2016; pp. 940–945.
85. Habibzadeh, H.; Qin, Z.; Soyata, T.; Kantarci, B. Large Scale Distributed Dedicated- and Non-Dedicated Smart City Sensing Systems. *IEEE Sens. J.* **2017**, *17*, 7649–7658, doi:10.1109/JSEN.2017.2725638.
86. Xu, L.; Hao, X.; Lane, N.D.; Liu, X.; Moscibroda, T. More with Less: Lowering User Burden in Mobile Crowdsourcing through Compressive Sensing. In Proceedings of the 2015 ACM International Joint Conference on Pervasive and Ubiquitous Computing, Osaka, Japan, 7–11 September 2015; ACM: New York, NY, USA, 2015; pp. 659–670.
87. Pouryazdan, M.; Kantarci, B. The Smart Citizen Factor in Trustworthy Smart City Crowdsensing. *IT Prof.* **2016**, *18*, 26–33.

88. Pouryazdan, M.; Kantarci, B.; Soyata, T.; Song, H. Anchor-Assisted and Vote-Based Trustworthiness Assurance in Smart City Crowdsensing. *IEEE Access* **2016**, *4*, 529–541.
89. Farahani, B.; Firouzi, F.; Chang, V.; Badaroglu, M.; Constant, N.; Mankodiya, K. Towards fog-driven IoT eHealth: Promises and challenges of IoT in medicine and healthcare. *Future Gener. Comput. Syst.* **2018**, *78*, 659–676.
90. Kleesiek, J.; Urban, G.; Hubert, A.; Schwarz, D.; Maier-Hein, K.; Bendszus, M.; Biller, A. Deep MRI brain extraction: A 3D convolutional neural network for skull stripping. *Neuroimage* **2016**, *129*, 460–469.
91. Fritscher, K.; Raudaschl, P.; Zaffino, P.; Spadea, M.F.; Sharp, G.C.; Schubert, R. Deep Neural Networks for Fast Segmentation of 3D Medical Images. In *Medical Image Computing and Computer-Assisted Intervention—MICCAI*; Springer: Cham, Switzerland, 2016.
92. Fakoor, R.; Ladhak, F.; Nazi, A.; Huber, M. Using deep learning to enhance cancer diagnosis and classification. In Proceedings of the 30th International Conference on Machine Learning, Atlanta, GA, USA, 16–21 June 2013.
93. Khademi, M.; Nedialkov, N.S. Probabilistic Graphical Models and Deep Belief Networks for Prognosis of Breast Cancer. In Proceedings of the 2015 IEEE 14th International Conference on Machine Learning and Applications, Miami, FL, USA, 9–11 December 2016; pp. 727–732.
94. Angermueller, C.; Lee, H.J.; Reik, W.; Stegle, O. DeepCpG: Accurate prediction of single-cell DNA methylation states using deep learning. *Genome Biol.* **2017**, *18*, 67.
95. Tian, K.; Shao, M.; Wang, Y.; Guan, J.; Zhou, S. Boosting Compound-Protein Interaction Prediction by Deep Learning. *Methods* **2016**, *110*, 64–72.
96. Che, Z.; Purushotham, S.; Khemani, R.; Liu, Y. Distilling Knowledge from Deep Networks with Applications to Healthcare Domain. *Ann. Chirurgie* **2015**, *40*, 529–532.
97. Lipton, Z.C.; Kale, D.C.; Elkan, C.; Wetzell, R. Learning to Diagnose with LSTM Recurrent Neural Networks. In Proceedings of the International Conference on Learning Representations (ICLR 2016), San Juan, Puerto Rico, 2–4 May 2016.
98. Liang, Z.; Zhang, G.; Huang, J.X.; Hu, Q.V. Deep learning for healthcare decision making with EMRs. In Proceedings of the IEEE International Conference on Bioinformatics and Biomedicine, Belfast, UK, 2–5 November 2014; pp. 556–559.

7

An Experimental Evaluation of the Reliability of LoRa Long-Range Low-Power Wireless Communication

Marco Cattani *, Carlo Alberto Boano and Kay Römer

Institute for Technical Informatics, Graz University of Technology, Graz 8010, Austria; cboano@tugraz.at (C.A.B.); roemer@tugraz.at (K.R.)

* Correspondence: m.cattani@tugraz.at

Abstract: Recent technological innovations allow compact radios to transmit over long distances with minimal energy consumption and could drastically affect the way Internet of Things (IoT) technologies communicate in the near future. By extending the communication range of links, it is indeed possible to reduce the network diameter to a point that each node can communicate with almost every other node in the network directly. This drastically simplifies communication, removing the need of routing, and significantly reduces the overhead of data collection. Long-range low-power wireless technology, however, is still at its infancy, and it is yet unclear (i) whether it is sufficiently reliable to complement existing short-range and cellular technologies and (ii) which radio settings can sustain a high delivery rate while maximizing energy-efficiency. To shed light on this matter, this paper presents an extensive experimental study of the reliability of LoRa , one of the most promising long-range low-power wireless technologies to date. We focus our evaluation on the impact of physical layer settings on the effective data rate and energy efficiency of communications. Our results show that it is often not worth tuning parameters, thereby reducing the data rate in order to maximize the probability of successful reception, especially on links at the edge of their communication range. Furthermore, we study the impact of environmental factors on the performance of LoRa, and show that higher temperatures significantly decrease the received signal strength and may drastically affect packet reception.

Keywords: LoRa; long-range technology; environmental impact; temperature; link quality; outdoor; underground; indoor; energy-efficiency; reliability

---

1. Introduction

An increasing number of radio technologies enabling low-power wireless communication over long distances has emerged in the past years. Ultra-narrowband technologies such as Sigfox (Labège, France) and Weightless-N [1] (Cambridge, UK), as well as spread-spectrum technologies such as LoRa [2] (San Ramon, CA, USA), allow for communicating up to few kilometers, and to build up low-power wide area networks (LPWANs) that do not require the construction and maintenance of complex multi-hop topologies [3,4].

A key characteristic of LPWAN technologies is indeed the ability to *trade* throughput for range and vice versa, i.e., one has the ability to fine-tune physical layer (PHY) settings to select a more sensitive (but slow) configuration that allows communication over a longer distance. This flexibility makes LPWAN technologies particularly appealing to developers of Internet of Things (IoT) applications requiring long-range communications with relatively low data rates. At the same time, however, the ability to fine-tune PHY settings requires *a thorough understanding* of their impact on network performance, especially on the reliability and energy-efficiency of communications [5].

The research community has recently devoted significant attention to the role of PHY settings in the context of LPWANs [5–7], especially LoRa technology. Out of the existing LPWAN technologies,

LoRa has especially attracted a large body of work due to the availability of commercial off-the-shelf radio transceiver and platforms [8–10], as well as its ability to operate in an infrastructure-free manner and to build up ad hoc mesh networks [5]). LoRa-based networks have been deployed in several settings, ranging from indoor [4] and urban [7] environments, to maritime [11] and mountain scenarios [12]. These deployments have shown the impact of PHY settings on connectivity range and sensitivity [5,13], as well as having given a first impression of the packet reception ratio that can be achieved at different distances with different hardware platforms and physical layer configurations. Bor et al. [14] have also shown through simulation that the choice of the PHY settings affects the number of LoRa nodes that can concurrently access the channel, which has an impact on the scalability of LoRa networks. Furthermore, Bor and Roedig [4] have presented the results of systematic indoor experiments showing that the set of LoRa settings leading to the most energy-efficient operation dynamically changes over time. Based on these results, the authors proposed a protocol that periodically probes different settings and that dynamically picks the ones minimizing energy consumption at run-time.

Interplay between PHY settings and link quality. Although the aforementioned works started to shed light on how to carry out an optimal selection of LoRa's PHY settings, they all share a common assumption: the best performance is obtained in the presence of highly reliable links. Most works, indeed, specifically target PHY settings maximizing the link quality, i.e., focus on selecting physical layer configurations that allow to sustain a packet reception ratio of 90% or higher [4,12,14] This practice is likely influenced from the behavior of non opportunistic low-power wireless data collection protocols for IEEE 802.15.4 radios, which favor high-quality links to intermediate and lossy ones [15,16]. However, adjusting the PHY settings of the radio to maximize the link quality has important implications w.r.t. energy efficiency when using long-range low-power wireless technologies such as LoRa. Maximizing the link quality, indeed, typically implies an increase in the transmission power and data overhead, and the selection of a more sensitive (and hence slow) physical layer configuration. As a result, one increases not only the likelihood to receive packets, but also the energy consumption of the radio, due to the higher transmission power, and the radio-on time, due to larger PHY layer overhead. This observation raises a yet unanswered question: *is it worth selecting PHY settings to reduce the data rate in order to increase the link quality?* This question is particularly relevant when two nodes are at the edge of the communication range: should one select a setting that reduces the data rate to increase the robustness of communication (and aim for a link achieving a high packet reception ratio) or rather accept having a link of intermediate quality (i.e., experiencing some packet loss), but with high data rate, and implement a re-transmission scheme on top? How this choice affects the energy-efficiency of the network still needs to be investigated.

Impact of environmental conditions on communication performance. The characteristics of LPWANs make them suitable for outdoor deployments on a large scale, and it is hence important to study in detail the impact of *environmental effects* such as changes in meteorological conditions, as well as variations in temperature and humidity on network performance. Unfortunately, to date, there is still little understanding about the impact of the environment on the reliability of LoRa communication, especially for links that are at the edge of their communication range. Iova et al. [12] have reported the vulnerability of LoRa communications to environmental factors such as presence of vegetation and temperature variations, but without quantifying their impact. Other works in the low-power wireless community have shown that some IEEE 802.15.4 radios are particularly vulnerable to changes in temperature, and that even the daily fluctuations recorded outdoors can render a good link useless [17–20]. However, these results are platform-specific and cannot be generalized to LoRa transceivers. Therefore, *if and how much temperature affects LoRa's communication performance* is yet to be answered.

Our contributions. In this paper, we carry out an experimental evaluation of the reliability of LoRa in different settings and provide an answer to the aforementioned open questions. First, we study how PHY settings and environmental factors affect the reliability of LoRa communications through an extensive experimental campaign indoor, outdoor, and underground. In line with earlier works [7], our experiments show that PHY settings have a significant impact on packet reception rate and that indoor environments are more challenging for LoRa communications. Our results also suggest that it is better to use faster (but more fragile) settings together with a re-transmission mechanism rather then selecting resilient and slower settings, maximizing packet reception rate and link quality. A detailed study of the overhead of each PHY setting in relation to its improvement on packet reception rate indeed shows that setting a maximizing data rate and minimizing range should be preferred.

Furthermore, our experimental results show a clear correlation between temperature, humidity, packet reception rate, and received signal strength. We hence analyze in depth how environmental factors such as temperature variations affect the reliability of LoRa communications by performing a series of systematic experiments in controlled settings on different hardware platforms. These experiments show that the reliability of LoRa drastically decreases at high temperatures. On the one hand, the signal strength of received packets decreases linearly when temperature increases, as was also observed for a number of IEEE 802.15.4 radios [17,18]. On the other hand, the decrease in signal strength can significantly affect LoRa links that are at the edge of the communication range, increasing packet corruption and loss up to a point in which a link is totally compromised.

The contributions of this paper are hence threefold:

- We study how PHY settings and environmental factors affect the reliability of LoRa through an extensive experimental campaign indoor, outdoor, and underground;
- We analyze the impact of LoRa's PHY settings on the effective data rate and energy efficiency of communications, highlighting that it is not worth selecting settings to reduce the data rate in order to increase the link quality;
- We systematically study the impact of temperature on the reliability of LoRa communications and show that high temperatures decrease the received signal strength and drastically increase packet loss and corruption for nodes at the edge of the communication range.

The paper proceeds as follows. In the next section, we introduce the reader to long-range technologies and to the LoRa physical layer settings that can be configured to fine-tune the operations of LoRa transceivers. Section 3 highlights the yet open questions with respect to LoRa's reliability as a function of PHY settings and environmental conditions. In Section 4, we describe our experiments indoor, outdoor, and underground, highlighting the strong impact of the chosen PHY settings and environmental conditions on the reliability of communications. Thereafter, we investigate in detail the interplay between PHY settings and link quality in Section 5 and carry out experiments in controlled settings to quantify the impact of temperature on LoRa's communication performance in Section 6. We finally summarize our contributions in Section 7, along with a discussion of future work.

2. Primer on LPWANs and LoRa

Low-power wide area networks complement short range wireless technologies such as Wi-Fi, Bluetooth Low Energy, and IEEE 802.15.4, and represent an interesting alternative to cellular technologies for urban-scale IoT applications. The success of LPWANs is due to their ability of providing long-range communication to thousands of devices at minimal cost and limited energy expenditure. Longer communication ranges allow for drastically simplifying duty cycling and networking protocol, as LPWANs can form star topologies where the low-power end devices are able to directly communicate with a more powerful orchestrator. This also allows for designing asymmetric communication schemes and to shift the load to the more powerful central device.

In order to increase the communication range, LPWAN technologies must improve the signal-to-noise ratio (SNR) at the receiver, either by narrowing down the receiver's bandwidth (reducing the receiver's noise-floor) or by spreading the energy of the signal over a wider freuency band (effectively reducing the spectral power density of the signal) [5]. NB-IoT [21] and Weightless-P [22], for example, encode the signal in low bandwidth (<25 kHz) to reduce the noise level and keep the transceiver design as simple and cheap as possible. Sigfox [23] and Weightless-N [24] further narrow the signal into ultra-narrow bands as narrow as 100 Hz, further reducing the perceived noise.

LoRa technology. Compared to these technologies, LoRa spreads the signal over a wider frequency band, and is more resilient to jamming and interference. LoRa is a proprietary LPWAN technology from Semtech (Camarillo, CA, USA) that recently attracted significant attention due to its ability to trade efficiently communication range against high data-rates, thus enabling IoT applications at an urban scale. The core of LoRa technology is its Chirp Spread Spectrum (CSS) modulation: the carrier signal of LoRa consists of *chirps*, signals whose frequency increases or decreases over time. LoRa's chirps allow the signal to travel long distances and to be demodulated even when its power is up to 20 dB lower than the noise floor. Because of this aspect, carrier sensing in LoRa is quite challenging: LoRa radios allow carrier detection via a *CAD mode*, a special reception state consuming half of the energy compared to the normal reception mode. However, the signals produced by different LoRa networks operating on different settings could create interference leading to false detections [7].

LoRa's communication performance can be fine-tuned by varying the selection of several PHY settings, including bandwidth, spreading factor, coding rate, transmission power, and carrier frequency, as summarized in Table 1. We explain next in detail the impact of each PHY parameters on data rate, receiver sensitivity (including resilience to interference), transmission range, and energy-efficiency [25].

Table 1. Summary of LoRa's configurable settings and their impact on communication performance.

Setting	Values	Effects
Bandwidth	125 . . . 500 kHz	Higher bandwidths allow for transmitting packets at higher data rates (1 kHz = 1 kcps), but reduce receiver sensitivity and communication range.
Spreading Factor	$2^6 \ldots 2^{12} \frac{\text{chips}}{\text{symbol}}$	Bigger spreading factors increase the signal-to-noise ratio and hence radio sensitivity, augmenting the communication range at the cost of longer packets and hence a higher energy expenditure.
Coding Rate	4/5 . . . 4/8	Larger coding rates increase the resilience to interference bursts and decoding errors at the cost of longer packets and a higher energy expenditure.
Transmission Power	−4 . . . 20 dBm	Higher transmission powers reduce the signal-to-noise ratio at the cost of an increase in the energy consumption of the transmitter.

Bandwidth (BW). Varying the range of frequencies (bandwidth) over which LoRa chirp spread allows for trading radio air time against radio sensitivity, thus energy efficiency against communication range and robustness. The higher is the bandwidth, the shorter is the air time and the lower is the sensitivity. A lower bandwidth also requires a more accurate crystal in order to minimize problems related to the clock drift. Given a bandwidth BW, typically in the range of 125. . . 500 kHz, LoRa's chip-rate R_C is computed as:

$$R_C = BW \quad chips/s.$$

Spreading Factor (SF). To transmit information, LoRa "spreads" each symbol over several chips (*spreading factor*) to increase the receiver's sensitivity even more. LoRa's spreading factor SF can be selected between 6 and 12, resulting in a spreading rate ranging from 2^6 to 2^{12} chips/symbol and a symbol-rate R_S that can be computed as:

$$R_S = \frac{R_C}{2^{SF}} = \frac{BW}{2^{SF}} \quad symbols/s,$$

and resulting in a modulation bit-rate that can be expressed as:

$$R_M = SF \cdot R_S = SF \cdot \frac{BW}{2^{SF}} \quad bits/s.$$

Note that, in LoRa, packets transmitted with different spreading factors are orthogonal with each other and do not cause collisions if transmitted concurrently.

Coding Rate (CR). To increase the resilience to corrupted bits, LoRa supports forward error correction techniques with a variable number CR of redundant bits, ranging from 1 to 4. The resulting bit-rate BR of LoRa becomes:

$$BR = R_M \cdot \frac{4}{4+CR} = SF \cdot \frac{BW}{2^{SF}} \cdot \frac{4}{4+CR} \quad bits/s.$$

The more interference bursts are expected, the higher the coding rate that should be used to maximize the probability of successful packet reception. Note that LoRa radios with different coding rates can still communicate, since the packet header (transmitted using the maximum coding rate of 4/8) can include the code rate used for the payload.

Transmission Power (TP). As most wireless radios, LoRa transceivers also allow for adjusting the transmission power, drastically changing the energy required to transmit a packet. By switching the transmission power, for example, from −4 to +20 dBm, the power consumption increases from 66 mW to 396 mW when using the RFM95 transceiver (HopeRF, Shenzhen, China) [26]. Note also that, for transmission powers higher than +17 dBm, hardware limitations and legal regulations limit the radio duty cycle to a maximum of 1%.

Carrier Frequency (CF). LoRa transceivers use sub-GHz frequencies for their communication: among others, the 433 MHz, 868 MHz (Europe), and 915 MHz (North America) industrial, scientific and medical (ISM) radio bands. Common LoRa modules such as the Semtech SX1272 [27] and HopeRF RFM95 [26] support communication in the frequency range [860–1020] MHz and are programmable in steps of 61 Hz. Ten channels with different bandwidths can be used to communicate using LoRa in the European 868 MHz ISM band.

3. Related Work

We now summarize the body of works characterizing the performance of LoRa communications and the effects of environmental conditions on its operations.

Characterization of LoRa performance. Because LoRa technology is closed-source, only a few details about its operations are actually available—mostly derived from Semtech's patent describing the modulation technology or from application notes written to help application designers fine-tune the performance of the transceiver to their needs. Many researchers found this information too limited and started benchmarking and reverse-engineering [28,29] the technology to better understand its mechanism and characteristics.

The first experiments focused on the range of reliable links and on the receiver sensitivity [14]—LoRa's core characteristics. In [11], LoRa has been evaluated in urban and maritime scenarios, and a signal attenuation model was derived. In [5], instead, experiments focused on testing LoRa's communication range on a set of diverse scenarios (from underground to overground, with and without line of sight) in order to provide a set of deployment guidelines.

Interestingly, in the evaluation process, different studies found that results were contradicting Semtech's claims on LoRa performance. In [13], researchers were not able to observe an improved sensitivity with increasing spreading factors. Bor et al. [14] found that LoRa's ability of penetrating buildings is rather limited compared to what was originally claimed. Similarly, the results that we present in this work show that communication in indoor scenarios with no line of sight are among the most challenging conditions for LoRa. Another challenge is represented by vegetation, as found by Iova et al. [12]. Finally, in [14,30,31], the authors model LoRa self-interference and channel utilization, concluding that LoRa's scalability is worse than what was originally promised. Other works focus on a more detailed characterization of LoRa, in particular on packet loss [32], on the ability of receiving packets from concurrent transmissions [7], and on the energy consumption at different transmission powers [4].

Different from previous works, this paper analyzes LoRa's PHY settings from a multi-objective perspective, with the goal of finding the best trade-off between data rate, packet reception rate, and energy efficiency.

Environmental effects on low-power radios. A large body of works has studied the impact of environmental conditions on network performance in low-power wireless radios, especially on IEEE 802.15.4-compliant radio transceivers. Several authors report the impact of meteorological conditions on packet reception, including the impact of weather conditions [20,33,34], and humidity [35], as well as the presence of vegetation [36]. One of the most comprehensive studies on wireless nodes deployed outdoors was carried out by Wennerström et al. [20], who have highlighted that packet reception ratio and received signal strength correlate the most with temperature, whereas the correlation with other factors such as absolute humidity and precipitation is less pronounced.

The strong impact of temperature on communication performance has been confirmed by several other works, also almost entirely focused on IEEE 802.15.4 transceivers. Bannister et al. [19] have shown the correlation between temperature and signal strength in a deployment in the Sonoran desert, and identified in a temperature-controlled chamber that the received signal strength of the TI CC2420 radio attenuates at high temperatures due to the impact of temperature on the radio's low-noise and power amplifiers. Based upon this work, Boano et al. [17,18] have confirmed these findings also on other platforms such as the TI CC1020 and CC2520, and also highlighted how this can cause a complete disruption of a wireless link. The authors have also shown how the impact of temperature cannot be neglected when designing duty-cycled medium access control protocols for low-power wireless radios [37,38]. To facilitate the study of how temperature affects the operation of low-power wireless protocols on a larger scale than in a temperature-controlled chamber, several low-cost testbed infrastructures have been proposed, the most popular being TempLab and HotBox [39,40].

The impact of environmental conditions on LPWAN radios, instead, has not yet been investigated in detail. Iova et al. [12] have deployed a number of LoRa networks in urban and mountain environments, and reported that environmental factors such as the presence of vegetation and temperature variations can negatively affect communication performance. The authors, however, did not quantify the impact of these environmental factors and their work does not yet clarify whether high temperatures degrade the quality of LoRa links in a similar way as observed on several IEEE 802.15.4 transceiver platforms.

In the remainder of this paper, we conduct a number of experiments to complement the body of aforementioned related works and answer two key questions that are yet open: (i) how does the selection of LoRa's PHY settings affect the efficiency of links, including the ones of intermediate quality? and (ii) how does temperature affect the performance of LoRa? To answer these questions, we start by carrying out experiments indoor, outdoor, and underground, and by analyzing how PHY settings and environmental factors affect LoRa's communication performance.

4. Evaluating the Performance of LoRa

To study the reliability and energy-efficiency of LoRa communications as a function of the PHY settings described in Section 2 and as a function of environmental factors, we conduct a series of small-scale deployments.

Experimental setup. All of our experiments are carried out at the Graz University of Technology, Austria: the exact location of the nodes is shown in Figure 1. We fix the senders at three given locations (S) and place three receivers at different distances (1, 2, 3) for three scenarios: *indoor* with obstacles (i), *outdoor* with direct line of sight (o), and *underground* covered by a metal manhole (u). Each transmitter sends a packet with a 5-byte payload every 3 s at transmission power +20 dBm, emulating a timely report of a typical IoT sensor for urban monitoring. Every six minutes, transmitter and receivers reboot and switch to a different setting according to a set of hard-coded combinations shown in Table 2. For each of the three scenarios (indoor, outdoor, and underground), we test each setting configuration sequentially every six minutes for a duration of 24 h, hence resulting in a total of 1600 packets exchanged per setting.

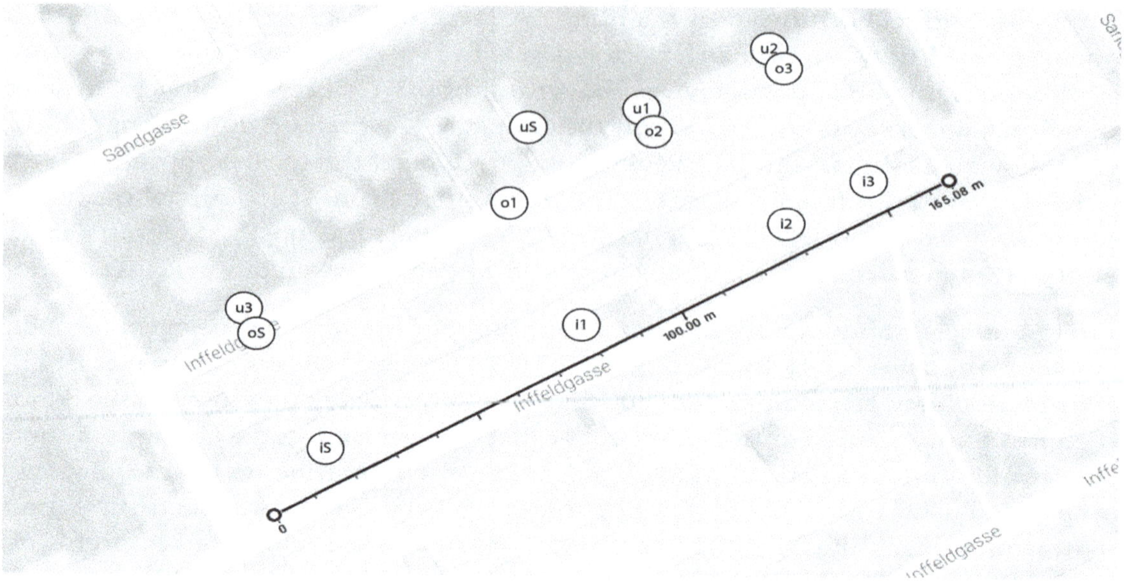

Figure 1. Deployment map for our experiments indoor (**i**), outdoor (**o**), and underground (**u**). The sender node for each scenario is indicated with iS, oS, and uS, respectively.

Table 2. LoRa settings used in our experiments: spreading factor (SF), code rate (CR), bandwidth (BW), and rit-rate (BR) Note that settings are ordered by decreasing bit-rate.

Setting ID	1	2	3	4	5	6	7	8	9	10	11	12	13	14	15	16	17	18
SF	7	7	7	9	7	7	9	9	7	9	9	12	9	12	12	12	12	12
CR	4/5	4/8	4/5	4/5	4/8	4/5	4/8	4/5	4/8	4/8	4/5	4/5	4/8	4/8	4/5	4/8	4/5	4/8
BW (kHz)	500	500	250	500	250	125	500	250	125	125	125	500	125	500	250	250	125	125
BR (kb/s)	21.87	13.62	10.93	7.03	6.83	5.47	4.39	3.51	3.41	2.2	1.76	1.16	1.09	0.72	0.58	0.37	0.30	0.18

Figure 2. Custom-built LoRa platform based on the Moteino MEGA (LowPowerLab, Canton, MI, USA) inside a water-proof enclosure (top removed) [41].

Hardware. The experiments are conducted using a custom-built platform (see Figure 2) based on the Moteino MEGA (LowPowerLab, Canton, MI, USA) [42]. The latter is equipped with an ATMega1284P microcontroller, and a HopeRF RFM95 LoRa transceiver operating at 868 MHz [26]. The device is powered by a 3.7 V Li-Ion battery with a capacity of 3.4 Ah that can be charged via a dedicated circuit. Without duty cycling the radio, this battery can sustain the device operation for more than 24 h, the maximum duration of our experiments. The platform we have built also embeds sensors to measure changes in the surrounding environment. In particular, temperature and humidity are read from a Bosch BME280 sensor (Gerlingen, Germany) via the I2C interface. For persistent storage, an SD card logs each received packet together with its sequence number, the sensed environmental conditions, as well as the time-stamp provided by a Maxim DS3231 real-time clock (San Jose, CA,

USA). We also save the presence of cyclic redundancy check (CRC) errors in the received packets in our traces. This hardware setup was used in our experiments both for senders and receivers.

Metrics. For each 6-min experiment, we compute the packet reception ratio (*prr*) and the receiver sensitivity, i.e., the lowest signal strength among successfully received packets. We then check the correlation of the computed *prr* with the employed PHY settings, as well as with the measured temperature, humidity, and received signal strength values.

4.1. Reliability of LoRa as a Function of PHY Settings

Figure 3 shows the packet reception ratio (i.e., the percentage of packets sent that were correctly received) indoor, outdoor and underground for a number of different radio settings (see Table 2). Figure 3 plots all 6-min experiments grouped by setting ID.

Horizontal red lines represent the median, while blue boxes represent the 25th and 75th percentiles. The remaining results are enclosed by vertical dashed black lines while statistical outliers are represented by red crosses. Note that these results were previously presented in [41].

Our results show that LoRa setting ID 11 (i.e., BW = 125, SF = 9, and CR = 4/5) achieves a packet reception ratio above 95% regardless of the scenario and distance between nodes. Nevertheless, setting ID 2 also performs remarkably well: although it sustains a lower *prr*, it sends packets using a bit-rate that is almost eight times faster than the one used by setting ID 11. This observation will be the starting point of our analysis in Section 5 answering the question of whether it is worth selecting PHY settings that reduce the data rate in order to maximize the link quality.

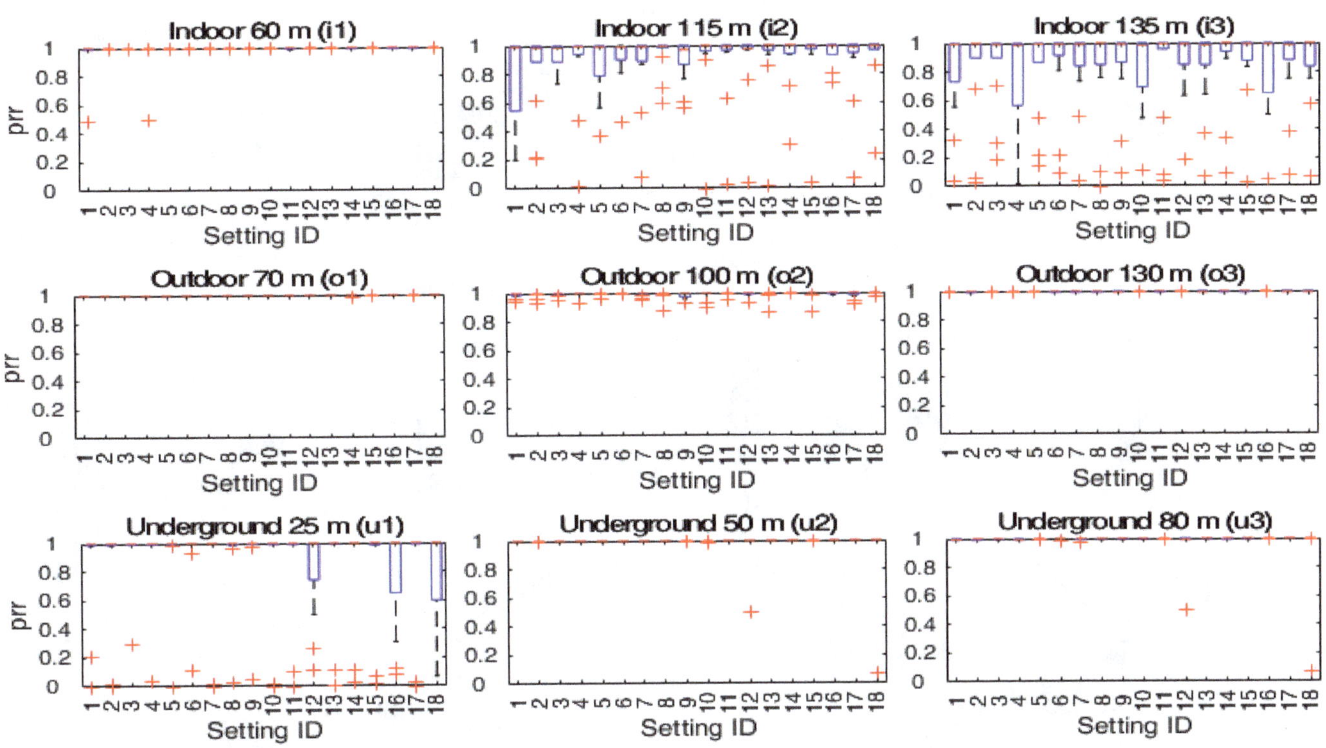

Figure 3. Packet reception ratio *prr*, i.e. the fraction of transmitted packets that are successfully received, for different distances, scenarios, and physical (PHY) settings (setting ID). Horizontal red lines represent the median, while blue boxes represent the 25th and 75th percentiles. The remaining results are enclosed by vertical dashed black lines while statistical outliers are represented by red crosses. Each node is positioned according to Figure 1.

Differently from previous works, in Figure 3, we also show the distribution of the experiment results rather than just the median and mean values. This allows us to make three observations. First, while the median *prr* is close to 1 for most settings, the quartiles and minima are not. Second, due to the lower multi-path and fading effects outdoor and underground, LoRa communications are more reliable in these scenarios rather than indoors (in line with what is observed in [7]), with packet reception ratios above 97% for almost all setting IDs. Third, the range of the LoRa radios is consistent throughout the different settings: even though the reception rate changes, all settings are able to deliver packets in similar conditions.

4.2. Factors Affecting LoRa Reliability

We explore next which environmental factors affect the reliability of LoRa. Towards this goal, we use the traces collected in the previous experiments and focus on the correlation between the packet reception rate (*prr*), setting ID (*set*), temperature (*temp*), humidity (*hum*), spreading factor (*sf*), coding rate (*cr*), bandwidth (*bw*), receiver sensitivity (*sens*), receiver signal strength (*rss*), and hour of the day (*hour*). In particular, we plot the Pearson correlation of each pair of parameters for different experimental scenarios in Figure 4: a value close to 1 (black) means that the two parameters are linearly correlated, whereas a value of 0 (white) implies that the two parameters are independent.

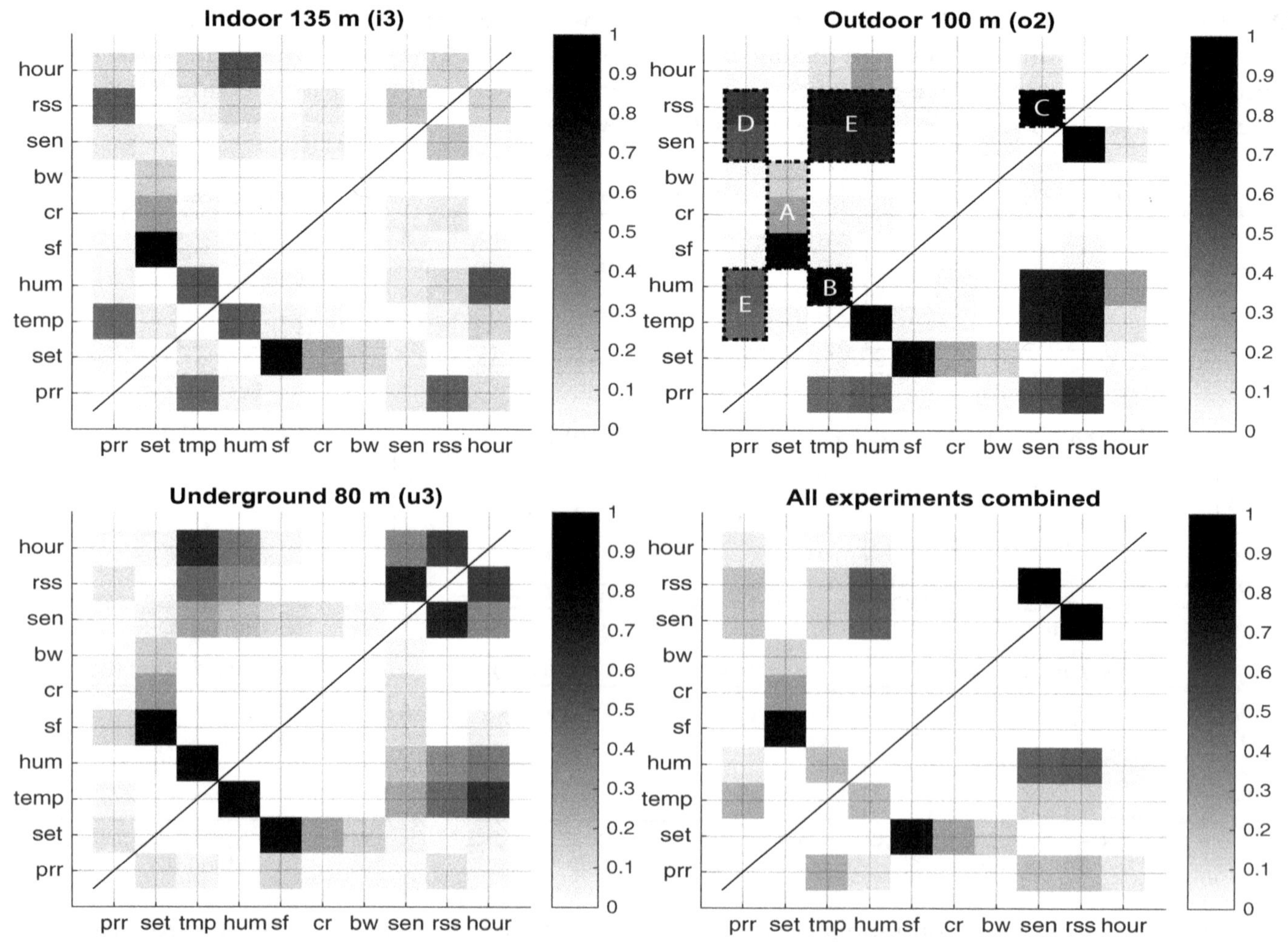

Figure 4. Correlation matrix for different LoRa settings indoor, outdoor, and underground. The plot on the bottom right combines all nine of the settings from Figure 3.

Figure 4 shows that, in all three scenarios (indoor, outdoor and underground), there are some obvious correlations. First (**A**), the setting ID depends on the bandwidth, coding rate (*cr*), and spreading factor (*sf*). This is to be expected, because the setting ID unequivocally describes a combination of these three PHY parameters. Second (**B**), temperature (*temp*) is highly correlated with humidity (*hum*) and both are correlated with the time of the day (*hour*). This is also an expected correlation, as these environmental factors are highly dependent on the sun exposure. Third (**C**), the radio sensitivity (*sen*) is correlated to the received signal strength (*rss*), since the former is defined as the minimum of the latter. Furthermore, one can also note in Figure 4 that the received signal strength (*rss*) is correlated with the packet reception ratio (*prr*) (**D**), as the LoRa radio is able to successfully decode packets that are above a certain signal-to-noise ratio.

Figure 4 also shows that temperature is tightly correlated with the received signal strength (*rss*) and the packet reception ratio (*prr*). This seem to hint that temperature variations may affect the operation of the employed LoRa radio in a similar way as observed on some IEEE 802.15.4 transceivers (see Section 3). When analyzing the figure in detail, one can actually observe a correlation cluster (**E**) between temperature, humidity, time of the day, packet reception ratio, and received signal strength: the strength of these correlations varies depending on the scenario and is stronger outdoors. To better understand the inter-dependency between the reliability performance (*rss* and *prr*) and environmental factors (temp), we carry out experiments in controlled settings in Section 6.

5. The Efficiency of LoRa as a Function of PHY Settings

The experimental campaign presented in the previous section has shown that it is possible to improve the reliability of LoRa by carefully choosing the PHY settings, i.e., some of the settings allow for sustaining a higher *prr*. In this section, we analyze the costs of such improvement in terms of energy efficiency and analyze in detail the trade-off between packet delivery rate and setting's bandwidth, providing an answer to the question: *is it more efficient to use resilient and slow settings or to use faster (but more fragile) configurations together with a re-transmission mechanism?*

To answer this question, we focus on the most challenging scenario in our experimental campaign, i.e., indoor, no line of sight, and with a distance between two devices of 115 m. Figure 5a shows the distribution of packet reception ratios as a function of setting ID. Averages are represented by '*', while median, quartiles, and extreme values are enclosed by a blue box and two black bars (outliers are indicated with crosses). PHY settings are ordered by decreasing bit-rate, from faster and more lightweight settings on the left, to slower settings increasing the transceiver's on-time on the right. As we can see, using the fastest setting (setting ID 1 and 2), the average *prr* is 80% with a worst case scenario where *prr* is as low as 20%. As expected, by selecting a PHY configuration that reduces the bit-rate (i.e., by decreasing the bandwidth and increasing the bit redundancy), the packet reception ratio improves, as well as its distribution.

Nevertheless, one can argue that it is more energy-efficient to re-transmit a packet using the faster settings available rather than employing PHY settings that reduce the bit-rate. To prove our point, we compute the expected number of re-transmissions (*ETX*) as:

$$ETX = \frac{1}{prr}$$

and compare them against each setting's original bit-rate (*BR*). As we can see in Figure 5b, the expected number of re-transmissions (squares) does not directly depend on the settings' bit-rate (triangles), suggesting that not all settings are worth their overhead.

In order to give an indication on how efficiently LoRa settings trade communication efficiency against reliability, we compute the effective bit-rate (*EBR*) as:

$$EBR[kb/s] = \frac{BR}{ETX} = BR \cdot prr$$

and show both mean and distribution in Figure 5c.

The EBR shows the expected bit-rate of each setting in the case packets are re-transmitted back-to-back until one is successfully received. As we can see from Figure 5c, the mean EBR is by far the highest when the fastest LoRa setting is used (setting ID 1). Setting ID 2, on the other hand, is more consistent, showing a lower variance and the highest minimum (crosses represent outliers).

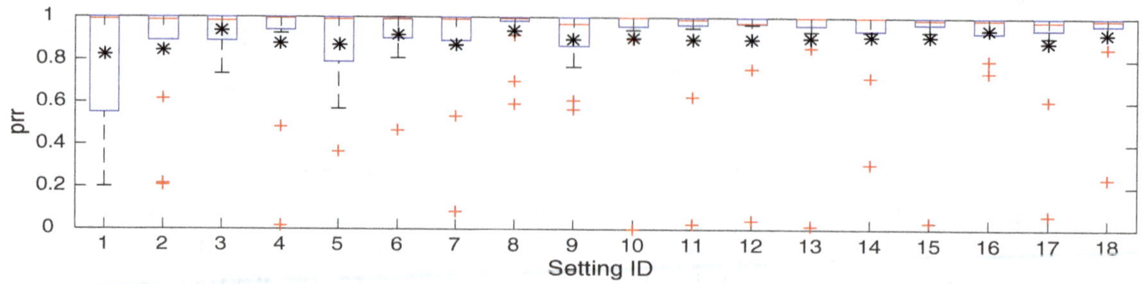

(a) Packet reception rate *(prr)* for different LoRa settings. The higher the *(prr)*, the better.

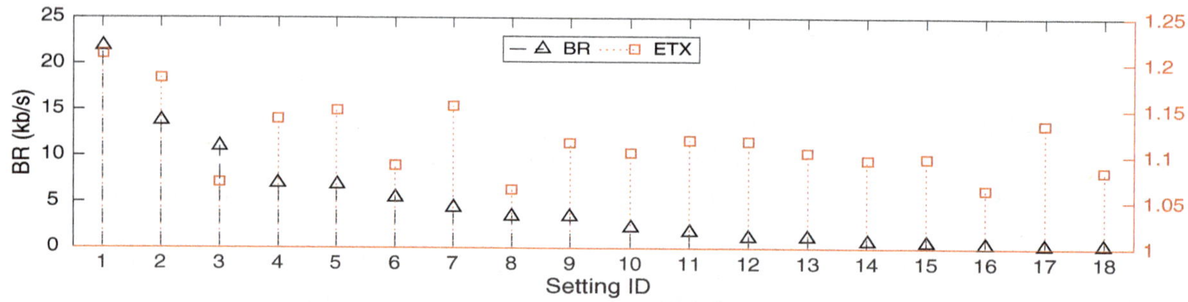

(b) Bit-rate (BR) versus expected number of retransmissions (ETX) computed as $\frac{1}{\text{prr}}$.

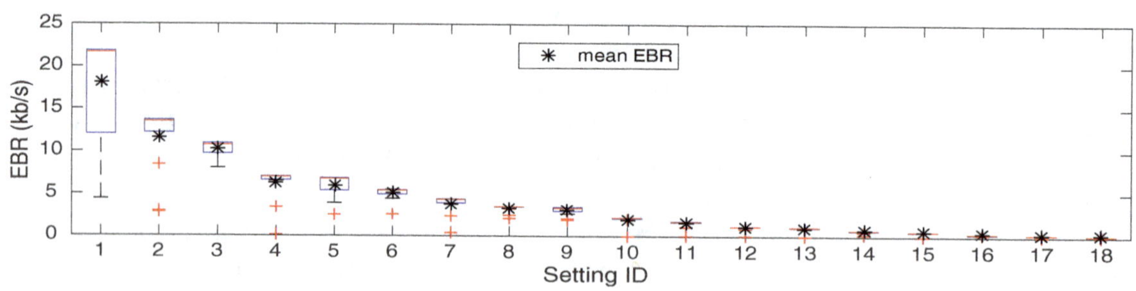

(c) Expected bit-rate computed as $\frac{\text{BR}}{\text{ETX}}$.

Figure 5. LoRa performance as a function of PHY settings in an indoor scenario without line of sight at a distance of 115 m. For (**a**,**c**), asterisks represent mean values, horizontal red lines represent the median, while blue boxes represent the 25th and 75th percentiles. The remaining results are enclosed by vertical dashed black lines while statistical outliers are represented by red crosses.

Even though these results can heavily depend on the surrounding environment, we argue that, *in order to maximize the effective bit-rate, one should opt for a re-transmission mechanism and use the settings with sufficient prr (e.g., >0.2) and the highest bit-rate possible.*

To test the validity of this claim, we additionally compute the EBR for 12 experiments run by independent researchers [5] and present the results in Figure 6. The experiment was run on several Libelium Waspmote LoRa motes; therefore, the setting ID shown in the figure (mode ID) is enumerated according to the Libelium application programming interface (API). As for Figure 5, we conveniently order the PHY settings starting from the one using the highest bit-rate on the left, to the one employing the lowest bit-rate on the right. As we can see, this set of experiments also confirms our observation: *faster settings result in higher bit-rates, even though the quality of the link (i.e., the (prr)) is lower*.

Next, we extend this analysis to different transmission power levels and explicitly also evaluate the energy efficiency of LoRa transmissions.

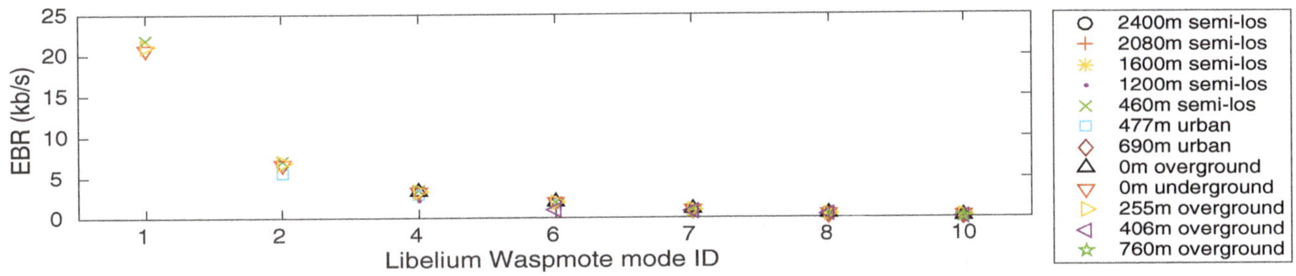

Figure 6. Effective bit-rate for 12 experiments run by independent researchers and presented in [5].

Energy efficiency of LoRa transmissions. We evaluate the energy efficiency of different PHY settings by repeating the indoor experiments without line of sight at a distance of 115 m. This time, we slightly vary the experimental setup as follows: we position the two nodes at the edge of transmission range and send each packet with a different power level, sequentially selected from the set {+20, +17, +15, +11, +7, +5} dBm. We first check if using the fastest PHY setting leads to the highest energy efficiency both in challenging and non challenging conditions (the lower the transmission power, the more challenging the communication). We then compute the most energy-efficient setting by computing the effective energy consumed to send a kilobit of data *EKB* as follows:

$$EKB[J/kb] = \frac{P}{EBR},$$

where P is the power consumption of the radio in watts and EBR is the effective bit rate.

Figure 7a shows the packet reception rate when using three different power levels: +20, +15, and +11 dBm. As expected, changing the transmission power drastically affects the *prr*, since the nodes are intentionally placed on the edge of the communication range. In agreement with the results presented previously, Figure 7b shows that the fastest settings are the ones with highest effective bit-rate *EBR*—*independently of the employed transmission power*. As lower transmission powers imply lower energy expenditures, we still need to answer *which transmission power configuration results in the highest energy efficiency*.

Figure 7c shows the energy required by each PHY setting to transmit a kilobit of data (*EKB*), including the cost of the re-transmissions. As we can see, the most efficient transmission power configuration (i.e., leading to the lowest EKB) is the highest, i.e., +20 dBm. Therefore, our experimental results suggest that, *together with the fastest setting, the highest transmission power should be preferred*: this combination provides the highest bit-rate *EBR* and the lowest energy consumption *EKB*. It is worth highlighting that, in less challenging scenarios in which several transmission powers achieve a $prr = 1$, the lowest one should be used, as the higher transmission powers may increase the energy consumption without any additional benefit.

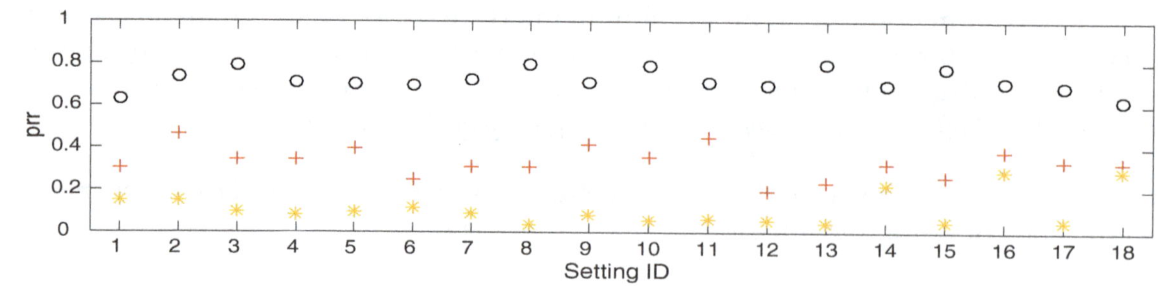

(a) Packet reception rate ((*prr*)) for different LoRa settings. The higher the (*prr*), the better.

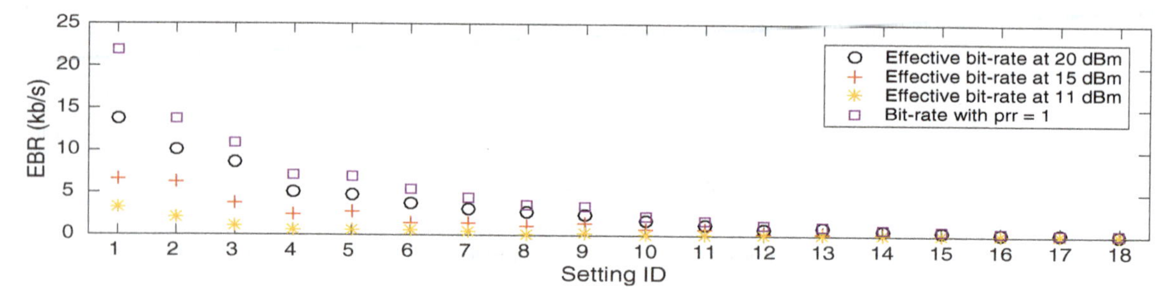

(b) Expected bit-rate computed as $\frac{BR}{ETX}$.

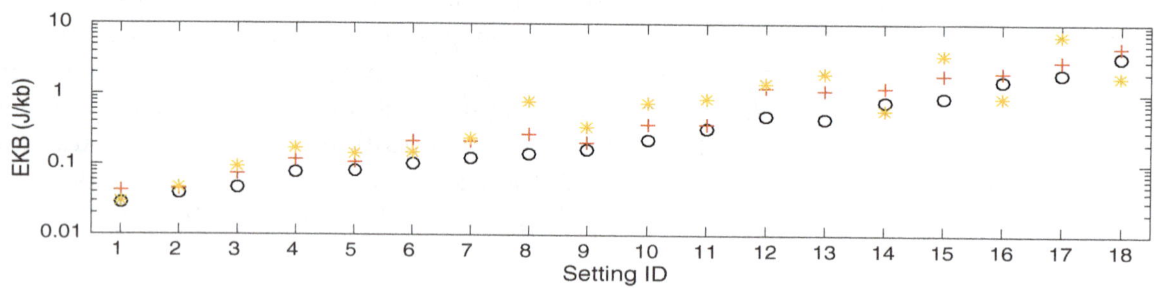

(c) Expected energy per kilobit (EKB). The lower EKB, the more energy efficient is LoRa.

Figure 7. LoRa performance for different settings and transmission powers.

6. Impact of Temperature on LoRa Transceivers

Our experimental campaign presented in Section 4 has shown that there is a strong correlation between temperature, packet reception rate, and received signal strength. To quantify this correlation and shed light on the impact of temperature on LoRa communications, we carry out a deeper investigation in controlled settings.

Experimental setup. We use the TempLab testbed [39] to expose a number of LoRa nodes to repeatable temperature variations as shown in Figure 8. The TempLab testbed available at Graz University of Technology has two different types of nodes [38]: LO nodes only heating the sensor nodes above room temperature and PE nodes having the capability to also cool down the node's temperature below zero degrees thanks to enclosures made of hard Polystyrene foam and ATA-050-24 Peltier air-to-air assembly modules (Custom Thermoelectric, Bishopville, MD, USA). Both LO and PE nodes are remotely controlled using an Aeon Z-Wave Stick Series 2 (Aeon Labs LLC, El Cerrito, CA, USA) sending commands to (i) Vesternet EVR_AD1422 Z-Wave Everspring wireless dimmers (Smartech Holdings Ltd., Manchester, UK) connected to Philips E27 infra-red 100 W light bulbs (Philips, Eindhoven, The Netherlands) and (ii) to Vesternet EVR_AN1572 Z-Wave Everspring on–off wireless switches connected to the Peltier modules.

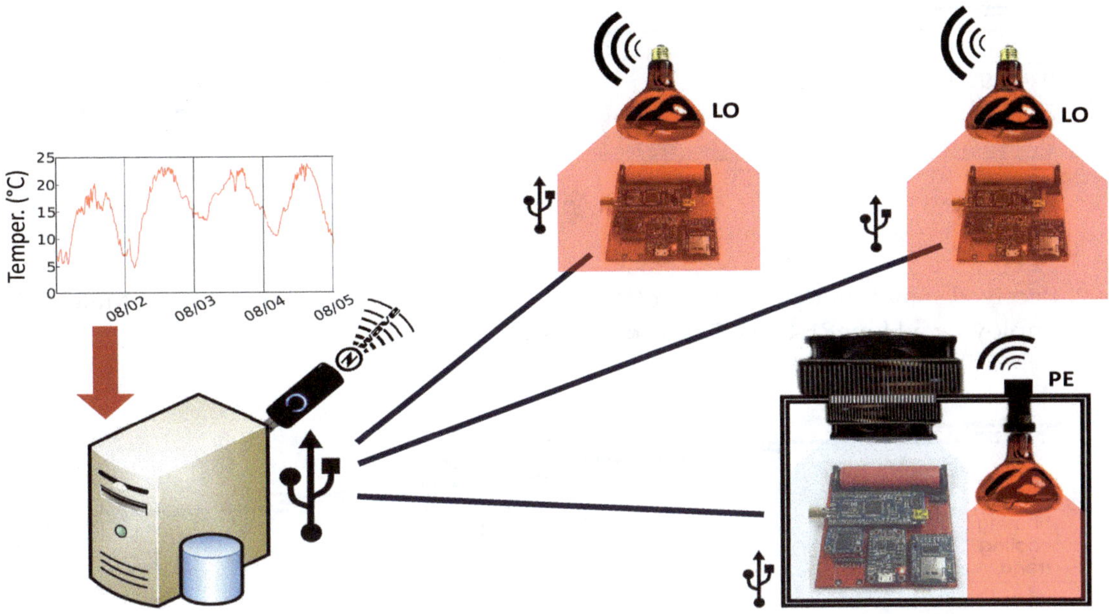

Figure 8. Sketch of the employed TempLab's setup to control the temperature of LoRa nodes.

We place the LoRa nodes without a connected SubMiniature version A (SMA) antenna inside PE nodes and let the nodes transmit packets as fast as possible without any radio duty cycling while temperature varies in the range of 0–60 °C. In particular, we scripted TempLab to first slowly increase temperature from 0 to 60 °C and then to quickly cool down to 0 °C. Each test has a duration of five hours and was repeated for different PHY settings and hardware platforms.

Impact on received signal strength. We plot the relationship between the received signal strength indicator (RSSI) of received packets and the median temperature measured by the two nodes for different hardware platforms. Figure 9 shows the curve recorded when using a Moteino MEGA board [42] equipped with a HopeRF RFM95 LoRa radio (the same platform used for the experiments described in Section 4). Each dot represents the median of the RSSI over 40 received packets. Similarly to what was reported in [18], the RSSI decreases linearly in discrete steps for a total of about 6 dB in the temperature range of 0–60 °C. This is because, for a given voltage, a higher temperature increases the resistance of conductors, while reducing the pass-trough current. For radio transceivers, this implies that higher temperatures reduce the received signal strength and signal-to-noise ratio.

Also according to [18], we observe an hysteresis in the relationship between RSSI and temperature when comparing the curves obtained when heating and when cooling the LoRa nodes. Similarly, Figure 10 shows the relationship between RSSI and temperature recorded when exposing ST Nucleo L073RZ boards (STMicroelectronics, Geneva, Switzerland) [9] equipped with a Semtech SX1272 radio to temperature variations. Also using this hardware, we observe a linear decrease of about 6 dB in the RSSI at high temperatures. We attribute the spikes recorded on the experiment of Figure 10 (when temperature varies between 30 and 40 °C) to a temporary multi-path fading effect of the environment. Note that Figures 9 and 10 refer to different experiments, both carried out using setting ID 6 (i.e., CR = 4/5, SF = 7, BW = 125 MHz).

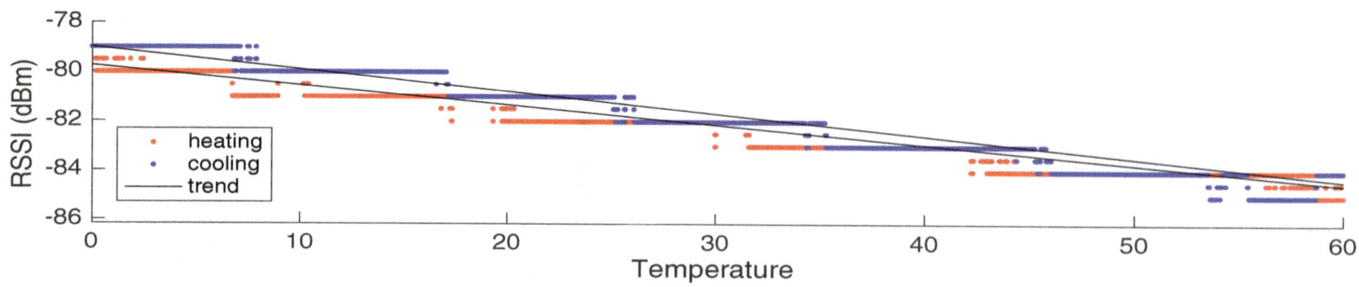

Figure 9. Received signal strength indicator (RSSI) as a function of temperature on the Moteino MEGA platform employing a HopeRF RFM95 transceiver [42].

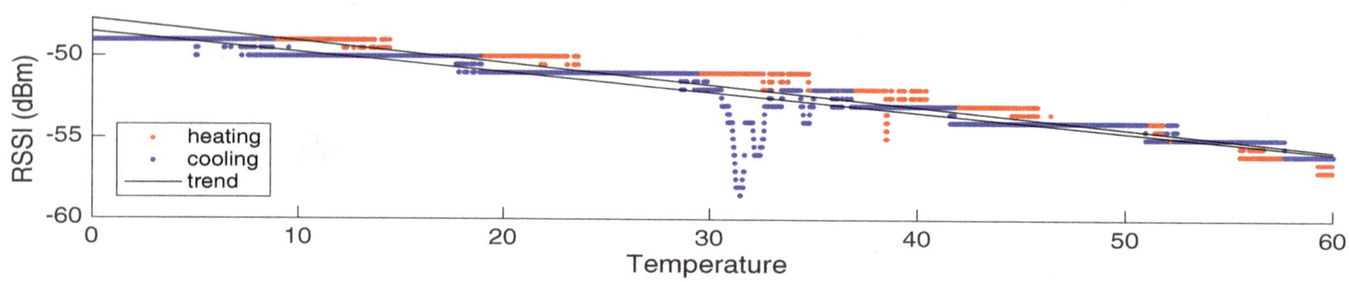

Figure 10. RSSI as a function of temperature on the ST Microelectronics Nucleo L073RZ platform employing a Semtech SX1272 transceiver [9].

Remarkably, the attenuation of received signal strength caused by an increase of temperature in the range [0–60] °C is comparable to the change in sensitivity that can be observed when switching from the fastest to the slowest PHY setting [14]. Therefore, in case of cold temperatures, it may even be possible to avoid using extremely slow radio settings by carefully deploying the LoRa devices in locations that are not directly exposed to sunlight.

Impact on packet reception ratio. We further analyze the effects of temperature variations on nodes that are at the edge of their communication range. We intentionally place two nodes at the limits of their communicating range and slowly change the temperature of the transmitter from 15 to 60 °C and quickly back to 15 °C. Figure 11 shows the distribution of lost, corrupted and successfully received packets for every minute in a 75-min experiment. We can observe that what was a perfect link at minute 0 (100% *prr* at 15 °C) slowly becomes unusable at higher temperatures. As soon as temperature (red line) starts to increase, either packets are received, but their content is corrupted, or the radio was unable to receive the packet at all. Once temperature starts decreasing again, the link is restored and sustains a high delivery rate. These experiments confirm results from previous studies on specific IEEE 802.15.4 radios [18,19], and show that temperature can drastically affect packet delivery. An important takeaway message is that LoRa nodes employing the radio transceivers used in our experiments should be deployed during the warmest time of the day or year, to ensure that network performance is sufficient throughout the system lifetime, and that nodes should be shielded from sunlight if possible.

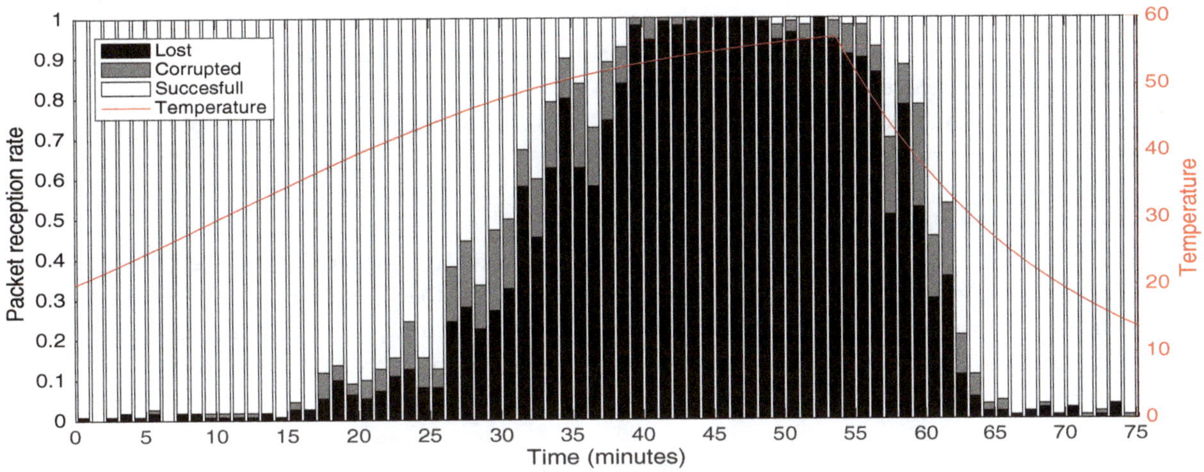

Figure 11. Increase of packet corruption and loss at higher temperature on a LoRa link at the edge of the communication range.

7. Conclusions

This paper presents an analysis of the performance of LoRa as a function of different PHY settings and environmental conditions. We first study the effects of different LoRa settings on the effective bit-rate that can be achieved (i.e., on the amount of information that LoRa is able to *successfully* deliver during a given period). Our experimental results suggest that, when nodes are at the edge of their communication range, using the fastest PHY setting and the highest transmission power is more efficient than selecting slower settings that maximize the link quality. Even though, for example, the fastest PHY setting in our experiments yields an *average* packet reception rate that is 10% lower than the slowest setting, the former's *effective bitrate* is 100× faster than the latter's. Compared to the slower settings, the efficiency of the fastest PHY setting is so high that even in its worst case scenario—when the *minimum prr* reaches 20%—the effective bitrate is faster than twelve of the slowest PHY settings (settings 7–18, from 1× to 25× better). Second, we analyze the external factors affecting the reliability of LoRa. Our outdoor experiments show a clear correlation between temperature, humidity, packet reception rate, and received signal strength. A deeper investigation in controlled settings shows that the signal strength of received packets decreases linearly when temperature increases in two different LoRa transceivers. Different LoRa radios have shown that, over a range of 60 °C, the received signal strength is consistently reduced by 6 dBm (1 dBm/10 °C). This decrease in signal strength can significantly affect LoRa links that are at the edge of the communication range, increasing packet corruption and loss, and rendering a perfectly good link (100% *prr* at 15 °C) completely unusable (0% *prr* at 60 °C).

As a future work, we plan to quantify the impact of other environmental factors on LoRa performance, e.g., humidity and radio interference. Our ultimate goal is to design and implement an environmental-aware MAC protocol tailored to LoRa that can sustain reliable and energy-efficient operations regardless of changes in the surrounding environmental conditions.

Acknowledgments: This work has been supported by the Sino Austrian Electronic Technology Innovation Center. This work was also partially performed within the LEAD-Project "Dependable Internet of Things in Adverse Environments", funded by Graz University of Technology.

Author Contributions: Marco Cattani has conceived and designed the experiments evaluating the interplay between LoRa's PHY settings and link quality described in Sections 4 and 5, as well as taken care of the write-up of the whole manuscript. Carlo Alberto Boano has conceived and designed the experiments evaluating the impact of temperature on LoRa transceivers described in Section 6 and taken care of the write-up of the whole manuscript. Kay Römer has participated to the general discussions and help revising the write-up of the manuscript.

References

1. Real Wireless Ltd. *A Comparison of UNB and Spread Spectrum Wireless Technologies as Used in LPWA M2M Applications*; Real Wireless: West Sussex, UK, 2015.
2. LoRa Alliance. LoRa: Wide Area Networks for IoT, 2017. Available online: http://www.lora-alliance.org/What-Is-LoRa/Technology (accessed on 22 May 2017).
3. Raza, U.; Kulkarni, P.; Sooriyabandara, M. Low Power Wide Area Networks: An Overview. *IEEE Commun. Surv. Tutor.* **2017**, *19*, 855–873.
4. Bor, M.; Roedig, U. LoRa Transmission Parameter Selection. In Proceedings of the 13th IEEE International Conference on Distributed Computing in Sensor Systems (DCOSS), Ottawa, ON, Canada, 5–7 June 2017.
5. Gnawali, O.; Fonseca, R.; Jamieson, K.; Moss, D.; Levis, P. Demystifying Low-Power Wide-Area Communications for City IoT Applications. In Proceedings of the 10th ACM Workshop on Wireless Network Testbeds, Experimental Evaluation, and Characterization (WiNTECH), New York, NY, USA, 3 October 2016; pp. 2–8.
6. Baños-Gonzalez, V.; Afaqui, M.S.; López-Aguilera, E.; Villegas, E.G. Throughput and Range Characterization of IEEE 802.11ah. *arXiv* **2016**, arXiv:1604.08625.
7. Bor, M.; Vidler, J.; Roedig, U. LoRa for the Internet of Things. In Proceedings of the 1st International Workshop on New Wireless Communication Paradigms for the Internet of Things (MadCom), Graz, Austria, 15–17 February 2016; pp. 361–366.
8. Libelium. *Waspmote-LoRa-868MHz-915MHz-SX1272 Networking Guide, v7.0*; Libelium: Zaragoza, Spain, 2017.
9. ST Microelectronics. *STM32 Nucleo Pack for LoRa Technology (P-NUCLEO-LRWAN1), DocID029505 Rev. 2*; ST Microelectronics: Geneva, Switzerland, 2016.
10. NetBlocks Embedded Networking. XRange SX1272 LoRa RF module. Available online: http://www.netblocks.eu/ (accessed on 22 May 2017).
11. Petäjäjärvi, J.; Mikhaylov, K.; Roivainen, A.; Hänninen, T.; Pettissalo, M. On the Coverage of LPWANs: Range Evaluation and Channel Attenuation Model for LoRa Technology. In Proceedings of the 14th IEEE International Conference on ITS Telecommunications (ITST), Copenhagen, Denmark, 2–4 December 2015; pp. 55–59.
12. Iova, O.; Murphy, A.L.; Ghiro, L.; Molteni, D.; Ossi, F.; Cagnacci, F. LoRa from the City to the Mountains: Exploration of Hardware and Environmental Factors. In Proceedings of the 2nd International Workshop on New Wireless Communication Paradigms for the Internet of Things (MadCom), Uppsala, Sweden, 20–22 February 2017.
13. Augustin, A.; Yi, J.; Clausen, T.; Townsley, W.M. A Study of LoRa: Long Range & Low Power Networks for the Internet of Things. *Sensors* **2016**, *16*, 1466.
14. Bor, M.; Roedig, U.; Voigt, T.; Alonso, J.M. Do LoRa Low-Power Wide-Area Networks Scale? In Proceedings of the 19th ACM International Conference on Modeling, Analysis and Simulation of Wireless and Mobile Systems (MSWiM), Valletta, Malta, 13–17 November 2016; pp. 59–67.
15. Baccour, N.; Koubâa, A.; Mottola, L.; Youssef, H.; Zúñiga, M.A.; Boano, C.A.; Alves, M. Radio Link Quality Estimation in Wireless Sensor Networks: A Survey. *ACM Trans. Sensor Netw.* **2012**, *8*, 34.
16. Gnawali, O.; Fonseca, R.; Jamieson, K.; Moss, D.; Levis, P. Collection Tree Protocol. In Proceedings of the 7th International Conference on Embedded Networked Sensor Systems (SenSys), Berkeley, CA, USA, 4–6 November 2009; pp. 1–14.
17. Boano, C.A.; Brown, J.; Tsiftes, N.; Roedig, U.; Voigt, T. The Impact of Temperature on Outdoor Industrial Sensornet Applications. *IEEE Trans. Ind. Inform.* **2010**, *6*, 451–459.
18. Boano, C.A.; Wennerström, H.; Zúñiga, M.A.; Brown, J.; Keppitiyagama, C.; Oppermann, F.J.; Roedig, U.; Nordén, L.Å.; Voigt, T.; Römer, K. Hot Packets: A Systematic Evaluation of the Effect of Temperature on Low Power Wireless Transceivers. In Proceedings of the 5th Extreme Conference on Communication (ExtremeCom), Reykjavik, Iceland, 24–29 August 2013; pp. 7–12.
19. Bannister, K.; Giorgetti, G.; Gupta, S.K. Wireless Sensor Networking for Hot Applications: Effects of Temperature on Signal Strength, Data Collection and Localization. In Proceedings of the 5th International Workshop on Embedded Networked Sensors (HotEmNets), Charlottesville, VA, USA, 2–3 June 2008.

20. Wennerström, H.; Hermans, F.; Rensfelt, O.; Rohner, C.; Nordén, L.A. A Long-Term Study of Correlations between Meteorological Conditions and 802.15.4 Link Performance. In Proceedings of the 10th IEEE International Conference on Sensing, Communication, and Networking (SECON), New Orleans, LA, USA, 24–27 June 2013; pp. 221–229.
21. Ratasuk, R.; Vejlgaard, B.; Mangalvedhe, N.; Ghosh, A. NB-IoT system for M2M communication. In Proceedings of the IEEE Wireless Communications and Networking Conference (WCNC), Doha, Qatar, 3–6 April 2016; pp. 1–5.
22. Weightless SIG. Weightless-P Open Standard, 2017. Available online: http://www.weightless.org/about/weightlessp (accessed on 22 May 2017).
23. Sigfox. Sigfox Technology, 2017. Available online: http://www.sigfox.com (accessed on 22 May 2017).
24. Weightless SIG. Weightless-N Open Standard, 2017. Available online: http://www.weightless.org/about/weightlessn (accessed on 22 May 2017).
25. Semtech Corporation. *LoRa Modulation Basics—Application Note 1200.22, Revision 2*; Semtech Corporation: Camarillo, CA, USA, 2015.
26. Hope RF Microelectronics. *RFM95/96/97/98(W)—Low Power Long Range Transceiver Module, v1.0*; Hope RF Microelectronics: Shenzhen, China, 2016.
27. Semtech Corporation. *SX1272/73—860 MHz to 1020 MHz Low-Power Long-Range Transceiver, Revision 3.1*; Semtech Corporation: Camarillo, CA, USA, 2017.
28. Myriad-RF. LoRa-SDR, 2017. Available online: http://github.com/myriadrf/LoRa-SDR (accessed on 22 May 2017).
29. DecodingLora, 2017. Available online: http://revspace.nl/DecodingLora (accessed on 22 May 2017).
30. Georgiou, O.; Raza, U. Low Power Wide Area Network Analysis: Can LoRa Scale? *IEEE Wirel. Commun. Lett.* **2017**, *6*, 162–165.
31. Voigt, T.; Bor, M.; Roedig, U.; Alonso, J. Mitigating Inter-network Interference in LoRa Networks. In Proceedings of the 2nd International Workshop on New Wireless Communication Paradigms for the Internet of Things (MadCom), Uppsala, Sweden, 20–22 February 2017.
32. Marcelis, P.; Rao, V.; Prasad, R.V. DaRe: Data Recovery through Application Layer Coding for LoRaWAN. In Proceedings of the 2nd International Conference on Internet-of-Things Design and Implementation (IoTDI), Pittsburgh, PA, USA, 18–21 April 2017; pp. 97–108.
33. Anastasi, G.; Falchi, A.; Passarella, A.; Conti, M.; Gregori, E. Performance Measurements of Motes Sensor Networks. In Proceedings of the 7th ACM International Symposium on Modeling, Analysis and Simulation of Wireless and Mobile Systems (MSWiM), Venice, Italy, 4–6 October 2004; pp. 174–181.
34. Boano, C.A. Application Support Design for Wireless Sensor Networks. Master's Thesis, Politecnico di Torino, Turin, Italy; Kungliga Tekniska Högskolan, Stockholm, Sweden, 2009.
35. Thelen, J.; Goense, D.; Langendoen, K. Radio Wave Propagation in Potato Fields. In Proceedings of the 1st Workshop on Wireless Network Measurement (WiNMee), Garda, Italy, 1–5 April 2005.
36. Marfievici, R.; Murphy, A.L.; Picco, G.P.; Ossi, F.; Cagnacci, F. How Environmental Factors Impact Outdoor Wireless Sensor Networks: A Case Study. In Proceedings of the 10th IEEE International Conference on Mobile Ad-Hoc and Sensor Systems (MASS), Hangzhou, China, 14–16 October 2013; pp. 565–573.
37. Boano, C.A.; Römer, K.; Tsiftes, N. Mitigating the Adverse Effects of Temperature on Low-Power Wireless Protocols. In Proceedings of the 11th IEEE International Conference on Mobile Ad Hoc and Sensor Systems (MASS), Philadelphia, PA, USA, 27–30 October 2014; pp. 336–344.
38. Boano, C.A. Dependable Wireless Sensor Networks. Ph.D. Thesis, Graz University of Technology, Graz, Austria, 2014.
39. Boano, C.A.; Zúñiga, M.A.; Brown, J.; Roedig, U.; Keppitiyagama, C.; Römer, K. TempLab: A Testbed Infrastructure to Study the Impact of Temperature on Wireless Sensor Networks. In Proceedings of the 13th ACM/IEEE International Conference on Information Processing in Sensor Networks (IPSN), Berlin, Germany, 15–17 April 2014; pp. 95–106.
40. Schmidt, F.; Ceriotti, M.; Hauser, N.; Wehrle, K. If You Can't Take the Heat: Temperature Effects on Low-Power Wireless Networks and How to Mitigate Them. In Proceedings of the 12th European Conference on Wireless Sensor Networks (EWSN), Porto, Portugal, 9–11 February 2015.

41. Cattani, M.; Boano, C.A.; Steffelbauer, D.; Kaltenbacher, S.; Günther, M.; Römer, K.; Fuchs-Hanusch, D.; Horn, M. Adige: An Efficient Smart Water Network based on Long-Range Wireless Technology. In Proceedings of the 3rd International Workshop on Cyber-Physical Systems for Smart Water Networks (CySWATER), Pittsburgh, PA, USA, 18–21 April 2017.
42. LowPowerLab. Moteino MEGA LoRa, 2016. Available online: http://lowpowerlab.com/shop/product/119 (accessed on 22 May 2017).

Analyzing the Relationship between Human Behavior and Indoor Air Quality

Beiyu Lin [1,*], Yibo Huangfu [2], Nathan Lima [3], Bertram Jobson [2], Max Kirk [3], Patrick O'Keeffe [2], Shelley N. Pressley [2], Von Walden [2], Brian Lamb [2] and Diane J. Cook [1]

1. School of Electrical Engineering & Computer Science, Washington State University, Spokane Street, Pullman, WA 99163, USA; djcook@wsu.edu
2. Department of Civil & Environmental Engineering, Washington State University, 2001 Grimes Way, Pullman, WA 99163, USA; yibo.huangfu@wsu.edu (Y.H.); tjobson@wsu.edu (B.J.); pokeeffe@wsu.edu (P.O.); spressley@wsu.edu (S.N.P.); v.walden@wsu.edu (V.W.); blamb@wsu.edu (B.L.)
3. School of Design & Construction, Washington State University, Spokane Street, Pullman, WA 99163, USA; nathan.lima@wsu.edu (N.L.); mkirk@wsu.edu (M.K.)

* Correspondence: beiyu.lin@wsu.edu

Abstract: In the coming decades, as we experience global population growth and global aging issues, there will be corresponding concerns about the quality of the air we experience inside and outside buildings. Because we can anticipate that there will be behavioral changes that accompany population growth and aging, we examine the relationship between home occupant behavior and indoor air quality. To do this, we collect both sensor-based behavior data and chemical indoor air quality measurements in smart home environments. We introduce a novel machine learning-based approach to quantify the correlation between smart home features and chemical measurements of air quality, and evaluate the approach using two smart homes. The findings may help us understand the types of behavior that measurably impact indoor air quality. This information could help us plan for the future by developing an automated building system that would be used as part of a smart city.

Keywords: indoor air quality; smart home environment; machine learning; data mining

---

1. Introduction

With global population growth and global aging issues, there will be a corresponding concern about living environment changes that impact human health both inside and outside buildings. In this paper, we focus on indoor air quality (IAQ) and its relationship to human behavior. The National Human Activity Pattern Survey [1] reports that individuals spent an average of 87% of their time indoors, so understanding IAQ and its impacts are of critical importance. Indoor air quality tremendously affects human health, and is considered one of the top five environmental risks to public health [2]. According to the United States Environmental Protection Agency (EPA), indoor pollutant levels may be two to five times, and occasionally 100 times, higher than outdoor pollutant levels [2].

According to a report by the Institute of Medicine [3], three major factors are affecting indoor air pollution: the properties of pollutants, building characteristics, and human behavior. The behaviors of occupants in buildings, as one of the three top components, impact IAQ by affecting the production and persistence of pollutants [4]. Behaviors include routine activities such as cooking, which increase the levels of nitrogen dioxide and carbon monoxide and might lead to hazardous levels of these chemical components. Behaviors also include interactions with the physical environment such as opening or closing windows or doors, which impacts the air exchange rate, thus increasing or decreasing indoor pollution levels.

Many studies have investigated sources of IAQ and their effects on human health [5–7]. Researchers recently have started analyzing the relationship between IAQ components and specific IAQ-related human behaviors, such as opening windows [8]. Studies have shown that some human behaviors, such as tending the fire and cooking, increase the total suspended particulates and carbon monoxide (CO) emissions [9]. Based on self-reports, additional domestic behaviors have been included in the analysis, such as sleeping and taking showers. These have been related to CO, particulate matter 10 (PM_{10}) and carbon dioxide (CO_2) [10]. Still, other researchers have investigated factors that drive residents to open windows and doors, thus influencing air exchange rates as well as air quality [11]. So far, the relationship of human behavior patterns and IAQ has been studied via questionnaire surveys for activities of daily living (ADLs). However, human behaviors might change daily due to flexible schedules and external factors including weekdays/weekends, holidays, and weather events. Self-report information is notoriously susceptible to error and bias [12], which introduces potential inaccuracies for IAQ studies.

With the rapid advancement of technology to monitor activities in sensor-filled spaces, algorithms have recently been introduced and enhanced to automatically recognize these activities using machine learning techniques [13–16]. In our study, we combine smart home (SH) technologies with machine learning algorithms to achieve real-time tagging of sensor data with ADL activity labels. An earlier study that used smart environments to relate indoor behavior to IAQ changes had a similar goal [17]. However, the previous study only considered a single behavior parameter (total sensed movement in the environment) and a single IAQ parameter (carbon dioxide level). We expand on the earlier study to consider actual classes of activities that residents perform in the home, rather than just movement level. We also consider a large set of IAQ chemical variables based on the list of criteria air pollutants provided by EPA.

Since human behavior is one of the three major factors that have an influence on IAQ, which in turn has a dramatic impact on human health, it will be beneficial to automatically recognize ADLs using machine learning techniques by monitoring activities in sensor-filled spaces. We hypothesize that machine learning techniques can help us understand the relationship between in-home behavior and IAQ. The findings will help us recognize the types of behavior that significantly impact IAQ, and use this information to develop an automated system to anticipate, prevent and prepare for indoor pollution levels. Such a system could maintain healthier environments, and thus play a central role in the development of smart cities.

To investigate our hypothesis, we collected both sensor-based behavior data and chemical indoor air quality measurements in smart home environments for two houses. We accomplished the investigation by conducting two machine learning-driven analyses. First, we used machine learning algorithms to determine which IAQ variables were measurably impacted by SH features. Second, we identified the particular smart home-based attributes that had the greatest impact on the IAQ variables.

2. Indoor Air Quality

The quality of air indoors is affected by chemical pollutants from diverse sources. The most common indoor air pollutants are from three sources: outdoor pollutants' sources, indoor combustion/cooking sources, and indoor material and chemical sources.

First, there are two primarily outdoor pollutants' sources that get into the home: ozone (O_3) and particulate matter (PM). The pollutant O_3 is photochemically produced by chemical reactions between sunlight, and nitrogen oxides (NO_x), and volatile organic compounds (VOCs). Many studies have been evaluating the amounts of O_3 that have adverse effects on human health, such as airway hyperreactivity and lung inflammation [18]. In the case of inhalable PM, this category of pollutants includes solid particles and liquid droplets suspended in air, and may cause lung cancer, emphysema, and respiratory infections [19]. For example, in our data collection periods, the experiments were conducted during periods with destructive wildfires that caused heavy smoke and very high levels of

PM. The high level of PM would have a great impact on the indoor air quality, the residents' behaviors, and their health. In our study, we concentrated on the outdoor PM less than 2.5 micrometers ($PM_{2.5}$).

Next, we considered pollutants from indoor combustion/cooking, and the corresponding effects. Combustion is the main cause of indoor PM, CO, NO_x and VOCs [20,21]. These pollutants have tremendous health impacts on the residents, such as respiratory infections in young children, chronic lung diseases, and associated heart disease in adults [22]. To monitor indoor PM in our study, we measured the mass concentration of PM less than 2.5 micrometers, as well as the number of small particles ($\geq$1 mm) and large particles ($\geq$5 mm) [23]. VOCs refer to a group of organic chemicals, and each one has its own possible reason for causing distinct health problems. After hours or days of exposure to the high levels of VOCs from cooking/combustion, a resident may experience eye, nose, throat irritation, and worsening asthma symptoms [24]. Selected VOCs, including formaldehyde, acetaldehyde, acetonitrile, methanol, ethanol, acetone, benzene, toluene, xylenes, styrene, and monoterpenes, were measured continuously with a proton transfer reaction mass spectrometer (PTR-MS, Dylos Corporation, Riverside, CA, USA.) [25]. The PTR-MS drift tube was operated at 120 Td. The response of the instrument to different VOCs was calibrated using an external multicomponent compressed gas standard [26]. Due to sensor limitations, our instruments failed to record the values of CO and NO_x during the experiment periods, so we limit our analysis to indoor PM and VOCs.

With regard to indoor material and chemical sources, we considered VOCs from carpet, furniture, building materials, solvents, cleaning supplies, and personal hygiene products [24]. The common VOCs from those sources will have adverse health impacts on residents, such as damage to the respiratory system, headaches, and skin irritations [27,28]. In our collection and analysis, we included all the above chemical variables in both indoor and outdoor environments, as well as data reported by a weather station.

Our testbeds consisted of two houses outfitted with sensors to transform them into smart homes. Data were collected in the first smart home, referred to as IAQ_1, for 27 days (620 h); the residents were a couple in their sixties. We also collected data in a second smart home, referred to as IAQ_2, for six days (187 h); the residents were a family that includes a couple in their fifties and two children, one in their teens, and one in their twenties. This study was approved by the Washington State University Institutional Review Board. In each home, we monitored the chemical components of indoor air quality described in this section, using the instruments summarized in Table 1. The instruments were contained in two separate racks. An indoor rack was placed in the living room to measure selected pollutants, as shown in the Table 1. A larger rack, the master rack, was placed in the garage. The master rack instruments sampled both indoor and outdoor air, alternating sampling between indoors and outdoors every 30 minutes using a three-way valve. The master rack was placed in the garage and Teflon tubing ran from the rack to the top of the roof for outdoor air sampling. For IAQ_1, indoor air was sampled from the return ducting of the furnace; the furnace fan was always on to ensure circulation through the ducts. For IAQ_2, indoor air was sampled using a Teflon tube that ran from the rack through the house to a main hallway, as illustrated in Figure 1. A weather station was placed on the roof. A more detailed diagram for the locations of the indoor and master racks are illustrated in Figure 2.

We examined smart home-based behavior data and chemical variables at the time scale of a single hour. Because the chemical sensors collect higher frequency data, we computed and stored the median values of the indoor and outdoor chemical variables for the corresponding hour of data collection. Similarly, we captured and integrated weather station data for the corresponding hour. Furthermore, the indoor air quality data was collected from a single point within the home, rather than individual rooms in the home. The positioning of the chemical sensors with respect to individual rooms in the house may have had an impact on our results, which we will discuss separately.

Table 1. Instruments for indoor air quality (IAQ) chemical data collection.

Analyte	Instrument(s)	Precision	Accuracy
	Indoor Rack Instruments		
CO_2	LGR Model 915-0011	100 ppbv	1%
	LiCOR 840A	<1 ppmv	1%
H_2O	LGR Model 915-0011	35 ppmv	1%
	LiCOR 840A	<0.01‰	1.5%
CH_4	LGR Model 915-0011	0.6 ppbv	1%
O_3	2B Technology Model 205	1 ppbv	2%
PM	TSI 8530 DustTrak for $PM_{2.5}$ mass concentration	0.01%	10%
	Dylos Corp DC1100 for PM number density		
	Master Rack Instruments		
O_3	TECO 49 O_3	2 ppbv	2%
PM	TSI 8530 DustTrak	0.01%	10%
CO_2	LiCOR 840A	<1 ppmv	1%
H_2O	LiCOR 840A	<0.01‰	1.5%
VOCs	Ionicon Analytik PTR-MS	3–30%	7%
CO	Teledyne 300U	0.5%	1%
NO_X	Teledyne 200U	0.2 ppbv	1%
	Weather Station		
Wind speed	AIRMAR WX200	0.1 m/s	5%
Wind direction	AIRMAR WX200	0.1 deg	5 deg
Temp	AIRMAR WX200	0.1 °C	1.1 °C
Pressure	AIRMAR WX200	0.1 mbar	1 mbar

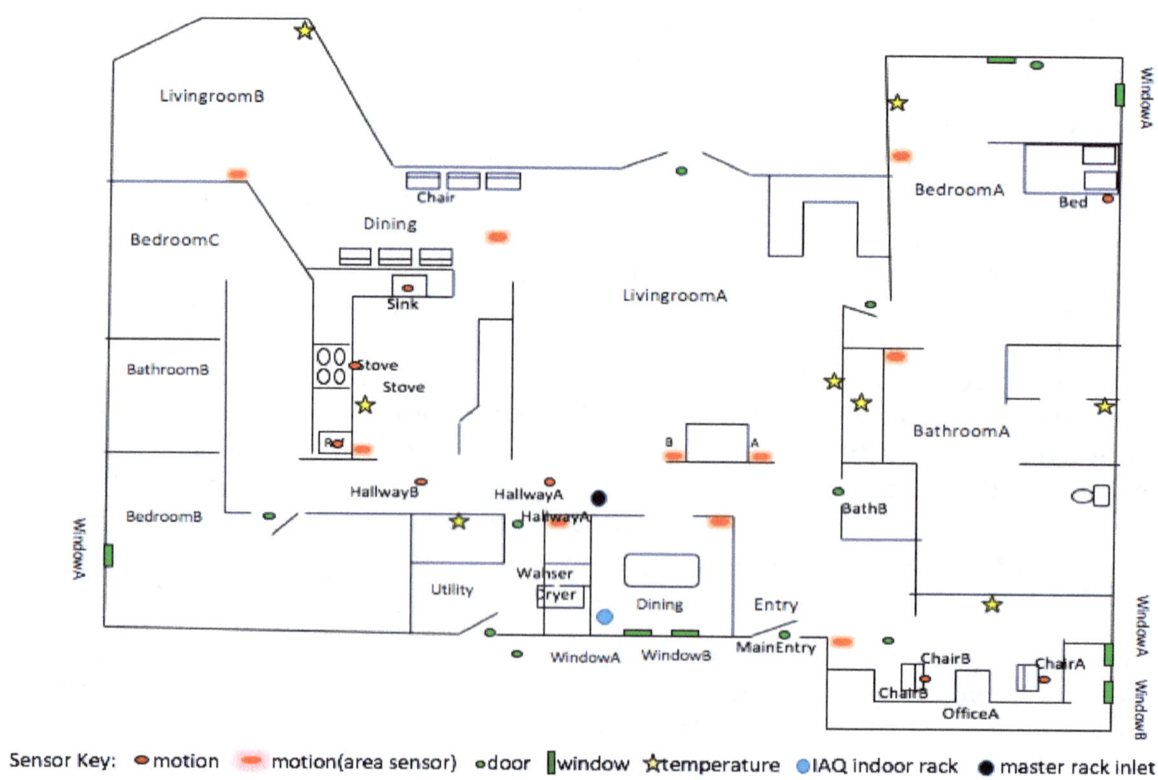

(a)

Figure 1. *Cont.*

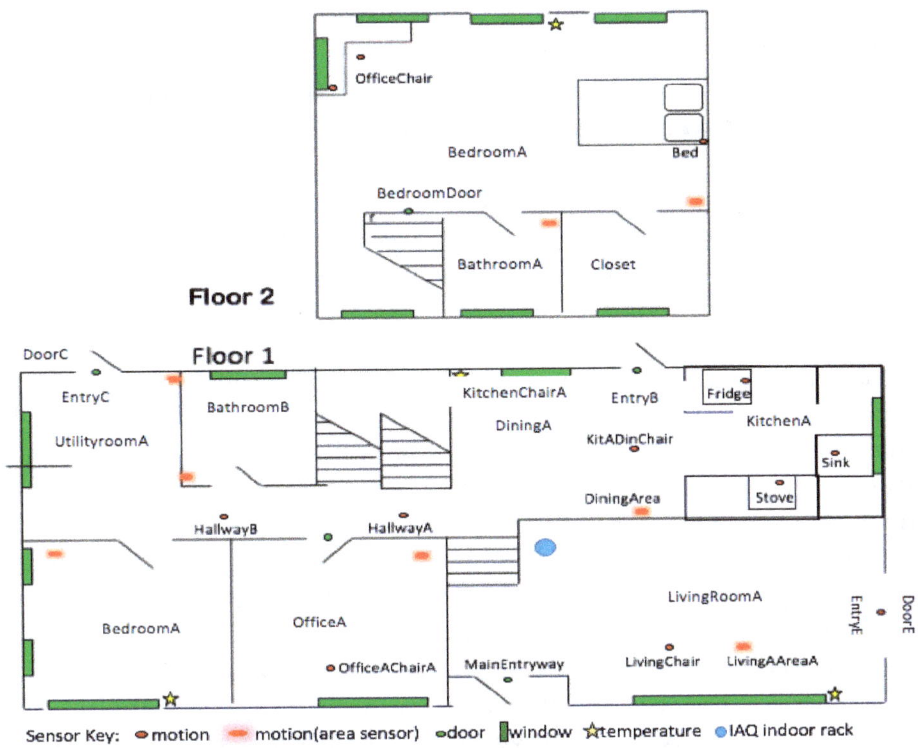

(b)

Figure 1. The floorplans and sensor layouts for the two smart homes. (**a**) The layout for IAQ_1; (**b**) The layout for IAQ_2.

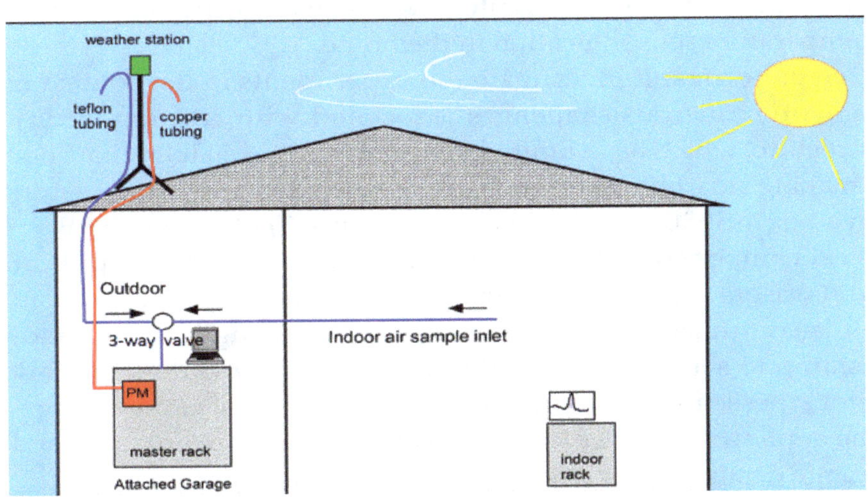

Figure 2. Locations of indoor and master racks.

3. Smart Home Houses

Our smart home testbeds for this study were located in the inland Pacific Northwest, and are maintained as part of the Center for Advanced Studies in Adaptive Systems (CASAS) smart home project. We performed our testing in two separate homes without automatic air exchange systems, each of which was a multiple-resident home. The physical layout and sensor placement for these two environments are shown in Figure 1. As shown in the figure, each smart house contained multiple bedrooms, bathrooms, offices and living areas. For convenience and consistency across all houses, we separated each type of room into two units: the main area of a particular category, and all secondary rooms of the same category aggregated together. For example, in the bedroom category, we collected features for the master bedroom and also collected features for the other bedrooms, which represented information aggregated from all of the other bedrooms in each house. Each of our smart homes had at least two bedrooms and bathrooms, so this approach provides fine-granularity feature specification, while also allowing generalization over multiple homes.

Each house was equipped with combination infrared motion/ambient light sensors and combination closure/temperature sensors that provided readings for the opening or closing of windows or doors, as well as the use of temperature-changing items such as showers and stoves. Based on conversations with IAQ experts and our previous studies [29], we identified four types of smart home features that are used to extract and correlate with chemical variables. These consist of the overall activity level (based on sensed movement), the duration of each automatically labeled activity, temperature, and the total area of the open doors and windows. Activity level is calculated as the number of motion sensor "ON" events in each room of the house. As with the chemical sensors, we captured this data for each hour during the continuous data collection period.

Because of the availability of activity recognition software, we could monitor activities that are performed in the home and capture the duration of each activity over the corresponding hour of data collection. We used machine learning techniques to tag the collected smart home sensor data (motion, door, light, temperature) with corresponding activity labels. Activity duration was then calculated as the time span of sensors' events during the hour labeled with the activity. Our machine learning techniques achieved an average of 95% accuracy for activity labeling based on threefold cross-validation [30]. The set of activities that we monitored for this study includes sleep, bed to toilet transition, relax, leave home, cook, eat, personal hygiene, bathe, enter the home, take medicine, wash dishes, and work.

To determine the area of open windows and doors throughout the house, we noted the size of each door or window and computed the product of the window/door size and the amount of time it was open during the hour. Finally, we computed the mean ambient temperature value sensed over one hour for each temperature sensor location in the home.

In this paper, we perform and investigate the experiments in the context of the CASAS smart home project. There are numerous challenges associated with creating a fully operational smart environment infrastructure, which have limited the number of available smart home houses. To assist with the process of making smart home technologies available in a variety of settings, CASAS initiated the "smart home in a box" (SHiB) project (shown in Figure 3) [31]. The SHiB architecture has three components: physical components, the middleware, and the software applications. The physical components include sensors and actuators that use a Zigbee "bridge" to communicate with the middleware, which is controlled by a publish/subscribe manager. The middleware is a process that adds the timestamp to sensor events and maintains sensor states. The middleware also uses a scribe bridge to store messages in a lasting archive, and an application bridge to share/exchange information with the applications. The SHiB architecture is easily maintained and expanded because of its lightweight bridge design (via application programming interfaces).

The SHiB sensor package includes infrared motion/ambient light sensors, magnetic doors/windows, and temperature sensors. They are attached using removable adhesive. All of these are ambient sensors that are only updated if there is a significant change in a state, for example, a door opening or closing. Narrow-area motion sensors are placed on the ceilings above some specific items in the house, including above the stove, entryway, and dining chairs. This is because narrow-area motion sensors can perceive motions that occur in a one-meter diameter area. As a complement of the narrow-area motion sensors, wide-area motion sensors are installed on the ceiling in large rooms such as the kitchen, living rooms, and bedrooms, and have a much wider coverage so as to recognize motions happening anywhere in the room. CardAccess magnetic contact sensors are used for external windows and doors, as well as for internal cabinets and doors in bathrooms and living rooms. CardAccess temperature sensors are placed in most of the rooms, including bathrooms and the kitchen, to both perceive key activities such as bathing and cooking, and to sense significant temperature changes at those points in each room.

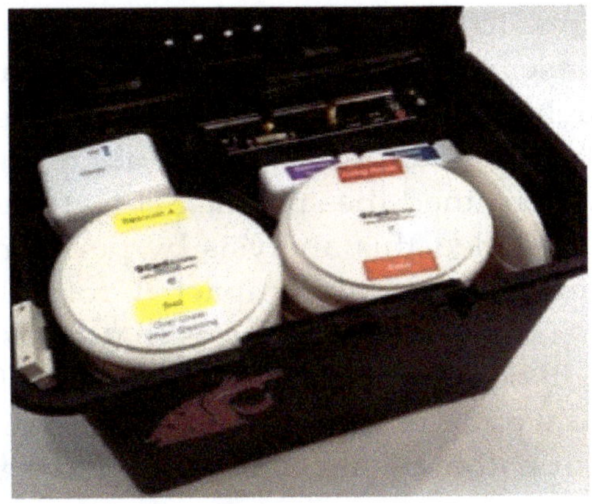

Figure 3. The Center for Advanced Studies in Adaptive Systems (CASAS) Smart Home in a Box (SHiB).

4. Activity Recognition

Activity recognition (AR) refers to mapping a sequence of perceived events onto an element from a group of predefined activity labels. Activity recognition is a well-researched area, and there is a large amount of prior work that introduces machine learning approaches to model the activities using techniques such as hidden Markov models (HMMs) [32] and segmented hierarchical infinite hidden Markov models (siHMMs) [33]. Methods are chosen according to the realism of the smart environment and the sensor technologies that are used for collecting the data. Our CASAS activity recognition algorithm is based on a sliding window method to perceive activities in a streaming fashion. The sensors that we use are ambient sensors triggered by a significant change in a state [30].

The necessary recognition steps in CASAS are gathering and performing preliminary processing on sensor data to handle missing or noisy data, separating it into feasibly sized subsequences by either supervised event segmentation or supervised window sliding approaches, and then pulling out subsequence features. As an alternative to traditional supervised learning-based segmentation, we employed an unsupervised change point detection and piecewise representation of the segments as separate activities. External annotators provide ground truth for training data. They look at a floor plan and the sensor data to provide an estimate of the corresponding activities, which is then used to learn a mapping from the extracted features to activity labels.

The experiments in this paper used the CASAS activity recognition algorithm to tag real-time activities on streaming data, as described in the last paragraph. The CASAS recognition algorithm is a generalization of activity models over several smart homes with no constrained circumstances related to pre-segmented data, single residents, or uninterrupted activities. To do this, we mapped a succession of the n latest sensor events to a label that indicated the activity. For example, this sequence of sensor events was mapped to a Sleep activity label:

2016-03-10 06:48:24.855293 BedroomABed ON Sleep
2016-03-10 06:48:29.727262 BedroomABed OFF Sleep
2016-03-10 06:48:30.479044 BedroomABed ON Sleep
2016-03-10 06:48:33.102565 BedroomABed OFF Sleep

5. Data Analysis

5.1. Experimental Setup

Global population growth and global aging issues will have a corresponding effect on behavioral changes and the quality of the air we experience inside and outside buildings. Here, we examine the relationship between occupant behavior and indoor air quality using machine learning techniques via monitoring human activities in sensor-filled spaces. We conducted two types of analyses on this data. In the first analysis, we performed three experiments to determine which IAQ variables were measurably impacted by SH features. To accomplish this goal, we used machine learning techniques to predict the value of each IAQ variable from the complete set of SH features (we refer to this experiment as AllSH_OneIAQ). We also highlighted the IAQ features that are most significantly impacted by smart home behavior, as indicated by the ability to predict the values using smart home sensor features.

In the second analysis, we determined the specific SH features that had the greatest influence on the IAQ variables. We accomplished this analysis by performing experiments to select a set of SH attributes that had the most significant impact (GroupSH_InIAQ). We then performed another experiment to select the individual SH features that measurably affect each IAQ variable (IndivSH_InIAQ). The findings will help us understand the types of behavior that have tremendous impacts on indoor air quality, and we can use this information to make suggestions to homeowners based on maximizing air quality, or automate the control of buildings.

5.2. Analysis 1: AllSH_OneIAQ

Our first analysis determined the IAQ variables that were measurably impacted by captured smart home-based behavior features (AllSH_OneIAQ). To validate the overall performance of SH features and IAQ variables, we used regression to estimate the value of each dependent variable (each IAQ variable), given the independent variables (SH features). There are many techniques that have been developed for regression analysis. In our project, we performed experiments based on three algorithms: random forest (RF), linear regression (LR) and support vector regression (SVR).

Decision tree learning is one of the most popular regression learning techniques. It can naturally handle data of mixed types and missing values, which occur in all of our datasets. We choose one of the best-known learning methods: random forest learning algorithm. Using random forest, a large set of decision trees are created, each using a different set of randomly selected feature inputs. Compared with other tree learning algorithms, RF improves the prediction accuracy and the stability when the data is changed a little. However, decision trees only map the feature vector to discrete target variables, so we also considered methods that are designed to handle numeric class values.

One model that deals with numeric variables is linear regression, where a single linear formula represents the mapping from input to class values. We used the linear regression learning algorithm as our second learning method. Since our data has a large number of features, we also used a third method, the support vector regression. It is a nonlinear regression technique, which complements the linear regression method.

We evaluated the performance of all three of the above algorithms by reporting the corresponding correlation coefficients (r). In our study, we did not consider the sign of the correlation coefficient, just the absolute value. This is because we wanted to determine whether a relationship exists between the smart home features and the chemical variable features, rather than analyze the type of direction of a relationship between these two complex models. We reported correlation coefficients that are moderate or large ($r \geq 0.3$). In addition, we evaluated the accuracy of our models based on 10-fold cross-validation by reporting the normalized root mean square error (NRMSE) as a performance measure.

In our project, we also report the statistical significance of the observed results. We set the null hypothesis as: there is no correlation between each dependent variable and the independent variables. The corresponding alternative hypothesis is set as: there is a correlation between each dependent variable and independent variables. We then choose the value of the first type error (probability of false rejection of a true null hypothesis) as 0.05, and the value of power (the probability of correctly rejecting a false null hypothesis) as 0.9. For these parameters, the sample size should be 113. Our sample sizes for IAQ_1 and IAQ_2 are 620 h and 187 h respectively, which are large enough to represent subjects where the probability of correctly rejecting a false null hypothesis is greater than 0.9.

To validate the hypothesis, we computed the correlations and NRMSE between the complete set of SH features and each predicted IAQ variable by performing the three regression learning algorithms (RF, LR, and SVR) on each house (IAQ_1 and IAQ_2), as well as on the aggregated dataset for both houses (denoted as IAQ_{1_2}). The results are summarized in Tables 2–4. The full set of results is provided online (http://eecs.wsu.edu/~blin).

Table 2. Overall smart home (SH) features used to predict the variables of the first smart home (IAQ_1). We report the classifier that was used, and the number of IAQ variables that are predicted with at least a moderate effect ($r \geq 0.3$).

Method	Number of $r \geq 0.3$	Total Number	Percentage	NRSME
Random Forest	48	51	94%	0.0961
Linear Regression	41	51	80%	0.2241
Support Vector Regression	42	51	82%	0.1415

Table 3. Overall SH features used to predict the variables of the second smart home (IAQ_2).

Method	Number of $r \geq 0.3$	Total Number	Percentage	NRSME
Random Forest	50	51	98%	0.1118
Linear Regression	39	51	76%	0.1314
Support Vector Regression	30	51	59%	0.1816

Table 4. Overall SH features predicted for the aggregated dataset of variables for both houses (IAQ_{1_2}).

Method	Number of $r \geq 0.3$	Total Number	Percentage	NRSME
Random Forest	50	51	98%	0.0798
Linear Regression	31	51	60%	0.2559
Support Vector Regression	27	51	53%	0.2591

As shown in Tables 2 and 3, the majority of the IAQ variables from both IAQ_1 and IAQ_2 exhibit a relationship with the SH features, because there are over 90% IAQ variables that are highly correlated with SH features, which results in an NRSME lower than 0.12 (using random forest). Further, based on the results shown in Table 4, we observed that the majority of IAQ variables from the aggregated dataset for both houses (IAQ_{1_2}) are also highly predictable from SH features (98% of the IAQ variables are highly correlated with SH features, and result in an NRSME of 0.0798 using random forest). According to this, we conclude that there is a generalized relationship between IAQ variables and SH features. Additionally, we list the correlation coefficients for IAQ variables from the aggregated dataset (IAQ_{1_2}) in Table 5.

In Table 5, we observe that there exists a relationship between human behavior and air quality inside and outside the homes. There are 16 indoor chemical variables (16 out of total 24 indoor chemical variables) that have higher correlation coefficients than those outside the house. Furthermore, there are five outdoor chemical variables (five out of 25 outdoor chemical variables) that have higher correlation coefficients than those inside the house. Thus, human behaviors have a greater impact on chemical variables measured indoors than those variables measured outdoors.

We are going to use three representative pollutants from both the indoor and outdoor categories to further interpret the results from Table 5. We chose $PM_{2.5}$, formaldehyde, and methanol as the representatives for outdoor pollutants, VOCs released from indoor materials, and VOCs released from occupant activities.

For $PM_{2.5}$, we observe that the correlation coefficient for the outdoor $PM_{2.5}$ is 0.5121. This indicates that there is a correlation between outdoor $PM_{2.5}$ and in-home human behaviors. Due to the wildfires, which caused heavy smoke with a large amount of outdoor $PM_{2.5}$ during the experimental period, residents closed windows and doors more often than usual, and stayed at home longer than usual. In the case of the indoor $PM_{2.5}$, the correlation coefficient is 0.4808, which shows that there exists a measurable relationship with human behavior, such as cooking and cleaning, and indoor $PM_{2.5}$.

Table 5. Each IAQ variable predicted by random forest (RF) in the aggregated dataset IAQ_{1_2}.

Higher Correlation Inside than Outside			Higher Correlation Outside than Inside		
Variable	**Correlation Inside**	**Correlation Outside**	**Variable**	**Correlation Inside**	**Correlation Outside**
C_3-benzenes	0.9554	0.3462	α-pinene fragment	0.6723	0.7495
C_2-benzenes	0.9537	0.5457	C_4-benzenes	0.5020	0.5299
temperature	0.9462	0.8830	particulate matter	0.4808	0.5121
methane	0.9334	NA	acetaldehyde	0.4313	0.5536
methanol	0.9265	0.5550	α-pinene	0.3225	0.6151
formaldehyde	0.9061	0.5407	wind speed	NA	0.7596
methyl ethyl ketone	0.8995	0.6076	wind direction	NA	0.7577
methyl vinyl ketone	0.8985	0.5954	pressure	NA	0.7330
styrene	0.8950	0.6155	relative humidity	NA	0.8420
toluene	0.8894	0.2180			
acetone	0.8779	0.5295			
benzene	0.8608	0.5598			
carbon dioxide	0.8465	0.8386			
isoprene	0.8338	0.5748			
water vapor	0.8276	0.6539			
ozone	0.8178	0.7971			
acetonitrile	0.7706	0.5988			
small particle count	0.4471	NA			
large particle count	0.4253	NA			

In Table 5, we observe that the correlation coefficient for the indoor formaldehyde is 0.9060. This large value indicates that there is a strong relationship between indoor formaldehyde and human behaviors. This is because indoor formaldehyde is mainly from indoor carpet, pressed wood products, and furniture. Indoor formaldehyde is also positively correlated with both indoor temperature and indoor humidity [27]. Human behaviors, such as cooking, bathing, washing dishes, and opening/closing windows or doors, make a significant contribution to the temperature and humidity changes inside the house. Thus, the relationship between human behaviors and humidity generate a positive correlation with indoor formaldehyde as well. In addition, the correlation coefficient for the outdoor formaldehyde is 0.5407. Outdoor formaldehyde is mainly produced from industrial wood manufacturing [28]. Hence, it is reasonable that the correlation coefficient is 36% lower than that for the indoor formaldehyde.

With regards to methanol, this chemical occurs either naturally in humans, animals, food, and plants, or industrially based on its use as a solvent, pesticide, and alternative fuel source [27]. The correlation coefficient for the indoor methanol is 0.9265, which is 37% higher than that for the outdoor methanol. This makes sense, because the indoor human behaviors, such as eating, drinking, breath, and solvent, would highly impact the indoor methanol.

5.3. Analysis 2: GroupSH_InIAQ and IndivSH_InIAQ

The above regression analysis quantifies the generalized relationship between IAQ variables and SH features. After regression analysis, we performed a second analysis to determine the specific SH features that have the greatest influence both as a group and individually on the IAQ variables selected from the first analysis. Although in earlier regression analysis we validated that a generalized relationship exists between smart home features and indoor air quality chemical variables based on the aggregated dataset from the two houses, there is a tremendous diversity of specific human behaviors in each house that will affect individual IAQ variables. Thus, in this analysis, we only consider each house and do not include the aggregated dataset. Specifically, we utilize learning algorithms for three experiments (shown in Table 6) to perform the automated selections of SH features for IAQ variables based on their ability to predict IAQ values. These three algorithms employ machine learning algorithms that only handle nominal class values. Because our data is numeric, we employ equal frequency binning to discretize the target variables by dividing the numeric range into a predetermined number (here, $n = 4$) of bins.

Table 6. Three classification algorithms for the second type of analysis.

Experiment Number	Attribute Evaluator	Classifier	Search Method	Lookup Cache Size
Experiment 1	WrapperSubsetEval	Random Forest	Best First	3
Experiment 2	WrapperSubsetEval	J48	Best First	3
Experiment 3	InfoGainAttributeEval		Ranking	

We note that the learning algorithms used for this analysis are different from those used for the first analysis and its corresponding experiments. The classifiers in the first analysis were regression algorithms. In contrast, we now need to employ classifiers that map the feature vector to discrete-valued class labels. We utilize algorithms that are popular for feature selection, namely RF, J48 (a decision tree learner) and information gain (InfoGain). Even though decision trees are typically used for classification (as done in Analysis 1 in Section 5.2), we also use them for feature selection in the current analysis, so as to determine which of the behavior-based attributes are most indicative of indoor air quality, and therefore exhibit the strongest relationship with indoor air quality parameters. InfoGain is used as a measure of information gain on the class that the attribute gives, so as to determine the relevance of that attribute and hence allow the elimination of attributes that are less relevant. The relevance of each attribute is evaluated by assigning a score, which is calculated as the difference in entropy with and without that attribute; afterwards, feature selection can be performed based on the scores. Entropy here measures the impurity of the sample that tells us the average number of bits needed to encode the information in the sample. Further, for classifiers RF and J48, we employ WrapperSubsetEval as an attribute evaluator, which uses a classifier to evaluate alternative attribute sets. The accuracy of the classifier for each attribute set is estimated by cross-validation.

We first perform two experiments to identify subset groups of SH features that together have the most noticeable impact on each chemical variable, and narrow down the size of the subset group to at most 15. To extend the second analysis further, we then perform a similar experiment to select individual SH features.

To be consistent with the first analysis (Section 5.2 Analysis 1: AllSH_OneIAQ), we summarize the behavior features that show the greatest impact on the same three representative chemical variables for each house (outdoor $PM_{2.5}$, indoor formaldehyde, and indoor methanol). The feature selection summary is given in Tables 7–12, which are separated by the particular chemical variable we are analyzing. Explanations for the feature names are provided in Table 13. The full set of results is provided online.

Table 7. Selected SH attributes that as a group predict outdoor $PM_{2.5}$ in IAQ_1.

RF	J48
HLabelBed_Toilet_Transition	HLabelPersonalHygiene
HTMasterBathroom	HTMasterBathroom
HTMasterBathroomWindowA	HTDoorMasterLivingRoom
HTDoorFirstFloorToUpstair	HTKitchen
HTKitchen	HTMasterOfficeWindowA
HTKitchenWindowA	WDMasterBedroomWindowA
HTMasterDingRoom	WDDoorMasterLivingRoom
HTMasterLivingRoom	
WDMasterBedroomWindowA	
WDDoorUtility	
WDDoor1stFloor	

Table 8. Selected SH attributes that as a group predict outdoor $PM_{2.5}$ in IAQ_2.

RF	J48
ALevelMasterBathroom	ALevelMasterLivingRoom
HLabelEat	HTKitchen
HTKitchen	HTMainEntry
HTMainEntry	HTMasterBedroom
HTMasterBathroom	HTMasterLivingRoom
HTMasterBedroom	
HTMasterLivingRoom	
HTMasterOffice	
HTUtility	

Table 9. Selected SH attributes that as a group predict indoor formaldehyde in IAQ_1.

RF	J48
ALevelLivingroom	ALevelDiningroom
ALevelOtherOffice	ALevelKitchen
HLabelBed_Toilet_Transition	ALevelMasterBedroom
HTBedroomAWindowB	ALevelMasterOffice
HTToTheFirstFloorDoor	HTToTheFirstFloorDoor
HTKitchen	HTKitchenA
HTKitchenAWindowA	
WDMasterBedroomDoor	

Table 10. Selected SH attributes that as a group predict indoor formaldehyde in IAQ_2.

RF	J48
HLabelWashDishes	HTMainEntry
HTKitchen	WDMainDoor
HTMainEntry	
HTMasterBathroom	
HTMasterBedroom	
HTOtherLivingRoom	
WDMasterBedroomWindowB	
WDOtherBedroomWindowA	
WDDoorUtility	

Table 11. Selected SH attributes that as a group predict indoor methanol in IAQ_1.

RF	J48
ALevelLivingroom	HTMasterBathroom
HTMasterBathroom	HTUtilityDoor
HTKitchen	HTKitchen
HTKitchenAWindowA	HTMasterLiving

Table 12. Selected SH attributes that as a group predict indoor methanol in IAQ_2.

RF	J48
HTKitchen	ALevelDiningRoom
HTMasterBathroom	HLabelCook
HTMasterLivingRoom	HLabelSleep
HTMasterOffice	HTKitchen
HTOtherLivingRoom	HTMasterBathroom
WDMasterBedroomWindowA	HTOtherLivingRoom

Table 13. Summary of SH feature name explanation, organized by prefix.

SH Features with Prefix	Feature Names
H	Hourly Data
T	Temperature features
ALevel	Activity Level features
Label	Labeled Activity Durations
WD	Open/Closed area of window/door

In Table 7, we observe that for the outdoor $PM_{2.5}$ in IAQ_1, features such as temperature in the bathroom, dining room, and kitchen are highly related with outdoor $PM_{2.5}$ values. We also observe that the duration of both personal hygiene and bed-to-toilet transition are selected. This makes sense because the high-level outdoor $PM_{2.5}$ during the wildfires caused residents to stay at home longer than usual, and therefore more activities to be detected in the house than usual, especially in the bathroom, dining room, and kitchen. Similar results are found for the selected features in IAQ_2 (based on Table 8) for the same reasons. For IAQ_2, the selected features are the temperatures in the main entryway, kitchen, master bedroom, master living room, and master office.

In Table 9, we observe that for indoor formaldehyde in IAQ_1, the selected features are the temperatures in the master bedroom, kitchen, and stairs to the first floor, as well as the overall activity levels in the master bedroom, the secondary office, and the area of an open door in the master bedroom. This makes sense, because we know that carpet is the main source of indoor formaldehyde, and the places with carpets in IAQ_1 are the bedrooms and the secondary office, which is also located inside the master bedroom. Further, temperature and humidity in rooms with carpets have positive impacts on indoor formaldehyde levels.

In Table 10, we notice that for indoor formaldehyde in IAQ_2, the selected features are temperatures in the master bathroom, kitchen, and main entry, and the duration of washing dishes. The temperature in the master bathroom could be an indication of taking a shower or running hot/cold water. Those activities in the bathroom and the duration of washing dishes may have a great contribution to the indoor humidity. In addition, the temperature feature for the main entry door is selected in IAQ_2, but not in IAQ_1. This might be because of the humidity difference during the experimental periods for the two testbeds. According to the weather station reports, for IAQ_2, the average outdoor water vapor was 10,443 parts per million (ppm) compared to 9827 ppm for IAQ_1. That is, the average humidity during the IAQ_2 experimental period was 616 ppm higher than that during the IAQ_1 period. Then, for IAQ_2, opening/closing the main entry door might allow the outdoor humidity to influence the indoor humidity.

In Table 11, we notice that in IAQ_1, the SH features that impact indoor methanol are temperatures in the master bathroom, kitchen, living room, and utility room, and the overall activity level in the living room. This makes sense, because in the kitchen or living room, there are food, fruits, vegetables, and other foods that contain methanol [27]. Temperatures in these rooms and the overall activity levels in the living room may indicate food processing, eating, or drinking, especially with the overly ripe or near rotting fruits or vegetables, smoked food, diet foods, or drinks with aspartame. The temperature in the utility room may indicate that the resident had been doing laundry. The liquid laundry detergents used in this process contain methanol [28]. This also partly explains the selected SH features for indoor methanol in IAQ_2, based on Table 12.

In Table 12, the selected features include temperatures in the kitchen, master bathroom, and secondary living room, the overall activity levels in the dining room, and the duration of cooking and sleeping. The duration of sleeping is selected in IAQ_2 because human breath also makes a contribution to the indoor methanol. In IAQ_2, there are two adults and one child, whereas in IAQ_1 there are only two adults. The living habits of residents in these two testbeds are also different. This may be a reason that the duration of sleeping is selected in IAQ_2 instead of in IAQ_1.

After selecting subsets of SH features for each IAQ variable by RF and J48 experiments, we conducted the third experiment to find the individual SH feature that had the greatest influence on each IAQ variable. That was accomplished through utilizing attribute selection by ranking the SH attributes using their individual scores. Sample results of this analysis for the same three chemical variables are shown in Tables 14–19. The full set of results is provided online.

Table 14. InfoGain method predictions for outdoor $PM_{2.5}$ in IAQ_1.

Information Gain Value	SH Features	Information Gain Value	SH Features
0.3860	HTMasterBathroom	0.2690	HTMasterBathroomWind
0.3675	HTDoor1stFloor	0.2568	HTMasterBedroomWind
0.3624	HTDiningroom	0.2024	HTMasterOfficeWindowA
0.3461	HTKitchenA	0.2008	HTKitchenWindowA
0.3433	HTMasterBedroom	0.1694	HTOtherBathroom
0.3394	HTOtherBedroom	0.1636	HTDoorMasterBedroomToBalcony
0.3330	HTDoorDiningRoom	0.1239	HTMasterBedroomWind
0.3202	HTDoorUtility	0.1145	ALevelMasterBedroom
0.2993	HTMasterLivingroom	0.0945	ALevelLivingroom
0.2990	HTDoorMasterLivingroom	0.0813	ALevelMainEntry
0.2922	HTMainDoor		

Table 15. InfoGain method predictions for outdoor $PM_{2.5}$ in IAQ_2.

Information Gain Value	SH Features
0.4677	HTMainEntry
0.2438	HTKitchen
0.1699	HTMasterBedroom
0.1501	HTMasterBathroom
0.1188	HTMasterOffice
0.0972	HTUtility

Table 16. InfoGain method predictions for indoor formaldehyde in IAQ_1.

Information Gain Value	SH Features	Information Gain Value	SH Features
1.1547	HTKitchen	0.4995	HTKitchenAWindowA
0.6945	HTMainDoor	0.8700	HTMasterBedroom
1.1039	HTToTheFirstFloorDoor	0.4781	HTMasterBathroomWindow
0.6512	HTMasterLivingroomDoor	0.8386	HTDiningRoomDoor
1.0974	HTDiningRoom	0.4412	HTMasterBedroomWindowA
0.5538	HTMasterLiving	0.8334	HTUtilityDoor
1.0541	HTMasterBathroom	0.2731	HTOfficeAWindowA
0.5153	HTMasterBedroomDoor	0.7271	HTOtherBathroom
0.8964	HTOtherBedroom	0.2541	HTMasterBedroomWindowB

Table 17. InfoGain method predictions for indoor formaldehyde in IAQ_2.

Information Gain Value	SH Features
0.2950	HTMasterBathroom
0.1850	HTKitchen
0.1830	HTMainEntry
0.1640	HTOtherLivingRoom
0.1470	HTUtility
0.1300	HTMasterBedroom
0.1100	HTMasterOffice

Table 18. InfoGain method predictions for indoor methanol in IAQ_1.

Information Gain Value	SH Features	Information Gain Value	SH Features
1.1617	HTUtilityDoor	0.8965	HTOtherBedroom
1.1503	HTMasterBathroom	0.8561	HTMasterBathroomWindow
1.1308	HTMasterBedroom	0.8515	HTMainDoor
1.0434	HTDiningRoomDoor	0.7545	HTKitchenWindowA
1.0160	HTDiningRoom	0.7399	HTMasterBedroomWindowA
0.9976	HTOtherBathroom	0.7340	HTMasterBedroomDoor
0.9832	HTKitchenA	0.7064	HTMasterLiving
0.9663	HTToTheFirstFloorDoor	0.5024	HTOfficeAWindowA
0.9361	HTMasterLivingroomDoor	0.4523	HTMasterBedroomWindowB

Table 19. InfoGain method predictions for indoor methanol in IAQ_2.

Information Gain Value	SH Features
0.2460	HTOtherLivingRoom
0.2010	HTMasterBathroom
0.1810	HTMainEntry
0.1520	HTMasterOffice
0.1060	HTKitchen

In Table 14, we notice that the majority of selected features that are strongly related with outdoor $PM_{2.5}$ are temperature variables; the top features are temperatures in the master bathroom, dining room, and kitchen. This is consistent with the results from Analysis 1, as shown in Table 7. In addition, this experiment allows us to observe that for IAQ_1, the temperature in the master bathroom had the highest correlation with outdoor $PM_{2.5}$. This makes sense, because heavy smoke from wildfires contains elevated levels of $PM_{2.5}$. Thus, residents spend more time at home for less exposure to the outside environment.

In IAQ_2, based on Table 15, we notice that the SH features that have the greatest impact are temperatures in the main entry, kitchen, master bedroom, and master bathroom. Moreover, the temperature in the main entry has the highest correlation with outdoor $PM_{2.5}$. This makes sense, because the temperature in the main entry might indicate opening/closing of the main door. Due to the heavy outdoor smoke, residents might open/close the main door more quickly than usual to prevent the outdoor smoke from coming into the house.

In the case of indoor formaldehyde in IAQ_1, based on Table 16, we observe that temperature in the kitchen has the highest correlation with formaldehyde. This is because the temperature in the kitchen was very similar to temperatures throughout the whole house (in general, the difference is less than 1 Celsius, except during the cooking time), and formaldehyde is positively related to the temperature. For IAQ_2, based on Table 17, the temperature in the master bathroom had the highest correlation with indoor formaldehyde due to the positive correlation with humidity.

Considering indoor methanol in IAQ_1, based on Table 18, we notice that the temperature in the utility room has the highest correlation with methanol. This is because methanol is a component of the liquid laundry detergents and temperature in the utility room may indicate the residents had been doing laundry. But for IAQ_2, from Table 19, we notice that the temperature in the secondary living room had the highest correlation with indoor methanol. That is because food and drink in the secondary living room contained methanol. Additionally, residents whose breaths have a contribution to the methanol level may spend a great deal of time in the secondary living room. Those results in the third experiment are consistent with the results from the first two experiments.

6. Discussion

In this study, we noticed that the temperature features are more frequently selected than other specific activities. This might be because temperature is impacted by multiple activities, such as cooking and running hot water, rather than selecting one specific activity that would exclude other activities. In addition, the change in temperature caused by an activity may last longer than the activity itself, and so affect the IAQ even after the activity has ended. The fact that these results are consistent with previous studies helps to validate the methodology as a whole.

In the analyses, we assume that some human activities occur based on the top selected temperature features. Future studies of this type should include information from occupant interviews to help explain the observations and to validate the occurrence of these activities.

Further, the study is based on homes equipped with both multiple SH sensors in each room and air quality measurements in one location inside and outside the house. The use of a single location in each home to measure indoor air quality and represent the air quality throughout the entire house may have impacted our results. Thus, future studies can be improved by using IAQ measurements placed in each room to capture the air quality. In addition, although the locations of indoor air quality measurements in each home is based on the house architecture, the inconsistence with the locations of IAQ measurements (either in living room or dining room) could also have an impact on the results.

7. Conclusions

Our goal was to examine the relationship between in-home behavior and indoor air quality based on collected data from smart home sensors and chemical indoor air quality measurements. We fulfilled this goal by collecting data in two smart home testbeds. We analyzed both the impact of overall smart home behavior on indoor air quality, and the relationship between individual groups of smart home features and indoor air quality variables. We identified and adapted machine-learning classifiers that are appropriate for each analysis.

The results of our first analysis indicated that there is a strong relationship between in-home human behavior and air quality. By examining an aggregated dataset, we also observed that this predictive relationship could be generalized across multiple smart homes. In our second analysis, the specific SH attributes that are most indicative of indoor air quality were found for each testbed. Based on the findings, it would be a reasonable suggestion for the resident to consider airing the rooms frequently.

In future work, we will design methods of automating ventilation control to improve indoor air quality based on sensed activities and other smart home features. For example, we will provide viable suggestions as to how to improve indoor air quality (e.g., turning on ventilation systems only at certain times of the day). These types of analyses can help us recognize the types of behavior that significantly impact IAQ and use this information to anticipate, prevent and prepare for indoor pollution, maintain better healthy environments, and plan for our changing future by developing an automated system for maintaining good indoor air quality.

Acknowledgments: This research was supported by both the US Department of Energy grants RD—83575601 and by the US Environmental Protection Agency Science To Achieve Results grants RD—83575601. The views expressed in this paper are those of the authors and do not necessarily reflect the views or policies of the US Department of Energy nor the US Environmental Protection Agency.

Author Contributions: Beiyu Lin, Brian Lamb and Diane J. Cook conceived and designed the experiments; Beiyu Lin performed the experiments; Beiyu Lin and Yibo Huangfu analyzed the data; Nathan Lima, Bertram Jobson, Max Kirk, Patrick O'Keeffe, Shelley N. Pressley and Von Walden contributed reagents/materials/analysis tools; Beiyu Lin, Brian Lamb, Bertram Jobson and Diane Cook wrote the paper.

References

1. Klepeis, N.E.; Nelson, W.C.; Ott, W.R.; Robinson, J.P.; Tsang, A.M.; Switzer, P.; Behar, J.V.; Hern, S.C.; Engelmann, W.H. The National Human Activity Pattern Survey (NHAPS): A resource for assessing exposure to environmental pollutants. *J. Expo. Sci. Environ. Epidemiol.* **2001**, *11*, 231–252. [CrossRef] [PubMed]
2. U.S. Environmental Protection Agency (EPA) and U.S. Consumer Product Safety Commission. *The Inside Story: A Guide to Indoor Air Quality*; National Service Center for Environmental Publications (NSCEP), EPA Document # 402-K-93-007; National Service Center for Environmental Publications: Cincinnati, OH, USA, 1995.

3. Institute of Medicine. *Climate Change, the Indoor Environment, and Health*; The National Academies Press: Washington, DC, USA, 2011. [CrossRef]
4. Field, R.W. *Climate Change and Indoor Air Quality*; Report to the U.S. Office of Radiation and Indoor Air; Environmental Protection Agency: San Francisco, CA, USA, 2010.
5. Tucker, W.G. Air Pollution, Indoor Air Pollution, and Control. In *Kirk-Othmer. Encyclopedia Chemical Technology*; John Wiley & Sons, Inc.: Hoboken, NJ, USA, 2003.
6. Berglund, B.; Brunekreef, B.; Knöppe, H.; Lindvall, T.; Maroni, M.; Mølhave, L.; Skov, P. Effects of Indoor Air Pollution on Human Health. *Int. J. Indoor Environ. Health* **1992**, *2*, 2–25. [CrossRef]
7. Francisco, P.W.; Jacobs, D.E.; Jacobs, L.; Jacobs, S.L.; Jacobs, J.; Rose, W.; Cali, S. Ventilation, Indoor Air Quality, and Health in Homes Undergoing Weatherization. *Int. J. Indoor Environ. Health* **2016**, *2*, 463–477. [CrossRef] [PubMed]
8. Fabi, V.; Andersen, R.V.; Corgnati, S.; Olesen, B.W. Occupants' window opening behaviour: A literature review of factors influencing occupant behaviour and models. *Build. Environ.* **2012**, *58*, 188–198. [CrossRef]
9. Barnes, B.R. Behavioral Change, Indoor Air Pollution, and Child Respiratory Health in Developing Countries: A Review. *Int. J. Environ. Res. Public Health* **2014**, *11*, 4607–4618. [CrossRef] [PubMed]
10. Chiang, C.M.; Chou, P.C.; Wang, W.A.; Chao, N.T. A study of the impacts of outdoor air and living behavior patterns on indoor air quality case studies of apartments in Taiwan. *Indoor Air* **1996**, *96*, 735–740.
11. Andersen, R.V.; Toftum, J.; Andersen, K.K.; Olesen, B.W. Survey of occupant behaviour and control of indoor environment in Danish dwellings. *Energy Build.* **2009**, *41*, 11–16. [CrossRef]
12. Matthews, C.E.; Steven, C.M.; George, S.M.; Sampson, J.; Bowles, H.R. Improving self-reports of active and sedentary behaviors in large epidemiologic studies. *Exerc. Sport Sci. Rev.* **2012**, *40*, 118. [CrossRef] [PubMed]
13. Roy, N.; Misra, A.; Cook, D. Infrastructure-assisted smartphone-based ADL recognition in multi-inhabitant smart environments. In Proceedings of the 2013 IEEE International Conference on Pervasive Computing Communications (PerCom), San Diego, CA, USA, 18–22 March 2013; pp. 38–46. [CrossRef]
14. Roy, N.; Misra, A.; Cook, D. Ambient and smartphone sensor assisted ADL recognition in multi-inhabitant smart environments. *J. Ambient Intell. Humaniz. Comput.* **2016**, *7*, 1–19. [CrossRef] [PubMed]
15. Pires, I.M.; Garcia, N.M.; Pombo, N.; Flórez-Revuelta, F. From data acquisition to data fusion: A comprehensive review and a roadmap for the identification of activities of daily living using mobile devices. *Sensors* **2016**, *16*, 184. [CrossRef] [PubMed]
16. Nasreen, S.; Azam, M.A.; Naeem, U.; Ghazanfar, M.A.; Khalid, A. Recognition Framework for Inferring Activities of Daily Living Based on Pattern Mining. *Arab. J. Sci. Eng.* **2016**, *41*, 3113–3126. [CrossRef]
17. Deleawe, S.; Kusznir, J.; Lamb, B.; Cook, D.J. Predicting air quality in smart environments. *J. Ambient Intell. Smart Environ.* **2010**, *2*, 145–154. [PubMed]
18. Uysal, N.; Schapira, R.M. Effects of ozone on lung function and lung diseases. *Curr. Opin. Pulm. Med.* **2003**, *9*, 144–150. [CrossRef] [PubMed]
19. Occupational Safety and Health Administration. *Substance technical guidelines for formalin: Occupational Safety and Health Standards, Toxic and Hazardous Substances, Formaldehyde*; Occupational Safety and Health Administration: Washington, DC, USA, 2012.
20. Hildemann, L.M.; Markowski, G.R.; Cass, G.R. Chemical composition of emissions from urban sources of fine organic aerosol. *Environ. Sci. Technol.* **1991**, *25*, 744–759. [CrossRef]
21. Mugica, V.; Vega, E.; Chow, J.; Reyes, E.; Sanchez, G.; Arriaga, J.; Egami, R.; Watson, J. Speciated non-methane organic compounds emissions from food cooking in Mexico. *Atmos. Environ.* **2001**, *35*, 1729–1734. [CrossRef]
22. Smith, K.R. Fuel combustion, air pollution exposure, and health: The situation in developing countries. *Annu. Rev. Energy Environ.* **1993**, *18*, 529–566. [CrossRef]
23. User Manual for DC1100 Air Quality Monitor DYLOS Corporation, Page 5. Available online: https://www.sylvane.com/media/documents/products/dylos-dc-1100-laser-particle-counter-owner's-manual.pdf (accessed on 27 July 2017).
24. Minnesota Department of Health, Volatile Organic Compounds in Your Home. *Indoor Air Unit.* 2016. Available online: http://www.health.state.mn.us/divs/eh/indoorair/voc/ (accessed on 3 October 2016).
25. Hansel, A.; Jordan, A.; Holzinger, R.; Prazeller, P.; Vogel, W.; Lindinger, W. Proton transfer reaction mass spectrometry: On-line trace gas analysis at the ppb level. *Int. J. Mass Spectrom. Ion Process.* **1995**, *149*, 609–619. [CrossRef]
26. Jobson, B.T.; McCoskey, J.K. Sample drying to improve HCHO measurements by PTR-MS instruments:

Laboratory and field measurements. *Atmos. Chem. Phys.* **2010**, *10*, 1821–1835. [CrossRef]
27. Odum, J.R.; Hoffmann, T.; Bowman, F.; Collins, D.; Flagan, R.C.; Seinfeld, J.H. Gas/particle partitioning, and secondary organic aerosol yields. *Environ. Sci. Technol.* **1996**, *30*, 2580–2585. [CrossRef]
28. Reitzig, M.; Mohr, S.; Heinzow, B.; Knöppel, H. VOC emissions after building renovations: Traditional and less common indoor air contaminants, potential sources, and reported health complaints. *Indoor Air* **1998**, *8*, 91–102. [CrossRef]
29. Dawadi, P.N.; Cook, D.J.; Schmitter-Edgecombe, M. Automated cognitive health assessment from smart home-based behavior data. *IEEE J. Biomed. Health Inf.* **2016**, *20*, 1188–1194. [CrossRef] [PubMed]
30. Krishnan, N.C.; Cook, D.J. Activity recognition on streaming sensor data. *Pervasive Mob. Comput.* **2014**, *10*, 138–154. [CrossRef] [PubMed]
31. Hu, Y.; Tilke, D.; Adams, T.; Crandall, A.S.; Cook, D.J.; Schmitter-Edgecombe, M. Smart home in a box: Usability study for a large scale self-installation of smart home technologies. *J. Reliab. Intell. Environ.* **2016**, *2*, 93–106. [CrossRef]
32. Trabelsi, D.; Mohammed, S.; Chamroukhi, F.; Oukhellou, L.; Amirat, Y. An unsupervised approach for automatic activity recognition based on Hidden Markov model regression. *IEEE Trans. Autom. Sci. Eng.* **2013**, *10*, 829–835. [CrossRef]
33. Saeedi, A.; Hoffman, M.; Johnson, M.; Adams, R. The segmented iHMM: A simple, efficient hierarchical infinite HMM. In Proceedings of the International Conference on Machine Learning, New York, NY, USA, 19–24 June 2016; pp. 2682–2691.

9

Integration of Sensor and Actuator Networks and the SCADA System to Promote the Migration of the Legacy Flexible Manufacturing System towards the Industry 4.0 Concept

Antonio José Calderón Godoy and Isaías González Pérez *

Department of Electrical Engineering, Electronics and Automation, University of Extremadura, Avenida de Elvas, s/n, 06006 Badajoz, Spain; ajcalde@unex.es

* Correspondence: igonzp@unex.es

Abstract: Networks of sensors and actuators in automated manufacturing processes are implemented using industrial fieldbuses, where automation units and supervisory systems are also connected to exchange operational information. In the context of the incoming fourth industrial revolution, called Industry 4.0, the management of legacy facilities is a paramount issue to deal with. This paper presents a solution to enhance the connectivity of a legacy Flexible Manufacturing System, which constitutes the first step in the adoption of the Industry 4.0 concept. Such a system includes the fieldbus PROcess FIeld BUS (PROFIBUS) around which sensors, actuators, and controllers are interconnected. In order to establish effective communication between the sensors and actuators network and a supervisory system, a hardware and software approach including Ethernet connectivity is implemented. This work is envisioned to contribute to the migration of legacy systems towards the challenging Industry 4.0 framework. The experimental results prove the proper operation of the FMS and the feasibility of the proposal.

Keywords: sensor and actuator network; fieldbuses; industrial communications; Ethernet; flexible manufacturing system; programmable logic controllers (PLC); supervisory control and data acquisition (SCADA); Industry 4.0; Industrial Internet of Things (IIoT)

---

1. Introduction

In industrial facilities, distributed Sensors and Actuators (S&A) are networked by means of the so-called fieldbuses. These digital communication networks fulfill necessities that process automation imposes like real-time and reliability [1]. Other nodes also integrated in the networks are controllers and supervisory systems, aiming to share the operational information compulsory for the proper behavior of the process.

Aiming at improving communication quality and costs in comparison to previous analog communication buses [2], fieldbuses have evolved since their inception from simple and proprietary approaches to heterogeneous communication infrastructures where wired and wireless networks coexist [3]. Different topologies, even combining diverse fieldbuses protocols, can be deployed, where several already standardized protocols contribute to enhance the interoperability handling [4]. Some examples of well-known fieldbuses are the Actuator Sensor Interface (AS-i), Modbus, Controller Area Network (CAN), PROcess FIeld BUS (PROFIBUS), PROcess FIeld NETwork (PROFINET), and Ethernet for Control Automation (EtherCAT).

Relating to Ethernet, when Internet technology became popular, a new wave of Ethernet-based networks was stimulated for automation [3]. Advances in Ethernet technology have made the medium more suited to industrial use, resulting in a trend towards Ethernet-based fieldbus protocols [5]. For years, Ethernet networks have become more and more popular due to advantages like speed, bandwidth, and easy integration with the Internet or the office network, among others. Ethernet is signalled as the basis for advances in industrial communications and standardization of industrial protocols [6]. In fact, real time Ethernet has become a standard in the industrial automation domain [3]. Improvements in industrial networking such as the incorporation of Ethernet technology have started to blur the line between industrial and commercial networks [5]. This also allows for easier interconnection between business and industrial networks in order to relay process and control information to interested parties [5].

These industry-focused networks also accommodate S&A for scopes out of the industrial domain, for instance, to implement remote laboratories [4,7], intelligent energy grids, i.e., Smart Grids (SGs) [8,9], or building automation systems [2,10].

Moreover, the increasing complexity of industrial systems results in a growing amount of data from signal sources that must be acquired, communicated, and evaluated [11]. This data is commonly managed by Supervisory Control and Data Acquisition (SCADA) systems. These are software applications responsible for gathering the data provided by S&A, as well as exchanging control parameters with the automation units, i.e., Programmable Logic Controllers (PLC). Numerical and/or graphical information of the plant behavior is displayed in real time to an operator, and the relevant variables are stored for further analysis [12].

Advanced functionalities are delivered as a consequence of the aforesaid complexity of industrial systems and the integration of Information and Communications Technologies (ICTs). In fact, a new industrial revolution is being predicted a-priori [13], the Industry 4.0 (I4.0). Industrial Internet of Things (IIoT) is an alternative term to refer to such a paradigm. I4.0 is considered the fourth industrial revolution in which digital factories are conceived as smart environments where machines, sensors, and actuators are interconnected to enable collaboration, monitoring, and control [14]. The implications of I4.0 are not only technological; it also involves economic, social, and ecological aspects [15]. Various pillars of I4.0 are proposed in the literature, among which common features are networked interconnection of the components, massive data gathering and analytics, wide-adoption of ICTs, interoperability management, smart S&A, collaborative robotics, new business models, and cloud computing exploitation, just to name a few [16].

I4.0 reshapes industry boundaries, creates entirely new industries, and exposes established manufacturing companies to new competitive challenges [15]. Regarding the latter issue, notwithstanding that I4.0 is envisioned to facilitate interconnection and computerization into the traditional industry [17], a big challenge facing I4.0 real-scale implementation is the legacy barrier. This is due to the fact that automation-devoted devices, mainly PLC, are expected to have a lifespan of decades, so the advents of innovative technology like those brought by I4.0, strike the legacy of already existing facilities. As it is evident, the investments associated with the modernization of software and hardware entities are a serious obstacle. Actually, most enterprises refuse a radical modernization of their entire automation system or simply cannot take the risk of quitting a running system.

Thus, the integration of the new advanced components within production systems requires a sound migration path from legacy production systems toward the improved production systems [18]. This way, aiming at facilitating migration to the I4.0, a step-by-step process must be followed. Instead of changing the whole system, it is necessary to extend capabilities of the hardware infrastructure that is in use to implement modern ways of information management [19]. Due to the networked operation, in I4.0, machinery and equipment must be provided with data sharing mechanisms [20]. In other words, connectivity is a key feature for the legacy of already existing infrastructures; therefore, a first stage to be I4.0-ready consists of adding network connectivity to current devices.

Concerning PLC, their capabilities are continuously increasing according to technological the Ethernet-based network to seamlessly link the S&A network with the SCADA system assures both heterogeneity management and the addition of advanced S&A.

Among the Research and Development (R&D) lines carried out in the Automation and Industrial Computation Laboratory of the University of Extremadura (Spain), I4.0 is receiving a lot of attention in order to study its implications from the perspective of automation and supervision. With this aim, an experimental advanced I4.0-compliant system is required to act as a benchmark to investigate the migration of legacy systems towards this challenging concept.

Within such a laboratory, a Flexible Manufacturing System (FMS) is found. It is composed of a set of stations interconnected mechanically by means of a transport conveyor, and logically through a fieldbus. Such a network acts as backbone where S&A, control units, and a supervisory system exchange data in real time. FMS has a modular architecture that allows the work stations to be reorganized for different processes and operations. As it is evident, FMS are ideal environments in which to research the I4.0 concept since they include characteristics like networked interconnection, data gathering, and distributed intelligence.

FMS are used nowadays in diverse R&D activities, as demonstrated by recently published papers. Girbea et al. [21] use an FMS as an environment to research production scheduling and systems integration. In [22], the application of the open source Arduino microcontroller in FMS is analysed. FMS as a scenario for a scheduling problem solved by means of Petri Nets is presented in [23]. An FMS serves as the application case of an I4.0-compliant architecture proposed by Pisching et al. [24]. In [25], a proposal of human-machine cooperation approach in the context of I4.0 is evaluated using an FMS. Many papers reporting the utilization of FMS scarcely provide details about the real and effective automation and management of all involved equipment. However, this information constitutes useful insights for researchers and practitioners aiming to implement advanced frameworks.

This paper presents the first step for the migration of a legacy system, an FMS, towards the I4.0 concept. Namely, an Ethernet-based communication solution has been developed in order to integrate the S&A network and the SCADA system. This approach is conceived as a middle layer involving hardware and software elements.

The motivation for this work arose when implementing the automation system of the FMS in order to be exploited for I4.0-related R&D. It was necessary to improve some functionality of the FMS, mainly enhancing its networked communication options and data management. Concretely, the installed PLC do not provide Ethernet connectivity by default, whereas the fieldbus PROFIBUS is natively supported, as well as a proprietary protocol of Siemens (Multi-Point Interface, MPI). A SCADA system could be linked to the S&A network via MPI connection or using a PROFIBUS adaption card coupled to the PC where the supervisory application runs. Though, in a certain sense, under the I4.0 vision, these options imply an isolated operation regarding modern devices which use widespread Ethernet communication and facilitate the inclusion of other I4.0 functionalities. This boundary was considered as an opportunity to study the legacy problem. Consequently, the presented proposal takes advantage of the already available components, fostering their orchestration following the I4.0 scheme.

The choice of an Ethernet-based solution is driven by the role that this communication medium plays in the I4.0 scenario. As asserted in [26], an Ethernet-Based network applicable to process industries is a key feature to achieving the concept of I4.0 for the next generation control systems. Within I4.0-ready environments, the heterogeneity of hardware and software entities must be handled to achieve an interoperable and standardized framework [11]. Under this viewpoint, the introduction of Ethernet connectivity is a step forward in the standardization and interoperability handling for legacy systems. From another perspective, various key requirements for I4.0-compliant systems are pointed out by Delsing [27], among which evolvability over time and technology generations have been considered in the present work. In I4.0, SCADA systems are required to exchange data through the network with new smart devices, i.e., modern S&A that are increasingly equipped with embedded Transmission Control Protocol/Internet Protocol (TCP/IP)-Ethernet ports. Therefore, using the Ethernet-based network to seamlessly link the S&A network with the SCADA system assures both heterogeneity management and the addition of advanced S&A.

This work constitutes a step forward in the direction of enabling I4.0 features for reliable legacy automation systems. Moreover, the experimental nature of the FMS affords an added value in the sense that it is not a theoretical proposal; instead, it is a real system effectively working.

The target group of this paper is researchers and practitioners in the scopes of industrial automation, S&A networks, and I4.0.

The remainder of this paper is structured as follows. Section 2 deals with the description of the used FMS. The proposed solution to integrate the S&A network with the SCADA system is reported in Section 3. The achieved results are expounded and discussed in Section 4. Finally, the main conclusions are provided.

2. Materials and Methods

Before addressing the description of the proposed approach to communicate the S&A network and the SCADA application, it is necessary to include, as a starting point, a general description of the FMS of the company SMC (Tokyo, Japan) [28]. The automated flexible manufacturing cell performs assembly and storage of a turning mechanism. To perform these tasks, the complete system consists of eight stations using components from different technologies (pneumatics, electrical engineering, robotics, etc.). The turning mechanism assembled in the different stations of the cell consists of the following elements: Body, Bearing, Shaft, Cap, and Screws. These constituents are shown in Figure 1a, whereas the pallet where the set is transported appears in Figure 1b.

(a)

(b)

Figure 1. Detail of pieces managed in the FMS: (**a**) Constituent elements of the turning mechanism; (**b**) Transport pallet.

Each of the stations is constituted by a table-like structure on which the different components of each process are arranged, such as robots, pneumatic cylinders, pneumatic distributors, motors, and sensors. On the front of each station, the electrical and electronic components that carry out the control of the station are arranged: switched voltage sources of 24 VC, protection elements, drivers for servomotors, PLC, etc., along with the buttons and indicators used for the manual operation of the station. Figure 2 shows the front panel and keypads of one of the stations.

Figure 2. Front panel and keypad of one of the stations.

Figure 3. Station FMS-201: Body supply.

The research group has three of the eight stations that make up the entire cell available, namely, station 1, station 7, and station 8, as well as an empty station, which is currently being equipped. The tasks of each station are now explained:

- Station 1 (FMS-201): Body supply. This station performs the processes of feeding the body, which serves as a support to the turning mechanism, and the verification of its correct position. This element is fed from a vertical container and can be found in two different positions. If the orientation of the body is correct, it is moved to the transport pallet located in the transfer system. If the base/body orientation is incorrect, the body is rejected (see Figure 3). The sensors involved in this station are: an inductive sensor to detect the presence of the base, a magnetic (Reed type) detector in-built in a pneumatic cylinder to detect the correct position of the base, and a vacuum pressure switch to guarantee the correct functioning of the vacuum pad. This station also consists of six pneumatic linear actuators.
- Station 7 (FMS-207): Robotized screwing. In this station, a widely diffused technology is used, such as robotics. This station consists of an ABB multipurpose industrial robot that weighs only 25 kg and can support a payload of 3 kg with a range of 580 mm and six degrees of freedom (see Figure 4). It should be noted that this station has two programing tasks:
    - Programming of robotic arm movements, contained in the ABB IRC5 controller.

- ○ Programming of the PLC that controls the work cycle of the station. The PLC commands the IRC5 controller to initiate the sequence that has been programmed for the robotic arm. The IR5 controller returns an end-of-cycle signal when all its instructions have been completed. It also provides information about the execution of the different phases of the cycle (screwing, changing the cover, etc.).

The robotized station is in charge of the adjustment of the four screws placed in the body of the turning device. Nevertheless, due to the fact that the research group does not own all the stations that compound the whole cell, the robot will carry out operations of assembly and disassembly of shafts and covers. Two different operating cycles have been programmed in the robot, whose execution depends on the coding of the transport pallet that supports the set. The so-called short cycle (code: 000) only implies that the cover is changed and the screwing is done. On the contrary, in the long cycle (code: 101), the cover and the shaft are changed and the screwing is conducted.

- Station 8 (FMS-208): Automatic warehouse. This station corresponds to an automated warehouse in the X-Y plane of finished parts. The storage of completed sets in a specific coordinate is done by means of two Mitsubishi servomotors that make up a Cartesian XY system (Figure 5). The collection and deposit of the completed set both in the transfer and in its corresponding position in the warehouse are solved through a vertical pneumatic axis (Z axis) equipped with a vacuum pad.
- Empty station: This one is intended as a reserve station for future extensions. Its objective is to close the modular transport system (transfer), so that the transport pallet can return from station 8 to station 1, and thus be able to carry out a cyclic and continuous execution of the assembly process.
- Modular Transfer: All the stations are interconnected by means of a transfer system, consisting of conveyor belts through which the set of pieces moves from one station to another. Consequently, the transport system makes the automatic performance of the process possible. Each of the stations includes an individual transfer section. In this way, multiple combinations of layouts can be developed, with the possibility of joining stations at 90° or 180°. Each individual transfer section includes the retainers and pallet lifters, electrical connections, air vents, and other elements required for its operation. The transfer system incorporates transport pallets in which the pieces are placed so that they can be transported from one station to another, as can be seen in Figure 6, where the aspect of the final piece mounted is appreciated.

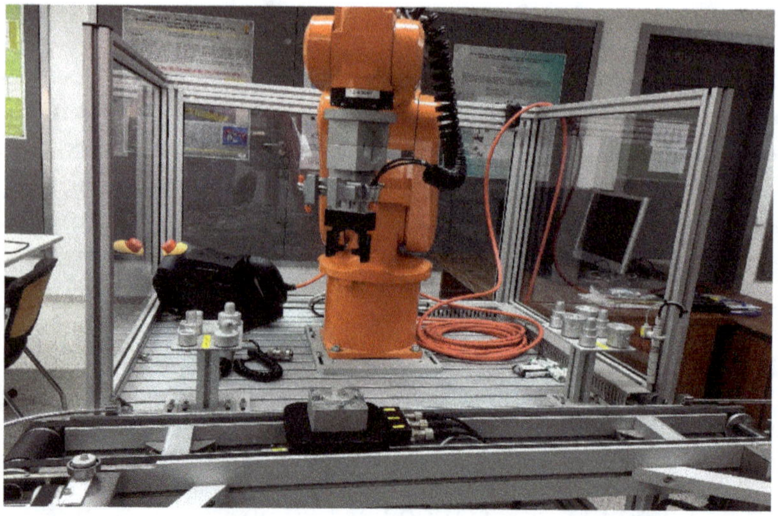

Figure 4. Station FMS-207: Robotized screwing.

Figure 5. Station FMS-208: Automatic warehouse.

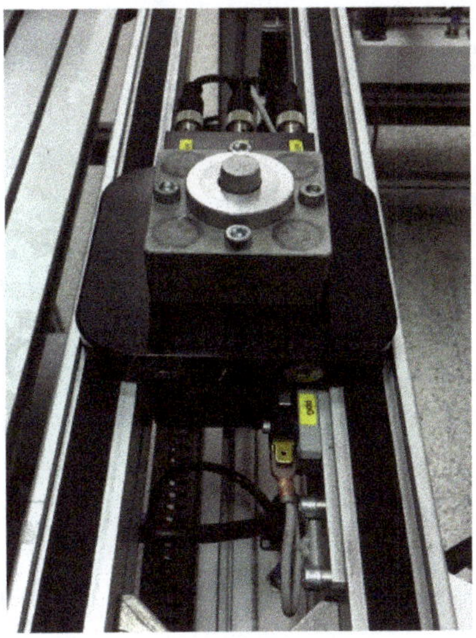

Figure 6. Transport pallet of the transfer system with a final piece.

Finally, for a better illustration of the FMS, Figure 7 depicts its layout through a block diagram, whereas Figure 8 shows a snapshot of the current four stations in the laboratory.

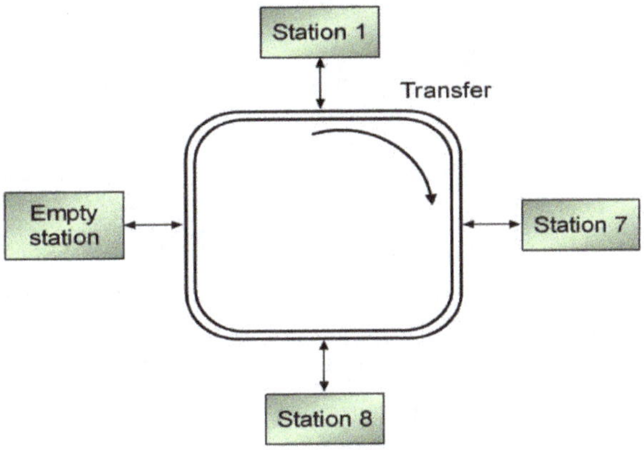

Figure 7. Block diagram of the layout of the stations.

Figure 8. Stations of the FMS in the laboratory.

3. Proposed System

The problem to tackle is the enhancement of the connectivity of the FMS automation system. The existent PLC support communication through the fieldbus PROFIBUS and the proprietary protocol MPI. These protocols are well-proved reliable and robust in industrial environments, but they lack the possibility of being directly connected to Ethernet networks. The Ethernet connectivity is a paramount requirement for I4.0, as discussed in the Introduction. Therefore, it was necessary to add such connectivity to start the migration process of the legacy system. The proposed solution to share data between the S&A network and the SCADA system is reported in this section. To begin with, it is necessary to describe the automation and supervision system that performs the operation of the FMS. After that, in Section 3.2, the abovementioned solution to integrate both systems is fully developed. To provide a whole perspective, Figure 9 depicts the block diagram of the proposal.

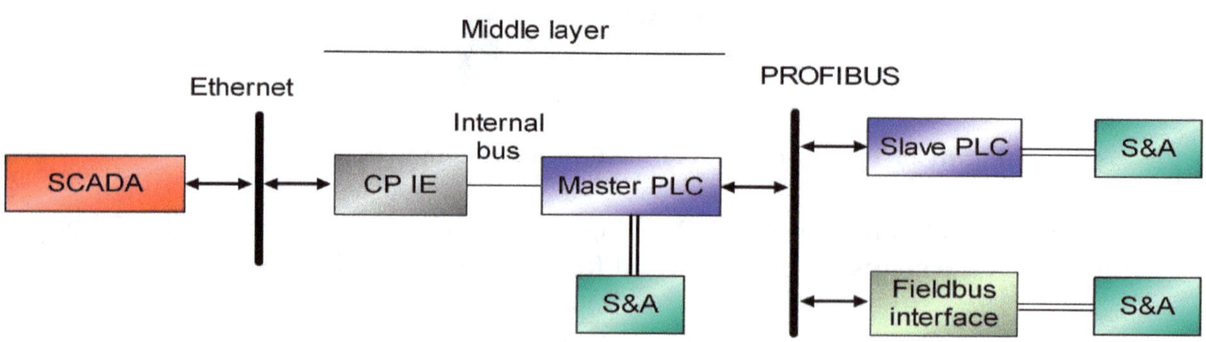

Figure 9. Block diagram of the proposed system.

From the point of view of the system operation, there are two network levels in the proposed system, namely: PROFIBUS and Industrial Ethernet (IE). Fieldbus PROFIBUS serves as a communication network among PLC (master and slave units) and distributed Input/Output (I/O) modules. Each of the aforementioned devices includes a communication port for this fieldbus, so, the physical layer of this configuration is already resolved. In this way, the master PLC has access to all the S&A of the manufacturing cell. The communication between the master PLC and the SCADA system is carried out by the standard communication network Ethernet. To this goal, a middle layer that acts as a bridge to link the fieldbus with the SCADA system enables the communication between the master PLC and the SCADA system. Such a middle layer is composed of a Communication Processor

(CP) and a software-defined structure to store operational information. Data of S&A are available in the fieldbus through PLC or dedicated interfaces. There are also S&A linked directly to the PLC of station 1.

3.1. Automation and Supervision System

The automation system is implemented by means of hardware and software subsystems, described below.

3.1.1. Hardware Subsystem

In this subsection, all the hardware elements that make up the entire cell will be exposed. The FMS, in its current configuration, is formed by the following elements:

- Four PLC Siemens S7-313C-2DP.
- One Ethernet CP Siemens CP 343-1.
- Two I/O modules Mitsubishi X8Y4.
- Two I/O modules Wago 750-343.
- One Industrial Robot Controller ABB IRC5.
- Two servomotors Mitsubishi MR-J2S-10 CL.

Figure 10 shows the connection diagram of the hardware elements that make up the automation and supervision system of the FMS.

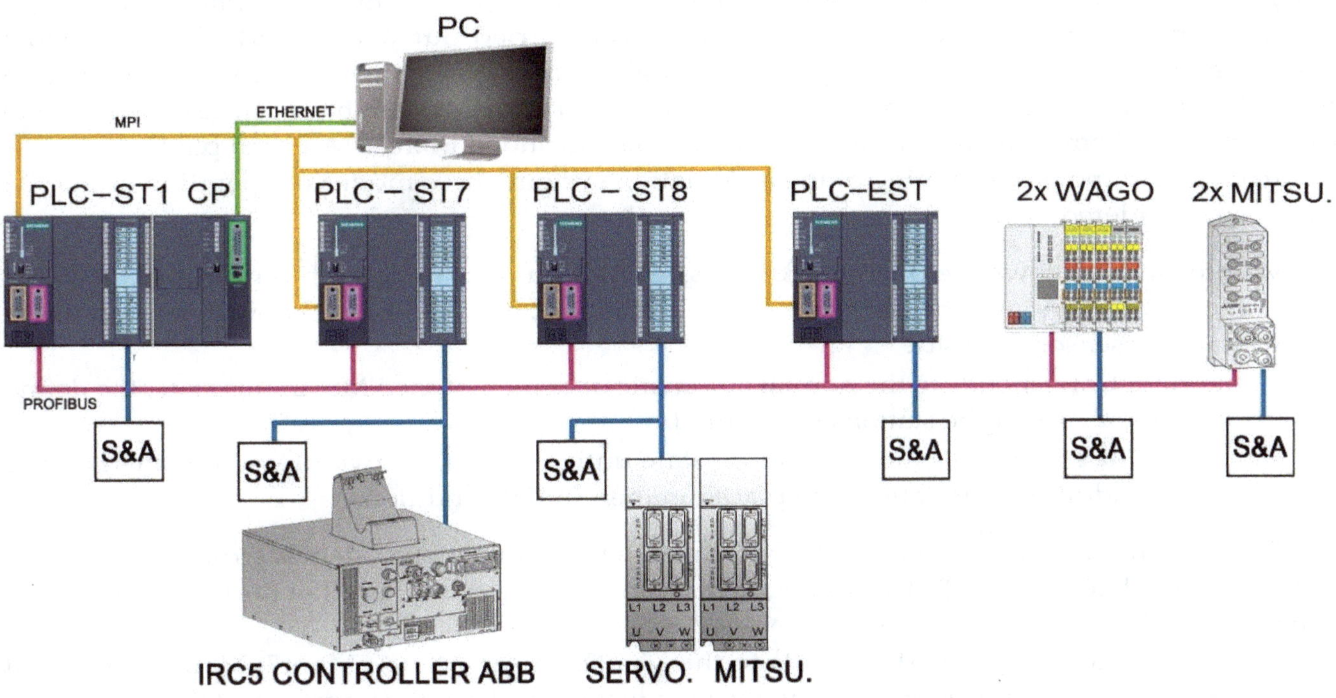

Figure 10. Block diagram of the automation and supervision system.

In the top of the automation and supervision system, a Master Terminal Unit (MTU) consisting of a PC-based SCADA system communicates with the PLC of station 1 (PLC-ST1 in Figure 10). This PLC plays the role of master PLC of the fieldbus PROFIBUS. So, the remaining PLC in the network (PLC-ST7, PLC-ST8, and PLC-EST in the same figure) act as slaves. To this aim, an Ethernet communication module, CP 343-1, is directly coupled to such master PLC through its internal bus. Details about the MTU-PLC communication are developed in Section 3.2.

The master PLC communicates with the rest of the components via PROFIBUS. These components are configured as slaves and correspond to the Mitsubishi and Wago distributed I/O modules and the remaining PLC. Eventually, for the initial programming and configuration tasks, the PC could also be connected to the PLC through an MPI connection, as indicated in Figure 9 in an orange color.

The PLC used for the automation of each station is located on the electrical panel mounted on a standardized Deutsches Institut für Normung (DIN) rail. The PLC incorporates 16 digital I/O, both working and load memory, PROFIBUS and MPI interface, and integrated counters and timers. Each slave PLC executes control tasks over the corresponding station and exchanges data with the others via PROFIBUS. In addition, it must be noted that each PLC acts as an interface to link the S&A with the fieldbus. In the case of station 1, S&A are connected directly to the master PLC.

The control of the modular transfer of stations 1 and 8 is carried out with the Mitsubishi distributed I/O modules. These modules, located below their corresponding station, are connected to the fieldbus and have eight inputs and four digital outputs for each one. In the same way, the Wago modules control the modular transfer segments of stations 7 and empty, also linked via PROFIBUS.

Regarding the robots, the control of the movements of the ABB IRB 120 robot is carried out by ABB's compact controller IRC5. Such a controller is connected to the digital I/O module of the station 7 PLC. On the other hand, the Cartesian robot of station 8 is controlled by two Mitsubishi servomotors, which are connected to the digital I/O module of the associated PLC.

3.1.2. Software Subsystem

The software platform used for the development of this project is Totally Integrated Automation Portal V13 SP1, which has been recently referred to as the foundation of digital factories [20]. The programming of the automation system has been carried out with the STEP 7 Professional package, an engineering software for programming and configuring Siemens controllers included in the TIA Portal platform. The programming of the supervisory system has been resolved with the WinCC package from Siemens (Munich, Germany), also included in the TIA Portal platform.

To carry out the programming of the automation system, three modes of operation of the FMS have been considered, namely:

- Maintenance: Allows the operator to access each input and force each output that makes up each of the stations and the transfer system. In this way, it is possible to know if any component (actuator or sensor) is faulty.
- Monitoring: Corresponds to the normal automatic operation of the FMS. In this state, the defined work cycle for each of the stations is executed.
- Stand-by: The stations remain in stand-by state until either of the two previous states has been selected. In addition, the stations are placed in their initial conditions.

The user program of each of the PLC that make up the automation system is structured in a series of blocks, called Functions (FC). The following diagram shows a flow chart that portrays the structure of the master PLC program (Figure 11).

From the Maintenance FC (FC1) calls to the MantTransfer, Mant1, Mant7, Mant8, and Mant9 functions are made. These FC correspond to the maintenance tasks of the transfer and the four stations, respectively. They are contained in the program of station 1. As aforementioned, this is the station which the SCADA system accesses to show the activation and deactivation status of the S&A of both the transfer system and the four stations.

The automatic operation mode of the FMS is programmed in the Monitoring FC (FC2). This mode is similar in the four stations, although the PLC program of station 1 (PROFIBUS master) includes some additional functions to control the other bus devices. This FC contains the following secondary FC:

- Stages: Conditions that mark the steps or stages that make up the processes of each station.
- PLC Outputs: Conditions that control the state of the actuators.

- Timers: Contains all timers that are used in programming.
- Work Bits: Conditions for the status of different important marks of each station, such as: initial conditions, cycle running, and base accepted, etc.
- Transfer Control: Conditions concerning the operation of the transfer.
- Slave control: Conditions concerning the operation of the slaves corresponding to stations 7, 8, and 9.
- SCADA: Where the variables related to the SCADA system are updated.
- Warnings: Contains the conditions that activate and deactivate the warnings defined by the developer (it only exists in the master PLC).

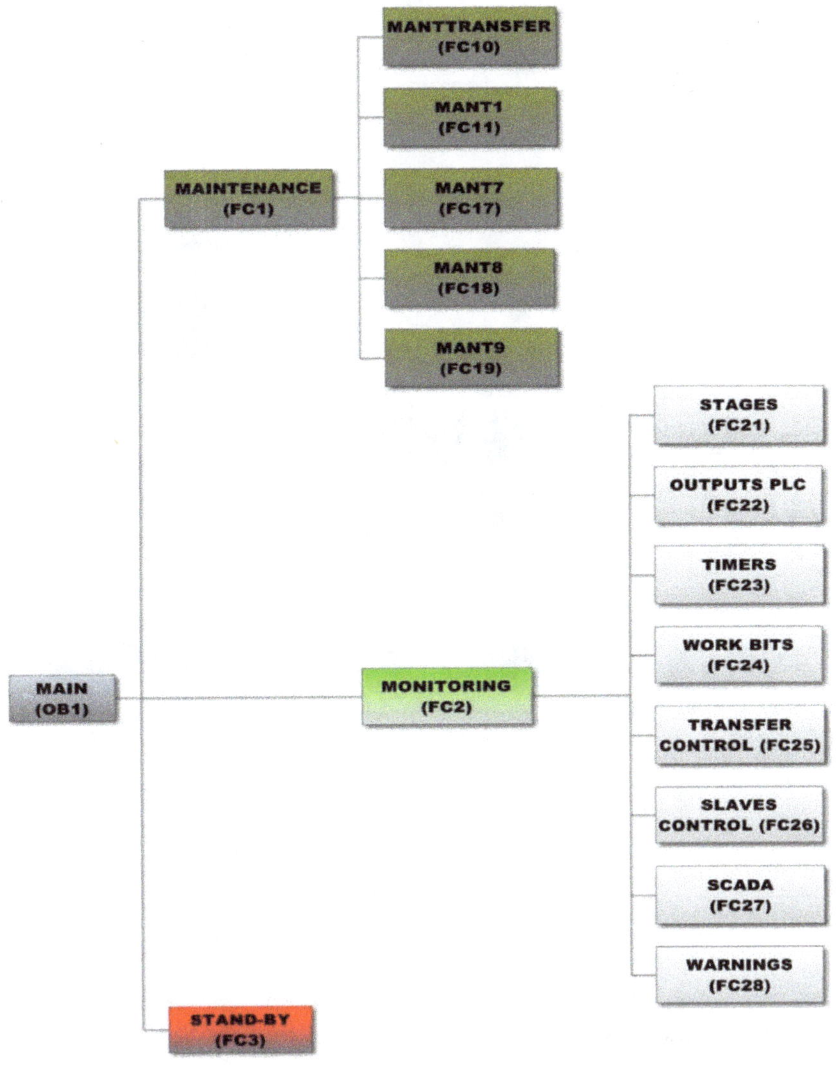

Figure 11. Block diagram of the master PLC program.

3.2. Integration S&A Network-Ethernet-SCADA

A middle layer composed of a CP and a software-defined structure to store operational data has been developed. On the one hand, the hardware linkage is implemented by a CP module that enables the integration of the master PLC in an Ethernet-based network. On the other hand, an array of Data Blocks (DB) in the master PLC makes the sharing of information between the SCADA and the S&A network independently of their location possible.

As previously commented upon, the CP module is responsible for establishing such a connection, both for the programming of the first station PLC and for communication with the SCADA system. The CP includes a RJ45 port to accommodate an Ethernet wire. Another capability is a customizable web page hosted by this module that allows web-enabled diagnostics via Hypertext Transfer Protocol (HTTP) clients. It should be noted that in the present work, this capability has not been exploited.

Concerning the software level, DB are memory positions of the PLC where measurements and signals are stored. These blocks can be of two types, global or instance DB. The access to the latter one is restricted to a particular Function Block (FB). The first type has been the selected one in order to have access to the hosted information by any part of the PLC program. Taking advantage of the Ethernet connectivity afforded by the CP, the data shared through the fieldbus and the DB-based storage, a seamless integration between the SCADA system and the S&A network has been developed and implemented.

To allow access to the data of all the stations and S&A, a software structure based on DB has been created in the master PLC of station 1 (Figure 12).

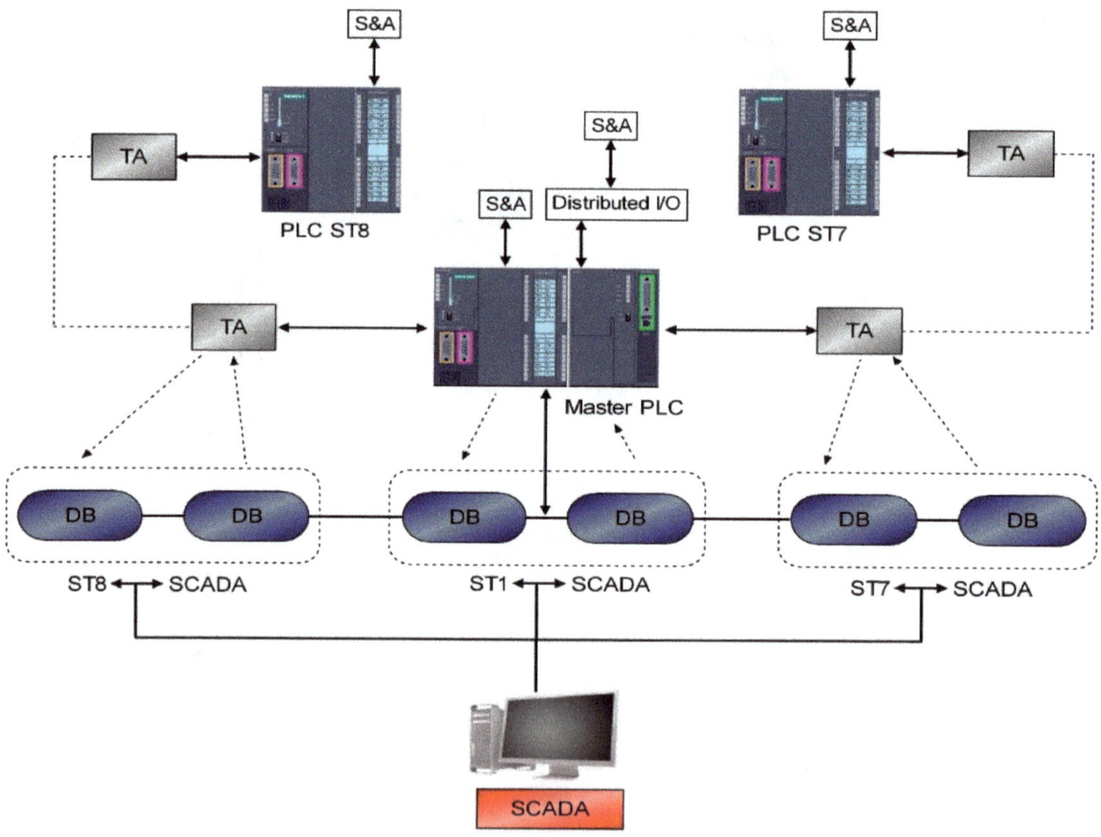

Figure 12. Connection scheme between signals of S&A and the SCADA system through DB blocks.

When parameterizing the PROFIBUS communication, a Transfer Area (TA) is defined to act as a buffer of the shared information. In the presented approach, two TA for each PLC have been defined: one for input data and the other for output data. As a consequence, two DB have been created for each PLC, where one DB is devoted to Read values and the other one hosts Write values. Read values are the signals from S&A, which constitute information to be read by the SCADA system. On the other hand, Write values correspond to command signals that the operator/user introduces through the supervisory interface. The empty station nowadays does not require any DB because there is no information to be shared. Hence, a total number of 6 DB are used, as can be observed in this figure.

This division of information facilitates modifications or maintenance tasks, both for current legacy equipment or future new devices.

In this way, the master unit gathers all the information of the FMS and makes it available for the SCADA. In other words, the information related to all S&A and internal parameters of PLC are "concentrated" in two DB for each station and can be shared with higher-level applications. Therefore, this information is available not only for supervisory interfaces, but also for other software applications devoted to the control of production, Enterprise Resource Planning (ERP), Manufacturing Execution Systems (MES), Computer-Aided Manufacturing (CAM), and so forth. Figure 12 depicts the information flows regardless of the physical medium where it occurs. A continuous data flow takes place between the different elements. For instance, signals from S&A of the transfer are interfaced with PROFIBUS via the distributed I/O modules, and exchanged with the master PLC through PROFIBUS. Once these stages are completed, the supervisory program accesses the signals by means of the Ethernet channel.

As a sample of this scheme, Figure 13 shows the aspect of the DB designed to exchange Write values between the SCADA system and the PLC of station 8, named DB80, with the master PLC acting as the intermediate layer.

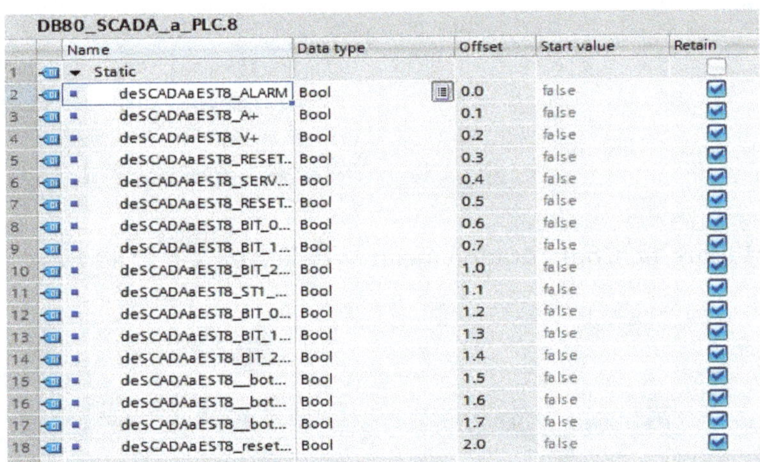

DB80_SCADA_a_PLC.8

	Name	Data type	Offset	Start value	Retain
1	Static				☐
2	deSCADAaEST8_ALARM	Bool	0.0	false	☑
3	deSCADAaEST8_A+	Bool	0.1	false	☑
4	deSCADAaEST8_V+	Bool	0.2	false	☑
5	deSCADAaEST8_RESET...	Bool	0.3	false	☑
6	deSCADAaEST8_SERV...	Bool	0.4	false	☑
7	deSCADAaEST8_RESET...	Bool	0.5	false	☑
8	deSCADAaEST8_BIT_0...	Bool	0.6	false	☑
9	deSCADAaEST8_BIT_1...	Bool	0.7	false	☑
10	deSCADAaEST8_BIT_2...	Bool	1.0	false	☑
11	deSCADAaEST8_ST1_...	Bool	1.1	false	☑
12	deSCADAaEST8_BIT_0...	Bool	1.2	false	☑
13	deSCADAaEST8_BIT_1...	Bool	1.3	false	☑
14	deSCADAaEST8_BIT_2...	Bool	1.4	false	☑
15	deSCADAaEST8__bot...	Bool	1.5	false	☑
16	deSCADAaEST8__bot...	Bool	1.6	false	☑
17	deSCADAaEST8__bot...	Bool	1.7	false	☑
18	deSCADAaEST8_reset...	Bool	2.0	false	☑

Figure 13. DB created to share data between SCADA and PLC of station 8.

Concerning the parameterization of the Ethernet-based communication, the master PLC and the MTU are connected in a Local Area Network (LAN). The configuration of this communication requires defining the IP address in both nodes. Despite the fact that the supervisory application is presented in the next section, it has been considered more convenient to expose here such a configuration for a better exposition of the developed solution.

The assignment of the IP addresses is made using the TIA Portal suite. Particularly, the IP address of the CP is defined in the Device configuration menu (Figure 14). The subnet mask must also be specified. Within this menu, the properties of the CP include the parameterization of the PROFINET interface, where the assignment of the IP address is performed. In this regard, it should be noted that this interface allows the use of fieldbus PROFINET or Industrial Ethernet. In the present case, the latter one has been applied as aforementioned. Other configurable features are related to the MPI network, but this connectivity has not been exploited for this application.

Regarding the MTU, the employed Ethernet interface is a common Peripheral Component Interconnect (PCI) Ethernet card. In WinCC, its IP address in established in an equivalent procedure to that followed for the CP. In addition, for a proper communication MTU–PLC, it is necessary to parameterize a connection within the WinCC, as shown in Figure 15. As can be seen, the addresses of both devices belong to the same range within the LAN.

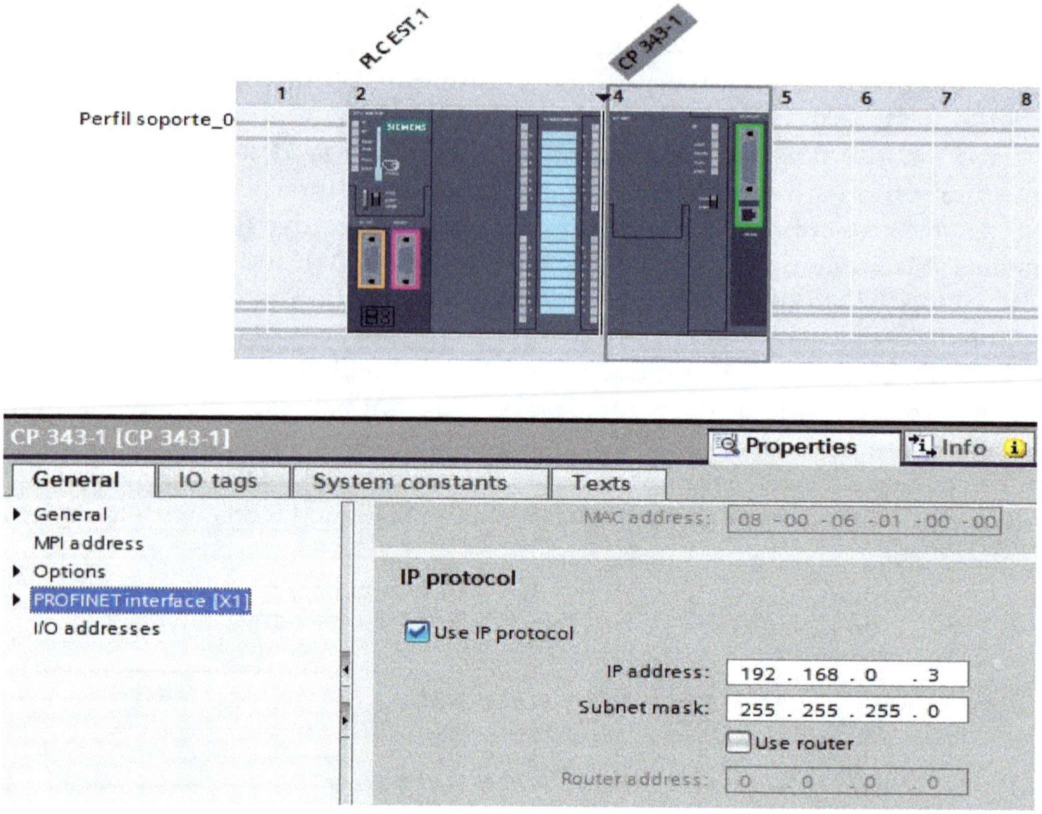

Figure 14. Configuration menu to define the CP IP address.

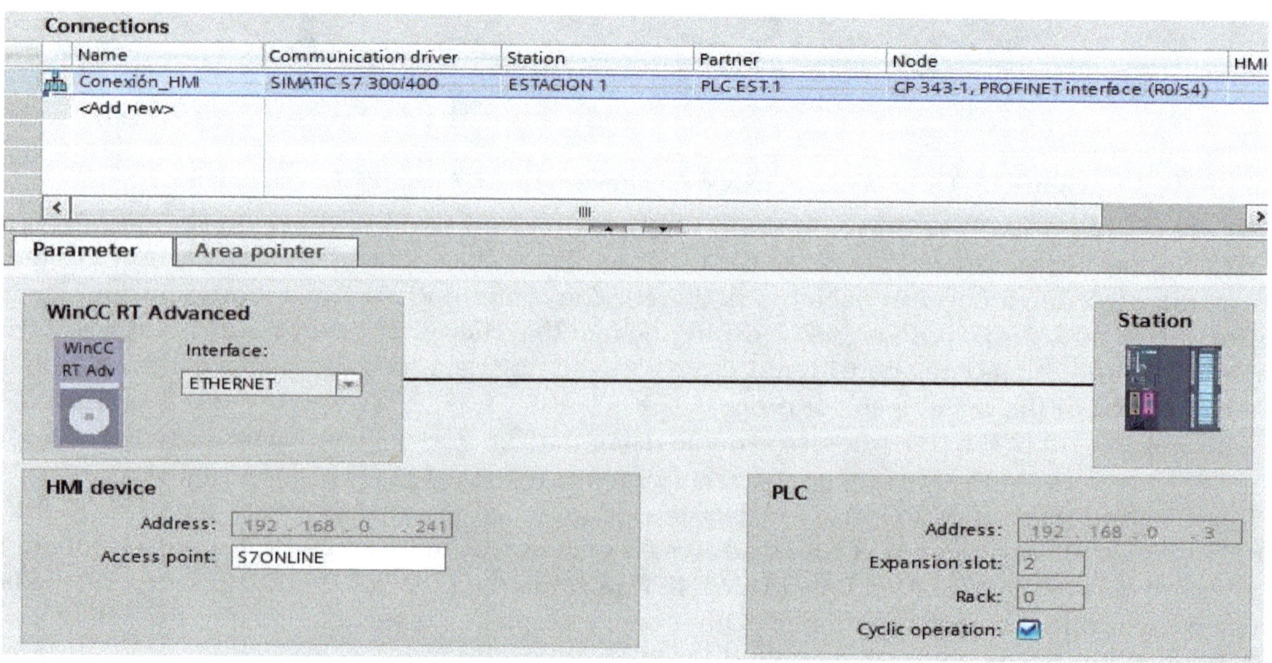

Figure 15. Configuration screen for the SCADA connection.

As a result, the Network View of TIA Portal depicts the configured network, including both the PROFIBUS segment and the Ethernet link (Figure 16). One can observe that both networks correspond to those depicted in Figure 10.

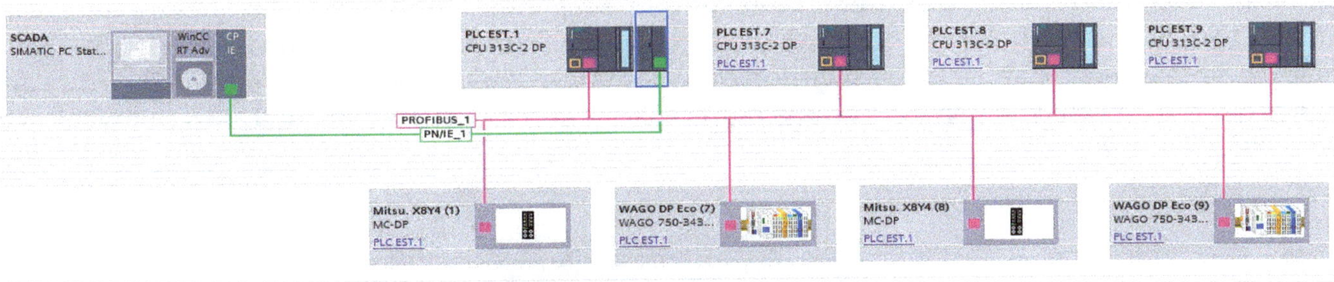

Figure 16. Network View of TIA Portal including PROFIBUS and Ethernet networks.

4. Results and Discussion

In this section, the achieved results are expounded and discussed. Once the Ethernet connection and the DB structure have been configured, all the operational information of the S&A network and of PLC is available to be monitored. To this aim, a SCADA system has been developed and deployed, demonstrating the feasibility of the proposed approach. Furthermore, a database has also been generated to afford this information to other software applications.

4.1. SCADA System

As briefly indicated in the Introduction, a SCADA system is a software application specially designed to work on computers in production control, providing communication with field devices (S&A, PLC, etc.) and controlling the process automatically from the screen. It also provides all the information generated in the production process to diverse users, both at the same hierarchical level as other supervisory users within the company (supervision, quality control, production control, data storage, etc.).

The program used for the design and implementation of the SCADA system is WinCC, included in the TIA Portal platform. Using the aforementioned program, the following functions are carried out: browsing through different screens, control of process variables, start-up of the system, user administration, warnings system, collection of process data and transfer to external files (historical data), scheduled tasks, etc.

The SCADA system is composed of 14 screens, through which the user can navigate to access all the options and is able to track the process of the FMS. The SCADA system has the following structure:

1. Maintenance: This option is focused on being able to carry out checks of each one of the S&A of the stations.
2. Monitoring: This second option makes it possible to carry out the monitoring of the productive process without the need to be in situ next to the stations thanks to the implemented Ethernet link. In this way, it is possible to know at each moment where the platform is located, what sensors are activated, and what action is being taken.
3. Data storage: This last option permits the access to an external database with the most important parameters of the work processes performed on each piece.

Once the SCADA runtime starts, the home screen containing the user identification by name and password, and access icons to the three operating modes of the system appear, as can be seen in

Figure 17. It should be noted that user administration is a paramount functionality in the exploitation of SCADA systems. With the establishment of access permissions, the protection of important data of the process is guaranteed and the execution of certain functions in runtime is regulated. Users and user groups are created and specific access rights called "authorizations" are assigned to them.

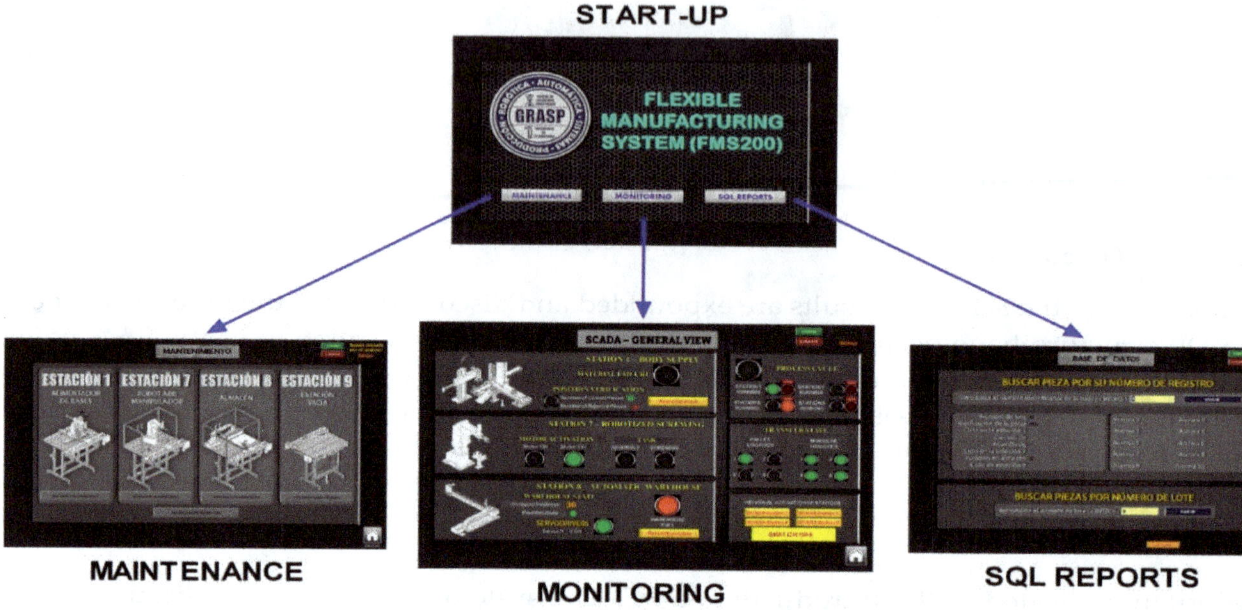

Figure 17. Navigation within the developed SCADA system.

The monitoring-devoted screens are described below. With these screens, the behavior of the FMS working in automatic mode can be supervised. This mode consists of a main view (Figure 17) in which the basic indicators appear to check the operation of the global system, as well as the buttons for set-up and to reset the counters of correct parts and of the full warehouse batch. Finally, from the main view, it is possible to access the warnings to visualize the operating states and the present faults and alarms in the installation. Obviously, the empty station has not been fully included in this screen since it does not perform operations over the piece, and only allows the cyclic transport of the pallet.

As a proof of concept, Figure 18 shows the main monitoring screen under real operation of the FMS. As can be seen, the working station is indicated in the top right corner, and in the presented case, station 1 is processing body pieces (see the corresponding green indicator). The pallet is placed in this station, which is shown by a green indicator in the section devoted to the transfer state. Concerning the automatic warehouse, an alarm state is reflected, showing that the warehouse is full. The maximum number of stored pieces (30) has been reached, so it is signaled by means of a red LED indicator in order to inform the operator. This human-requesting situation is solved by resetting the counter (button Reset Warehouse) once the positions have been manually released.

There are also screens to access the monitoring of each station separately. Through animations and Boolean indicators, the evolution of the process can be observed in real time. Other actions that can also be performed are resetting the counters of accepted and rejected parts, restarting the ABB robot system, or resetting the warehouse positions. For instance, Figure 19 shows the screen dedicated to monitor the first station. The reflected case corresponds to that commented on for the previous screen. This one provides illustrative information about the operation of the S&A of the first station. Namely, the pallet is detected, the correct position of the base is verified, and the vacuum pad works properly to place the base over the pallet.

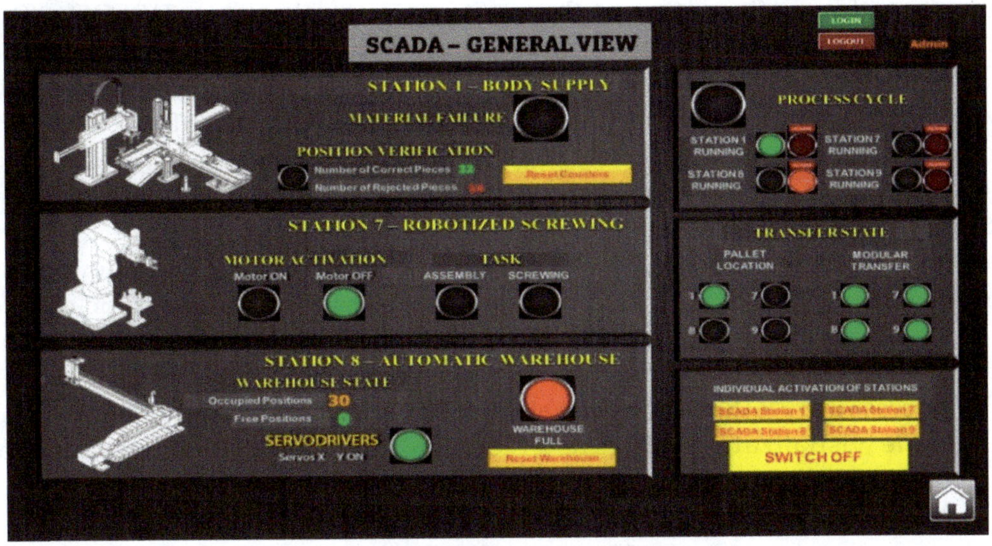

Figure 18. Main view of Monitoring.

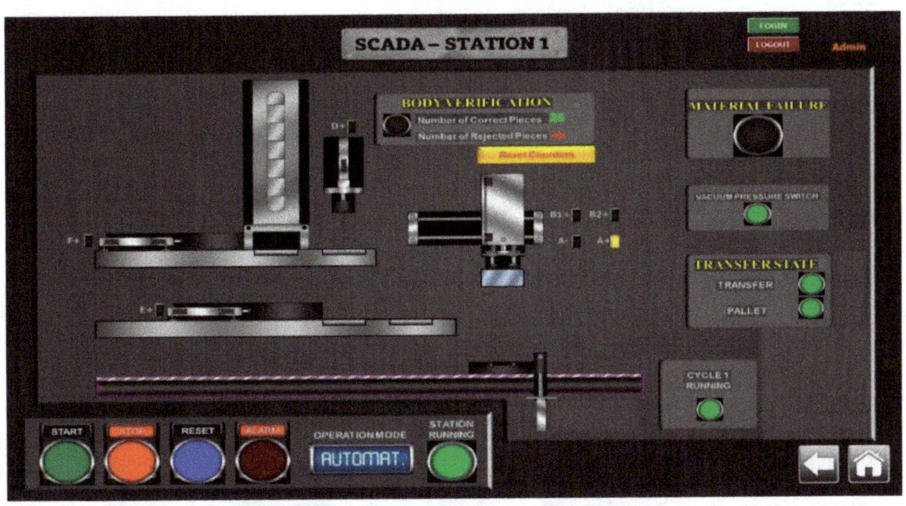

Figure 19. Monitoring screen of station 1.

4.2. Database Report

Concerning data storage, it must be remarked that the continuous recording of data allows its subsequent use and, therefore, also its graphic representation, comparison, creation of statistics, analysis, etc. Consequently, the registration of process data (historical) and its possible exploitation is a basic functionality of the supervisory system. For this purpose, the monitoring systems are linked to databases, usually external to them. In this way, the monitoring systems allow the historical tracking of the product (traceability), the comparison of campaigns, or their use as a virtual test bench for the training of operators without the need of direct connection to the process. The use of external databases allows access both from the monitoring environment and from other applications through standard languages, which results in the most convenient way to integrate industrial computer systems. Structured Query Language (SQL) is one of the most widespread languages and is adopted by most manufacturers and suppliers of industrial software. In this work, the process of archiving information of interest was solved through Visual Basic (VB) scripts that are generated by variable value change events.

To this aim, first, the configuration of the communication between WinCC software and SQL Server is addressed. The access to an SQL database in WinCC Runtime software is done through scripts. To access an SQL database, it must be previously created on the SQL server. WinCC Runtime software acts as an SQL client. A communication using the Open DataBase Connectivity (ODBC) standard with the SCADA application establishes data streaming towards the database. The goal of ODBC is to allow access to any data from any application, no matter what DataBase Management System (DBMS) stores the data.

Now, through scripts, the developer can create new registers for each new piece, update the information for the current task of the cycle, or allow the user to search a specific register by entering the identifier in the input field. Figure 20 shows the result obtained for the database using the aforementioned scripts. In this way, the information of the FMS operation is successfully stored and can also be accessed by other software applications like those devoted to ERP, MES, CAM, or web-based remote visualization. Figure 21 depicts, in a simplified way, this availability of database-supported information.

id	verification	end-cycle_1	cap/shaft	screwed	end-cycle_7	position	end-cycle_8	batch number
1	ACCEPTED	DONE	Only cap	CORRECT	DONE	1	DONE	1
2	ACCEPTED	DONE	Cap & shaft	CORRECT	DONE	2	DONE	1
3	REJECTED	DONE	No information	No information	No information	0	No information	1
4	REJECTED	DONE	No information	No information	No information	0	No information	1
5	ACCEPTED	DONE	Cap & shaft	CORRECT	DONE	3	DONE	1
6	ACCEPTED	DONE	Cap & shaft	CORRECT	DONE	4	DONE	1
7	ACCEPTED	DONE	Only cap	CORRECT	DONE	5	DONE	1
8	ACCEPTED	DONE	Only cap	CORRECT	DONE	6	DONE	1
9	ACCEPTED	DONE	Only cap	CORRECT	DONE	7	DONE	1
10	ACCEPTED	DONE	Cap & shaft	CORRECT	DONE	8	DONE	1

Figure 20. Table of the file obtained with VB scripts and SQL statements.

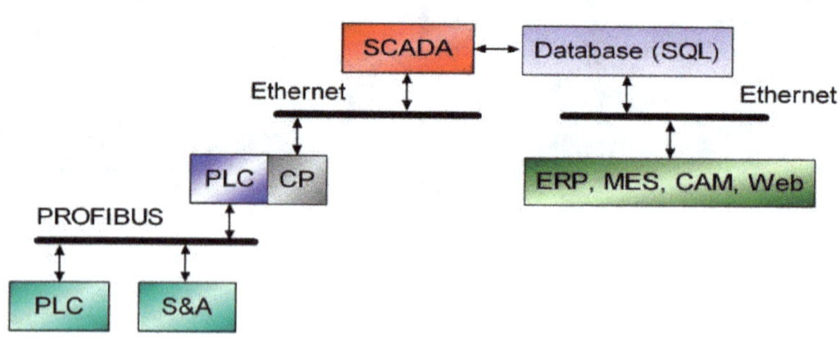

Figure 21. Block diagram of third-party applications to the SQL database.

4.3. Discussion

In view of the achieved results, the following discussion is conducted. This work constitutes the first step in the migration of a legacy system, namely an FMS, towards the I4.0 concept. As expounded in the previous subsections, the experimental results prove the feasibility of the proposed solution.

The presented development has enabled the integration of an S&A network with a SCADA system. To sum up, the legacy feature of the existing PLC has been overcome by means of an Ethernet link between the supervisory interface and the master PLC. Therefore, all data of the S&A network can be effectively retrieved by such a supervisory interface. The designed SCADA system affords the relevant information with user-friendly and easy-to-use features.

The proposed solution is generalist in the sense of being valid if other SCADA software is used. For instance, the widely known LabVIEW environment could also be used, so particularities of such software should have to be configured but the communication bridge would remain the same. In this sense, the communication between the master PLC and the SCADA application should be conducted through the standard Open Platform Communications (OPC), but the middle layer implemented in this work would be the same. Likely, information storage in databases is a widely supported functionality regardless of the specific supervisory software suite.

As more features of I4.0 are included within the FMS, it will be necessary to manage the heterogeneity of components (software and hardware). This issue will be approached by means of the aforementioned OPC protocol, namely the Unified Architecture (UA) specification, whose advantages are able to address the challenges introduced by the Industry 4.0 [19].

The implemented scheme is envisioned to facilitate the addition of other S&A that include embedded Ethernet connectivity, especially modern IoT-enabled S&A.

Remote access for online supervision is afforded by the remote connection options of the SCADA system. In the present case, WinCC offers web connections, as well as a Virtual Network Computing (VNC) desktop interface. In the developed system, this function has not been exploited in order to avoid unauthorized intrusions in the network of the university.

The time and effort devoted to implementing the proposed approach has required deep expertise related with automation to configure the PLC, whereas the deployment of the Ethernet link with the SCADA system has not required high level knowledge about networking over Ethernet. The structure of DB also implies a profound skill for managing distributed I/O signals in the PLC and the fieldbus PROFIBUS. It must be taken into account that the usage of industrial fieldbuses imposes detailed parameterization, whereas the Ethernet means does not require low level configurations. This issue is considered as a benefit in order to facilitate the migration towards I4.0.

In an exercise of self-criticism, the presented FMS still has a long way to go until it becomes a fully I4.0-compliant system. In fact, innovative trends can be incorporated like Radio Frequency IDentification (RFID), virtual/augmented reality, Cloud computing, cyber security means, or open source resources.

Regarding that last trend, a drawback was found when designing the SCADA system with the TIA Portal due to the associated license costs. This aspect is important because the available budgets are continuously decreasing. To solve this situation, a research line is focused on developing a supervisory system using open source software. In fact, open source hardware and software projects are key accelerators for the industry adoption of IoT [29]. In the same regard, the FMS is able to act as a benchmark to research the integration of PLC and open source hardware devices in the context of legacy systems. An example consists of connecting an Arduino microcontroller through an Ethernet shield or an inbuilt port, so a communication channel is established.

Virtualization of manufacturing processes by means of virtual or augmented reality is also a merging movement, envisioned to enhance human-machine interaction in the I4.0 context [30]. Ethernet connectivity will facilitate the sharing of information between the virtualized environment, for instance a 3D virtual world, and the experimental facility.

5. Conclusions

This paper has presented the design and implementation of a communication link between an S&A network and a SCADA system within a FMS. The proposal consists of a middle layer involving hardware and software elements. Such an FMS is used to investigate I4.0, so an enhancement of its connectivity capabilities according to the I4.0 main principles has been carried out.

A fully functional and experimental system has been deployed, which, in fact, is presently working. The reported results prove the feasibility of the proposed solution in order to seamlessly connect the S&A network and the SCADA system, taking advantage of the DB-based structure in the master PLC and the Ethernet CP. All the signals of the S&A are effectively shared and managed through the network, available for higher hierarchical level applications.

Consequently, this work aims to humbly contribute to the migration path of legacy systems towards the challenging I4.0 scenario. In this sense, authors would like to remark that the proposal has constituted a preliminary stage in order to make a legacy system a candidate to fulfil I4.0 requirements. Therefore, further efforts must be conducted to make the system fully I4.0-compliant.

Future works include the addition of more I4.0 features like Cloud storage and open source technologies under the Service-Oriented Architecture (SOA) paradigm. The inclusion of the OPC UA specification will also be approached to handle the heterogeneity of entities. Another further guideline consists of introducing other operational scenarios related to I4.0, taking advantage of the incorporated connectivity.

Author Contributions: A.J.C. conceived and developed the presented systems; I.G. and A.J.C. performed the experimental validation; I.G. analyzed the data; I.G. and A.J.C. wrote the paper. All authors have seen and approved the manuscript.

Acknowledgments: Authors wish to thank the anonymous reviewers for their valuable suggestions that improved this article.

References

1. Zand, P.; Chatterjea, S.; Das, K.; Havinga, P. Wireless industrial monitoring and control networks: The journey so far and the road ahead. *J. Sens. Actuator Netw.* **2012**, *1*, 123–152. [CrossRef]
2. Domingues, P.; Carreira, P.; Vieira, R.; Kastner, W. Building automation systems: Concepts and technology review. *Comput. Stand. Interfaces* **2016**, *45*, 1–12. [CrossRef]
3. Wollschlaeger, M.; Sauter, T.; Jasperneite, J. The future of industrial communication: Automation networks in the era of the internet of things and industry 4.0. *IEEE Ind. Electron. Mag.* **2017**, *11*, 17–27. [CrossRef]
4. González, I.; Calderón, A.J.; Mejías, A.; Andújar, J.M. Novel networked remote laboratory architecture for open connectivity based on PLC-OPC-LabVIEW-EJS integration. Application to remote fuzzy control and sensors data acquisition. *Sensors* **2016**, *16*, 1822. [CrossRef] [PubMed]
5. Galloway, B.; Hancke, G.P. Introduction to Industrial Control Networks. *IEEE Commun. Surv. Tutor.* **2013**, *15*, 860–880. [CrossRef]
6. Givehchi, O.; Landsdorf, K.; Simoens, P.; Colombo, A.W. Interoperability for industrial cyber-physical systems: An approach for legacy systems. *IEEE Trans. Ind. Inform.* **2017**, *13*, 3370–3378. [CrossRef]
7. Mejías, A.; Reyes, M.; Márquez, M.A.; Calderón, A.J.; González, I.; Andújar, J.M. Easy handling of sensors and actuators over TCP/IP Networks by Open Source Hardware/Software. *Sensors* **2017**, *17*, 94. [CrossRef] [PubMed]
8. González, I.; Calderón, A.J.; Andújar, J.M. Novel remote monitoring platform for RES-hydrogen based smart microgrid. *Energy Convers. Manag.* **2017**, *148*, 489–505. [CrossRef]
9. Sayed, K.; Gabbar, H.A. SCADA and smart energy grid control automation. In *Smart Energy Grid Engineering*; Gabbar, H.A., Ed.; Academic Press, Elsevier: New York, NY, USA, 2017; ISBN 978-0-12-805343-0.
10. Figueiredo, J.; Sá da Costa, J. A SCADA system for energy management in intelligent buildings. *Energy Build.* **2012**, *49*, 85–98. [CrossRef]
11. González, I.; Calderón, A.J.; Barragán, A.J.; Andújar, J.M. Integration of Sensors, Controllers and Instruments Using a Novel OPC Architecture. *Sensors* **2017**, *17*, 1512. [CrossRef] [PubMed]
12. Calderón, A.J.; González, I.; Calderón, M.; Segura, F.; Andújar, J.M. A new, scalable and low cost multi-channel monitoring system for polymer electrolyte fuel cells. *Sensors* **2016**, *16*, 349. [CrossRef] [PubMed]
13. Cohen, Y.; Faccio, M.; Gabriele, F.; Mora, C.; Pilati, F. Assembly system configuration through Industry 4.0 principles: The expected change in the actual paradigms. *IFAC PapersOnLine* **2017**, *50*, 14958–14963. [CrossRef]
14. Iglesias-Urkia, M.; Orive, A.; Barcelo, M.; Moran, A.; Bilbao, J.; Urbieta, A. Towards a lightweight protocol for Industry 4.0: An implementation based benchmark. In Proceedings of the IEEE International Workshop of Electronics, Control, Measurement, Signals and Their Application to Mechatronics, San Sebastian, Spain, 24–26 May 2017. [CrossRef]
15. Müller, J.M.; Kiel, D.; Voigt, K.-I. What Drives the Implementation of Industry 4.0? The Role of Opportunities

and Challenges in the Context of Sustainability. *Sustainability* **2018**, *10*, 247. [CrossRef]
16. González, I.; Calderón, A.J.; Figueiredo, J.; Sousa, J.M.C. Design of an educational platform for automation and supervision under the Industry 4.0 framework. In Proceedings of the 12th International Technology, Education and Development Conference (INTED), Valencia, Spain, 5–7 March 2018.
17. Lu, Y. Industry 4.0: A survey on technologies, applications and open research issues. *J. Ind. Inf. Integr.* **2017**, *6*, 1–10. [CrossRef]
18. Lüder, A.; Schleipen, M.; Schmidt, N.; Pfrommer, J.; Henben, R. One step towards an Industry 4.0 component. In Proceedings of the 13th IEEE Conference on Automation Science and Engineering (CASE), Xian, China, 20–23 August 2017. [CrossRef]
19. Hoffmann, M.; Büscher, C.; Meisen, T.; Jeschke, S. Continuous integration of field level production data into top-level information systems using the OPC interface standard. *Procedia CIRP* **2016**, *41*, 496–501. [CrossRef]
20. Chen, J.; Tai, K.C.; Chen, G.C. Application of programmable logic controller to build-up an intelligent industry 4.0 platform. *Procedia CIRP* **2017**, *63*, 150–155. [CrossRef]
21. Girbea, A.; Suciu, C.; Nechifor, S.; Sisak, F. Design and implementation of a service-oriented architecture for the optimization of industrial applications. *IEEE Trans. Ind. Inform.* **2014**, *10*, 185–196. [CrossRef]
22. García, M.V.; Irisarri, E.; Pérez, F.; Estévez, E.; Marcos, M. OPC-UA Communications Integration using a CPPS architecture. In Proceedings of the IEEE Ecuador Technical Chapters Meeting (ETCM), Guayaquil, Ecuador, 12–14 October 2016. [CrossRef]
23. Kammoun, M.A.; Ezzeddine, W.; Rezg, N.; Achour, Z. FMS scheduling under availability constraint with supervisor based on timed Petri nets. *Appl. Sci.* **2017**, *7*, 399. [CrossRef]
24. Pisching, M.A.; Pessoa, M.A.O.; Junqueira, F.; Filho, D.J.; Miyagi, P.E. An architecture based on RAMI 4.0 to discover equipment to process operations required by products. *Comput. Ind. Eng.* **2018**, in press. [CrossRef]
25. Pacaux-Lemoine, M.-P.; Trentesaux, D.; Zambrano, G.; Millot, P. Designing intelligent manufacturing systems through Human-Machine Cooperation principles: A human-centered approach. *Comput. Ind. Eng.* **2017**, *111*, 581–595. [CrossRef]
26. Rogoll, G.; Suzuki, T. Advance physical layer for digital communication in process industries. In Proceedings of the 56th Annual Conference of the Society of Instrument and Control Engineers of Japan (SICE), Kanazawa, Japan, 19–22 September 2017. [CrossRef]
27. Delsing, J. Local cloud internet of things automation: Technology and business model features of distributed internet of things automation solutions. *IEEE Ind. Electron. Mag.* **2017**, *11*, 8–21. [CrossRef]
28. SMC. Available online: https://www.smc.eu/portal_ssl/WebContent/main/index_restyling.jsp?lang=en&ctry=EU&is_main=yes&dfl_locale=yes (accessed on 10 April 2018).
29. Martinez, B.; Vilajosana, X.; Kim, I.H.; Zhou, J.; Tuset-Peiró, P.; Xhafa, A.; Poissonnier, D.; Lu, X. I3Mote: An open development platform for the intelligent industrial internet. *Sensors* **2017**, *17*, 986. [CrossRef] [PubMed]
30. Wittenberg, C. Human-CPS Interaction—Requirements and human-machine interaction methods for the Industry 4.0. *IFAC-PapersOnLine* **2016**, *49*, 420–425. [CrossRef]

10

Localization in Wireless Sensor Networks: A Survey on Algorithms, Measurement Techniques, Applications and Challenges

Anup Kumar Paul [1,2,*] and Takuro Sato [2]

[1] Department of Electronics and Communications Engineering, East West University, Dhaka 1212, Bangladesh
[2] Waseda University, Tokyo 169-8050, Japan; t-sato@waseda.jp
* Correspondence: anuppaul@ewubd.edu

Abstract: Localization is an important aspect in the field of wireless sensor networks (WSNs) that has developed significant research interest among academia and research community. Wireless sensor network is formed by a large number of tiny, low energy, limited processing capability and low-cost sensors that communicate with each other in ad-hoc fashion. The task of determining physical coordinates of sensor nodes in WSNs is known as localization or positioning and is a key factor in today's communication systems to estimate the place of origin of events. As the requirement of the positioning accuracy for different applications varies, different localization methods are used in different applications and there are several challenges in some special scenarios such as forest fire detection. In this paper, we survey different measurement techniques and strategies for range based and range free localization with an emphasis on the latter. Further, we discuss different localization-based applications, where the estimation of the location information is crucial. Finally, a comprehensive discussion of the challenges such as accuracy, cost, complexity, and scalability are given.

Keywords: localization; range free; survey; wireless sensor network; mobile anchor

---

1. Introduction

In the future generation of communications networks, real-time localization and position-based services are required that are accurate, low cost, energy efficient and reliable [1,2]. Nowadays, Wireless Sensor Networks (WSNs) can be applied in many applications, such as natural resources investigation, targets tracking, unapproachable places monitoring and so forth. In these applications, the information is collected and transferred by the sensor nodes. Various applications request these sensor nodes' location information. Moreover, the location information is also indispensable in geographic routing protocols and clustering [3,4]. All these mentioned above make localization algorithms become one of the most important issues in WSNs researches. Thus, locations of sensor nodes are important for operations in WSNs. Localization in WSNs has been intensively studied in recent years, with most of these studies relying on the condition that only a small proportion of sensor nodes, called anchor nodes, know their exact positions through GPS devices or manual configuration [5–7]. Other sensor nodes estimate their distances to anchor nodes and calculate positions with multi-lateration techniques. These methods provide satisfactory level of accuracy with a small proportion of anchor nodes in WSNs [8,9].

The sensor nodes are randomly deployed in inaccessible terrain by the vehicle robots or aircrafts to be used in many promising applications, such as health surveillance, battle field surveillance, environmental monitoring, coverage, routing, location service, target tracking, and rescue [10]. The Global Positioning System (GPS) or the standalone cellular systems are the most promising and accurate positioning technologies. Although they are widely accessible, the limitation of high cost and energy consuming of GPS system makes it impractical to install in every sensor node where the lifetime of a sensor node is very crucial. On the other hand, the cellular signals are interrupted in scenarios with deep shadowing effects [11]. In order to reduce the energy consumption and cost, only a few number of nodes which are called anchor or beacon nodes, contain the GPS modules. The other nodes could obtain their position information through localization method. Wireless sensor network is composed of a large number of inexpensive nodes that are densely deployed in a region of interests to measure certain phenomenon. The primary objective is to determine the location of the sensor node. Node self-localization can be classified into two categories: range based localization and range free localization. The former method uses the measured distance/angle to estimate the location. In addition, the latter method uses the connectivity or pattern matching method to estimate the location.

Various localization algorithms and methodologies have been proposed to deal with different problems in different applications. A combination of different range based techniques called hybrid positioning is a well known approach for localization that exhibits sufficient accuracy and coverage [12]. On the other hand, the localization algorithms based on hop distance and hop count based information between anchor nodes and sensor nodes are commonly known in the literature as connectivity-based or range-free algorithms. Depending on the process used to estimate the distances between the intermediate nodes, range-free algorithms may fall into two categories: heuristic, and analytical [13–33]. Also, range free localization algorithms are categorized based on the deployment scenarios. The categorization has been divided into four groups: (1) static sensor nodes and static anchor nodes [34,35]; (2) static sensor nodes and mobile anchor nodes [36,37]; (3) mobile sensor nodes and static anchor nodes [38,39]; and (4) mobile sensor nodes and mobile anchor nodes [40,41].

Although there are many localization techniques available to solve positioning problems in the WSNs, there are practical limits on the combination of these techniques as well as on the minimal number of anchor nodes that can be deployed in such scenarios. For example, in many situations, only one or two anchor nodes are able to communicate with the sensor nodes that need to be localized. Hence, new positioning techniques based on hybrid data fusion and/or heterogeneous access are proposed and analyzed [42].

In this paper, we present a detailed survey on recent localization techniques and concepts with their fundamental limits, challenges and applications. Although literature survey on localization techniques are available in [8,43–48], only a few papers exist that focus on range free localization techniques [49] without focusing on recent advanced techniques and applications. Thus, the survey in [45] is outdated, whereas [43] focuses only on ultrasonic positioning systems. The work in [8] describes relatively recent localization techniques but focuses only on the indoor localization techniques and briefly discusses about range free localization. The works of [46,47] review different technologies, such as Wireless Local Area Network (WLAN), used for indoor positioning. However, they do not discuss positioning neither from the perspective of energy efficiency nor from the requirement in recent applications, such as ambient assisted and health living applications. The survey in [48] provides notable categorization of various fingerprint-based outdoor positioning techniques, discussing how each method works. So, we intend to present a survey focused specially on range free techniques. Moreover, the rapid growth of various localization approaches in this field and the need for a complete and up-to-date survey of the techniques, applications and future trends, provide the motivation for this survey paper.

The remainder of this paper is organized as follows. Basic distance measurement techniques for localization in WSNs are described briefly in Section 2 with their common pitfalls and challenges. Different localization algorithms and their comparative analysis are discussed in Section 3. Section 4 describes various localization based applications. Section 5 presents various evaluation criteria for localization. Then we present perspective and challenges in range free localization algorithms in section 6. Finally, Section 7 concludes the paper.

2. Basic Measurement Techniques for Localization in WSNs

The localization algorithms for WSNs depend on various measurement techniques. There are many factors that affect the accuracy of the localization algorithms and consequently, the choice of the localization algorithms to be used in various applications. For example, network architecture, sensor density in an area, number of anchor nodes, geometric shape of the measurement area, sensor time synchronization, and the signaling bandwidth among the sensors are the key factors to be taken into account while designing a localization algorithm. However, it is the type of measurement and the corresponding precision that fundamentally determines the accuracy of localization algorithm.

Measurement techniques in WSNs localization can be broadly classified into three categories [50]. Angle of Arrival (AOA) measurements, distance related measurements and the Radio Signal Strength (RSS) profiling techniques. Figure 1 shows the classification. In the subsequent discussion, we briefly discuss these techniques along with their limitations in different WSNs applications.

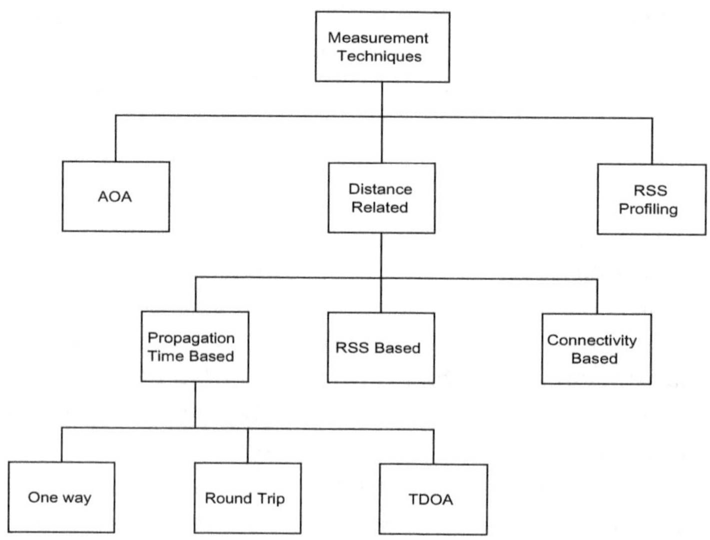

Figure 1. Classification of measurement techniques for localization algorithm.

2.1. *Angle of Arrival (AOA) Measurements*

The AOA measurement techniques are also known as the bearing measurements or the direction of arrival measurements. The AOA measurements can be obtained from two categories of techniques: one from the receiver antenna's amplitude response and another from the receiver antenna's phase response. These techniques calculate the angle at which the signal arrives from the anchor node to the unknown sensor nodes. Then, the region where the unknown sensor is located, is a line having a certain angle from the anchor node. In AOA measurement techniques, at least two anchor nodes are needed to calculate the position. The localization error could be large if there is a small error in measurement. The accuracy is depended on the directionality of the antenna and measurements are further complicated by the presence of shadowing and multipath effect of the measurement environment. A multipath component from the transmitted signal may appear as a signal coming from entirely different direction and consequently causes a very large error in measurement accuracy [50]. Thus, AOA technique is of limited interest in localization unless it is used with large antenna arrays [8]. As a result, for WSNs with tiny sensor nodes, this option is not energy efficient at all.

2.2. *Distance Related Measurement*

Distance related measurements can be further classified as propagation time measurements (one way, round trip and time difference of arrival (TDOA)), RSS based and connectivity based measurements.

2.2.1. Propagation Time Measurement

In **one way propagation time measurement**, the principle approach is to measure the difference between the sending time of the transmitting signal and the receiving time of the signal at the receiver. The distance between the transmitter and the receiver is then computed using this time difference and the propagation speed of the signal in the media. Time delay measurement is a relatively mature field. However, a major limitation in implementing the one way propagation time measurement is that, it requires the synchronization between the local time at the transmitter and the local time at the receiver. Any difference between the local times at the transmitter and the receiver will cause large error in estimating distance and consequently the position estimation error will be large. At the speed of light, a very small synchronization error of 1 ns will translate into a distance measurement error of 0.3 m [50]. The accurate synchronization requirement may add extra cost to the sensor nodes, by demanding a highly accurate clock or may add complexity to the sensor network by demanding a sophisticated synchronization algorithm. This disadvantage makes this option less attractive for WSNs localization.

Round trip propagation time measurement measures the difference between the times when a signal sent by a sensor node is returned from the second sensor node to the first sensor node. In this technique, there is no need for time synchronization, since the time difference is measured at the transmitting sensor node using the same local clock. The major source of error in this technique is the delay required in the second sensor node to handle the signal, process it and send back again. This internal delay is either known via a priori calibration or measured at the second sensor node and send back to the first sensor node where it is subtracted. In addition to the synchronization problem, both one way and round trip propagation time measurements are affected by noise, signal bandwidth, non line-of-sight and multipath environment. To overcome some of the limitations, Ultra Wide Band (UWB) signals have been used for accurate propagation time measurements [51]. UWB can achieve very high accuracy because its bandwidth is very large and therefore its pulse has a very short duration. This feature makes fine time resolution of UWB signals and therefore the separation of multipath signals possible.

Time difference of arrival measurement measures the difference between the arrival times of a transmitting signal at two separate receivers respectively, assuming the locations of the two receivers are known and they are perfectly synchronized. This technique requires three receivers to uniquely locate the transmitter location. The accuracy is affected by synchronization error and multipath. The accuracy improves when the distance between receivers are increased because this increases the difference between the times of arrival.

2.2.2. Received Signal Strength (RSS) Based Measurement

Received signal strength measurement estimates the distance between two sensor nodes from the received signal strength of the signal [52,53]. Most sensors have the capability to measure the RSS. The distance estimated from the RSS is a monotonically decreasing function. The relation is modeled by the following log-normal model:

$$P_r(d)[dBm] = P_0(d_0)[dBm] - 10n_p log_{10}\left(\frac{d}{d_0}\right) + X_\sigma \tag{1}$$

where $P_0(d_0)[dBm]$ is a reference power in dB milliwatts at a reference distance d_0 from the transmitter, n_p is the path loss exponent that measures the rate at which the received signal strength decreases with the distance, X_σ is a zero mean Gaussian random variable with standard deviation σ and it accounts for the random effect caused by shadowing. Both n_p and σ are environment dependent. Given the model and the model parameters, which are known via a priori measurements, the distance between two sensor nodes can be obtained from the RSS measurements. Localization algorithm can then be applied to use this distance and estimate the position using multilateration technique.

Another interesting technique to measure distance between an optical transmitter and an optical receiver is the lighthouse approach [54]. In this approach, the distance is measured by estimating the time duration that the receiver dwells in the optical beam. The advantage is that the optical receiver is of small size and low cost. However, it requires line of sight between the transmitter and the receiver.

2.2.3. Connectivity Based

Connectivity based measurement is the simplest form of all the measurement techniques we have discussed so far. In this technique, a sensor is connected to another sensor if it is within the radio transmission radius of each other. Such measurement technique is treated as the binary measurement. In this technique, a sensor node is connected to another sensor node (binary 1) or not connected directly if it is outside the radio transmission range (binary 0). From one sensor to anther sensor, the distance is thus represented as the hop count and various algorithms are applied to measure the average hop distance as accurately as possible [14]. This category of WSNs localization algorithm is popularly known as the range free localization algorithm.

2.3. RSS Profiling Measurement

RSS based measurement estimates the distance between sensor nodes as was discussed in the previous section. The localization algorithms then use this distance to calculate the position of the sensor nodes. However, the implementation of this kind of algorithm faces two major challenges: first, the wireless environments, especially the indoor wireless environments and the outdoor wireless environments with irregular objects inside the measurement area, make the distance estimation from RSS very difficult. In addition, second, the determination of model parameter is also a very difficult task. To overcome such difficulties, RSS profiling measurement techniques [55–58] that estimate sensor location from the map of RSS measurements are used to improve the accuracy.

The RSS profiling measurement works by first constructing a form of map of signal strength of anchor nodes at different locations of the measurement area. The map is obtained either offline via a priori measurements or online by deploying some sniffing devices [56] at some known locations. This kind of technique is mainly used for WLAN, but they would appear to be attractive for WSNs too [50].

In RSS profiling based localization systems, in addition to anchor nodes and unknown sensor nodes, a large number of sample points, e.g., sniffing devices [56] or reference points are distributed throughout the coverage area. At each sample point, the RSS signal strength is obtained from different anchor nodes, where n_{th} entry corresponds to the n_{th} anchor nodes. Obviously, different entries have different signal strength and many of them have zero values or near to zero values due to the large distance from the anchor nodes. The collection of all these points constitute the RSS map of the interested region and is a unique signature corresponding to the anchor locations and the wireless environment. The model is stored in a central location. The non anchor node estimates its position by referring to the RSS map. It calculates the signal strength of its current location and then match the position from the corresponding map whose signal strength is a closest match.

3. Localization Algorithms in WSNs

Based on the measurement of inter-sensor distance, localization algorithms in WSNs can be broadly classified into two categories: centralized and distributed [50]. In centralized localization technique, all the inter-sensor measured distances are sent to the central location where the positions of each and every sensor node are calculated. On the other hand, in distributed localization technique, the individual sensor nodes calculate their own position by utilizing the distance measurement from other anchor nodes. Major approaches for designing centralized algorithms are Multi Dimensional Scaling (MDS) [59], linear programming [60] and stochastic optimization algorithms [61,62]. Some well known distributed localization algorithms are DV-Hop [52], DV-Distance [14] and a number of other algorithms based on the above two algorithms [17,18,63].

Centralized and distributed localization algorithms are further subdivided into range based and range free algorithm. Moreover, fusing the information from different positioning systems with different physical principles can improve the accuracy and robustness of the overall system. This leads to the development of another category known as hybrid data fusion [8].

Range based localization technique utilizes the measurement techniques such as AOA, TOA, TDOA and RSSI as is discussed in the previous section to estimate the distance between sensor nodes and then calculates the position. Range based technique usually achieves high ranging accuracy but requires extra hardware and consumes more energy. In the following sub-section, we focus on range free localization and hybrid data fusion techniques.

3.1. *Range Free Localization Algorithm*

Range free localization technique, which is totally dependent on the contents of the received packet and is a much cheaper solution than many range based localization techniques [64] in WSNs. Range free schemes are simple, inexpensive and energy efficient where localization is performed using geometric interpretation, constraint minimization and resident area formation [65].

3.1.1. Hop Count Based

Almost all the range free localization techniques mainly use hop count based information to calculate the position. DV-Hop [52] and Centroid [36] are the pioneering approaches of this type. Centroid is designed for sensor nodes which have at least three neighbor anchor nodes. Assume that the sensor node N has three neighbor anchors A_1, A_2, A_3, whose coordinates are (x_1, y_1), (x_2, y_2), and (x_3, y_3), and all nodes have equal communication range. The principle of Centroid is to regard the central point Ncentroid of anchors as the estimated position. The position of Ncentroid, denoted as (xcentroid, ycentroid) could be calculated as $(xcentroid, ycentroid) = ((x_1 + x_2 + x_3)/3, (y_1 + y_2 + y_3)/3)$. Centroid has very low communication and computation cost, and can get relatively good accuracy when the distribution of anchors is regular. However, when the distribution of anchors is not even, the estimated position derived from the Centroid algorithm will be inaccurate. On the other hand, the hop count based method DV-Hop and hop-terrain [66] requires small number of anchors.

DV-Hop plays an essential role in many localization methods to give primal distance estimation from sensor nodes to anchor nodes. DV-Hop propagates distance estimation among anchor nodes represented by number of hops throughout a WSN. Anchor nodes can then estimate the average distance of each hop, with which each sensor node calculates its estimated distances to anchor nodes. By multilateration, the location is then calculated as follows:

Let (x, y) be the unknown node $D's$ location and (x_i, y_i) be the known location of the $i'th$ anchor node receiver. Let's say the $i'th$ anchor node distance to unknown nodes are d_i and the total number of anchors deployed in the network is n. Then, here is the following formula for calculating location in range free localization [63].

$$\begin{cases} \sqrt{(x-x_1)^2 + (y-y_1)^2} & = d_1 \\ \sqrt{(x-x_2)^2 + (y-y_2)^2} & = d_2 \\ \vdots & \vdots \\ \sqrt{(x-x_i)^2 + (y-y_i)^2} & = d_i \end{cases} \tag{2}$$

$$A = -2 \times \begin{pmatrix} x_1 - x_n & y_1 - y_n \\ x_2 - x_n & y_2 - y_n \\ \vdots & \vdots \\ x_{n-1} - x_n & y_{n-1} - y_n \end{pmatrix} \tag{3}$$

$$B = \begin{pmatrix} d_1^2 - d_n^2 - x_1^2 + x_n^2 - y_1^2 + y_n^2 \\ d_2^2 - d_n^2 - x_2^2 + x_n^2 - y_2^2 + y_n^2 \\ \vdots \\ d_{n-1}^2 - d_n^2 - x_{n-1}^2 + x_n^2 - y_{n-1}^2 + y_n^2 \end{pmatrix} \tag{4}$$

$$P = \begin{pmatrix} x \\ y \end{pmatrix} \tag{5}$$

where, $P = (A^T A)^{-1} A^T B$.

However, DV-Hop requires not only uniformly deployed WSNs but also the same attenuation of signal strength in all directions. To modify the disadvantage of existing DV-Hop localization algorithm, the relevant literature proposed many improved algorithms based on the following metric:

Improvement based on average hop distance: In the randomly deployed node density and connectivity of the network, there are many works that modified the average hop distance between anchor nodes to improve the position estimation accuracy [67–70]. Such as [67], it improved the location accuracy by modifying the network average hop distance based on minimum mean square error criteria as $HopSize_i^N = \frac{\sum_{j \neq i} h_j d_{ij}}{\sum_{j \neq i} h_j^2}$. Where d_{ij} is the straight line distance between anchor nodes i and j, h_j is the hop segment number between anchor nodes i and j. Another algorithm such as [68], it calculated the error e^{ij} as $e^{ij} = d_{est}^{i,j} - d_{true}^{i,j}$, where $d_{est}^{i,j}$ is the estimated distance between anchor nodes i and j, $d_{true}^{i,j}$ is the Euclidean distance between anchors i and j. Then finally adjusting the average hop distance by $HopSize_{eff}^{i,j}$=HopSize$_i - \frac{e^{i,j} + e^{i,m}}{h^{i,j} + h^{i,m}}$, where m is the closest anchor node to anchor node i and $HopSize_i$ is calculated as $HopSize_i = \frac{\sum_{j \neq i} \sqrt{(x_i - x_j)^2 + (y_i - y_j)^2}}{\sum_{j \neq i} h_{ij}}$, where (x_i, y_i) (x_j, y_j) are the coordinates of anchor nodes i and j and h_{ij} is the number of hops between anchors i and j. The algorithms [67,68], made improvements on distance estimation and consequently the accuracy of the DV-Hop algorithm.

Improvement based on node information and nearest anchors: There are still some disadvantages in the improved algorithms that are based on the average hop distance, such as no obvious improvement on localization accuracy, especially when the transmission route is not straight but detoured. These approaches are accurate insofar only when the topology is isotropic, i.e., shortest paths between anchors and sensors approximate to their Euclidean distances. However, there may be large errors in the distance estimates if the topology is not isotropic or contains a hole (aka anisotropic environment) [71]. Therefore, some modified methods were proposed using the anchor node information and the relationship between anchor node and sensor node or topological structure information to improve the DV-Hop localization method. In order to alleviate the influence of holes (obstacle shape), Shang et al. [72] suggest using only four nearest anchors, assuming that the shortest paths to the nearest anchors may be less affected by irregularities, and this does produce good results in some cases but with a drawback of the possibility to falsely discard some good anchors which can improve the localization accuracy.

3.1.2. Analytical Geometry Based

Most popular alternatives suitable for range free localization algorithms are based on analytical algorithms [7,9,29,33] which evaluate theoretically the average hop distance of the network using the statistical characteristics of the network deployment. The obtained average hop distance is locally computable at each sensor node and likewise other range free method, it has to be broadcasted to other sensor nodes.

To cope with the problem of anisotropy in a network, pattern driven localization scheme [29] is proposed. For anisotropic environment, this paper devised two methods to calculate the estimated distance between anchors and sensors based on whether the anchor is slightly detoured or strongly detoured from normal sensor nodes. For slightly detoured anchors, it utilizes the information from the nearest anchors (namely reference station) and this reference station must be within three or four hops away from normal sensor nodes. Which means that, the anchors distribution density must be very high. It devised one method to discard the strongly detoured anchors. However, no indication of how many anchors fall in the strongly detoured category because it may be impossible to accurately determine which anchors are slightly detoured and which are moderately or strongly detoured. The author in [7,9] deals with this problem by calculating the angle of the detoured path between anchor and sensor nodes. Another analytical algorithm [33] argues that average hop distance and number of hops between anchor and sensor nodes are not sufficient to calculate accurate position of the sensor nodes. It also depends on number of forwarding nodes (which forward any data between two nodes). By utilizing this information along with other information, the author in [33] showed that further accuracy can be achieved.

3.1.3. Mobile Anchor Based

In this technique, a mobile anchor with GPS capability moves into a sensing area and periodically broadcast its current geometric coordinates. The other sensor nodes collect the location coordinates of the mobile anchor node. Later, the sensor nodes choose three non-collinear coordinate points of the mobile anchor node and apply different mechanisms to estimate position. Based on this principle, several localization algorithms are devised [30,73–76].

The author in [73] proposed a geometric conjecture (perpendicular bisector of the chord of a virtual circle) based range free localization algorithm, where a mobile anchor traverses a sensing area and periodically broadcasts its current location coordinates. The neighboring sensor nodes keep track of entering and departing anchor coordinate points to construct a chord on its communication range. The sensor node repeats this process until it gets at least three coordinate points from the moving anchor node on its communication range. The line segments between these three selected coordinate points create two chords on its communication range. Later, the perpendicular bisector of the two cords gives the position estimates of the sensor nodes. To further improve the localization accuracy, the author in [74] proposed a geometric constraint based localization scheme. In this scheme, the selection process of the three anchor coordinate points on the communication range of the sensor node remains the same as in [73]. Initially, the intersection of the selected two anchor coordinate points determine the constraint area of the sensor node. This process is repeated with another two intersected points to further narrow down the constraint area of the sensor node. Finally, the average of all the intersection points give the position estimates of the sensor node.

Another approach [75] proposed a constraint area based localization using mobile anchor. In this approach, the specific type of moving anchor's trajectories create a specific type of constraint areas for the sensor node. To identify the potential location of the sensor node within different constraint areas, a number of intersections are created within different constraint areas until the final arrival of the coordinate points before the final departure of the anchor node. Each intersection further narrow downs the potential location of the sensor node within the overlapping constraint areas. However, the scheme shows high localization error when random waypoint mobility model is used for the moving anchor node. Also the scheme is computationally expensive due to multiple intersection computation. Another approach [76] proposed a curve fitting method along with a mobile anchor node to calculate the location of the sensor node. In this approach, the arrival and departed coordinate points of the moving anchor nodes are recorded and this is repeated as many times as the moving anchor re-enters the communication region of the sensor node. The localization begins through fitting a curve on the few selected coordinate points on communication range and iteratively refined through Gauss-Newton method. The center coordinates of the fitted curve define the position of the sensor node. Another mobile anchor based localization is proposed in [30], where the localization begins with approximation of the geometric arc parameters. The approximated arc parameters are used to generate the chord on the virtual circle. Later, the perpendicular bisector of the chords along with the approximated radius are used to estimate the position of the sensor node. The accuracy is improved for boundary nodes too.

Although several techniques are devised so far, a common pitfall to all mobile anchor based localization schemes arise when considering the longer periodic interval of the message send by the anchor node and the irregular radio propagation pattern.

3.2. Hybrid Data Fusion

Hybrid data fusion is based on the principle of fusing the information from different positioning systems with different physical measurement techniques in order to achieve higher accuracy as compared to other stand-alone localization techniques. Recently, research work has been focusing on two main approaches in hybrid data fusion: centralized and distributed. Iterative positioning [77–79] and cooperative link selection [80,81] are used with the distributed approach. In iterative multilateration, once the position is estimated for unknown nodes, this node is used as the anchor node for other unknown sensor nodes. Multiple iterations are needed to complete the localization process.

Another interesting work [82] utilizes the technique of combining angle based localization, map filtering, and pedestrian dead reckoning (PDR) where absolute position estimates are provided by the angle based localization techniques. Pedestrian dead reckoning provides accurate length and shape of the traversed route. Thus, the estimates obtained from angle based localization techniques and the PDR movement are merged together with a vector map built in a particle filter is used as the fusion filter. Hence, merging different information from different positioning techniques lead to higher positioning accuracy.

Hybrid data fusion is also used for the purpose of pedestrian tracking [83]. Usually, this hybrid technique merges inertial measurement and RSS information via a Kalman filter. Classic hybrid methods [84,85] were based on fingerprinting RSS method or map based method. On the other hand, another method [83] uses a channel modeling technique, where a propagation channel model gives a direct relation between the distance of two nodes and the RSS. Then, triangulation or multilateration is utilized to estimate the node position from a set of distances to some known anchor nodes. This approach has minimal calibration cost. Additionally, fusion between inertial measurements and channel based localization provides higher accuracy as compared to fingerprinting based methods.

Another hybrid data fusion system is achieved by merging the information from WLAN with the build-in camera on a smartphone for position estimation [86]. This approach utilizes visual markers pre-installed on the floor for the position correction. Visual information is combined also with the radio data to track a person wearing a tag using a mobile robot in indoor environments [87]. The author in [88] presented a method to integrate range-based sensors and ID sensors (i.e., infrared or ultrasound badge sensors) using a particle filter to track people in a networked sensor environment. As a result, their approach is able to track people and determine their identities owing to the advantages of both sensors.

Another method is based on the fusion of video and compass data acquired by the anchor node [89]. This method calculates the anchor node location by using a digital compass (magnetometer), an image taken by a video camera and the exact location data for some geographically-located referential objects (e.g., solitary trees, electricity transmission towers, furnace chimneys, etc.) situated in the deployment area. This method, due to the low price of digital compasses, is particularly suitable for video-based or multimedia-based WSNs, where the nodes already equipped with digital compasses may simply become anchor nodes or anytime the GPS receiver is not considered to be an appropriate solution. The author in [90] developed a hybrid localization system in WSNs, which is composed of coarse-grained localization system and fine-grained localization system. The coarse-grained localization system takes the wireless signal strength as the reference for distance and gets the rough region as the unknown node. The fine-grained localization system is in charge of location refinement that takes image to localize the unknown node with camera sensor nodes.

Hence, different kinds of information fusion lead to an improvement in the positioning accuracy, usually at the cost of additional complexity. For instance, data fusion occurs also with different types of RF sensors to improve the localization accuracy since different positioning systems may complement each other [91].

3.3. Comparative Performance of Centralized and Distributed Localization Algorithms

Centralized and distributed algorithms can be compared from several perspectives including, location estimation accuracy, implementation and computational complexity, and energy efficiency [50].

Distributed localization algorithms as compared to the centralized algorithms are considered to be more computationally efficient and can be easily implemented in a large scale WSN. However, in certain network types, where centralized information collection architecture already exists, such as health monitoring, precision agriculture monitoring, environment monitoring, road traffic control network etc., the measurement data from individual sensor node needs to be collected and processed centrally. In such a network, the individual sensor nodes have limited processing capability for saving energy; the localization related data can be piggybacked with other monitoring data and send back to the central processing node. Therefore, a centralized processing algorithm is more convenient in such situations than distributed algorithm with existing centralized architecture.

While considering the estimation accuracy of localization algorithms, centralized algorithms provide more accurate estimation results than distributed algorithms. One of the key reason behind this is that, centralized algorithms have global view of the network. However, centralized algorithms suffer from scalability problems and are not suitable at all for large scale sensor network. Other drawbacks of centralized algorithms as compared to distributed algorithms are their higher computational complexities, unreliability due to the inaccurate accumulated information (loss of information may occur over multihop) collected from multihop sensor nodes to the central node in WSNs.

On the other hand, while considering design complexity, distributed algorithms are more difficult to design than centralized algorithms, due to the complexity of local behavior and global behavior. That is, a distributed algorithm which works locally optimal may not behave equally optimal globally and is an open research problem. Error in distance estimation between sensor nodes propagated to other nodes which further deteriorate the estimation accuracy of the distributed algorithm. Moreover, distributed algorithms require a number of iterations to arrive at a stable solution. This may take longer time for a localization algorithm than the acceptable in some applications.

From the perspective of the energy consumption, the energy needed for specific type of operation (processing, transmitting, and receiving) in the specific hardware and the setting of the transmission range needs to be considered in centralized and distributed algorithms. Depending on the setting, it is seen that the energy required to transmit a single bit could be used to process 1000–2000 instructions [92]. Centralized algorithms require each sensor to send the localization related information over multihop to the central node whereas distributed algorithms require only local exchange of information within single hop (between neighboring nodes). However, in distributed algorithms, many such information exchanges (iterations) are required among sensor nodes to arrive at a stable solution. A comparative research about the energy efficiency of centralized and distributed algorithms are presented in [93], where the author concluded that in distributed algorithms, the number of iterations needed to arrive at a stable solution do not exceed the number of hops to the central processor, then distributed algorithms are more energy efficient as compared to the centralized algorithms.

It is worth noting that the differences between centralized and distributed algorithms are sometimes ambiguous. Any distributed algorithms can be applied to centralized manner. In addition, distributed versions of centralized algorithms can also be designed for certain applications. A typical way of designing distributed versions of centralized algorithms would be to divide the total network area into small areas, where in each area the centralized algorithms will be applied and then collecting the areas final result through the overlapping sensor nodes from each area and stitching these sensor nodes to obtain a global map [59,94,95]. Such algorithms may offer optimal tradeoff between the merits and demerits of centralized and distributed algorithms.

4. Localization Based Applications

Positioning and navigation for mobile devices is a booming market with expected size of 4 billion dollar in 2018 [8]. A reliable, user friendly, and accurate position information in navigation for mobile user might open the door for many promising applications and the creation of new business opportunities. It is thus considered to be a cornerstone in realization of Internet of Things (IoT) vision.

Location based services: Location based services provide spatial information to the end users through wireless networks and/or the Internet. Applications that provide location based services can offer the context and the connectivity needed to dynamically associate the position of a user to context sensitive information about current environments. Location based services send data by knowing the geographical location accessed by a mobile user. Thus, this service is very essential both in indoor and outdoor environment. For example, indoor applications with location based services can provide safety information, up to date cinemas, events or concerts in the vicinity. Moreover, application of this type include navigation application to direct the user to the place of interest. Location based services are also used for advertisement, billing, and for personal navigation to guide guests of trade-shows to the targeted booth. Also, it can be used in the bus or train stations to guide the passengers to the desired platform.

Ambient assisted living (AAL) and health applications: Indoor localization is one of the most important constituent for the AAL tools. AAL tools are advanced tools performing human-machine interactions. AAL tools aim to enhance the health status of the older adults by making them able to control their health conditions [96]. Such applications are used to track and monitor the elderly people. Some of the indoor localization systems based on the AAL applications are "Smart Floor Technology" to detect the presence of people and the "Passive Infrared Sensors" to notice the motion of people [97].

Other applications are based on ultra wide band (UWB) technology [98]. For example, orthopedic computer-aided surgery as well as its integration with smart surgical tools such as wireless probe for real-time bone morphing is implemented. UWB positioning system is proven to achieve a real time 3D dynamic accuracy of 5.24 mm–6.37 mm. Hence, this dynamic accuracy implies the potential for millimeter accuracy. This accuracy satisfies the requirement of 1 mm–2 mm 3D accuracy for orthopedic surgical navigation systems.

Robotics: Robotics is one of the main applications of localizations. Many researches and developments are conducted for implementing multi-robot system applications. The movement of robots in large indoor environments, where cooperation between them is required is a critical application of localization. For example, cooperation between robot teams enhances the mission outcomes in applications such as surveillance, unknown zone explorations, guiding or connectivity maintenance [8]. Ubiquitous Networking Robotics in Urban Settings (URUS) project [99] is an excellent example of using localization for evacuation in case of emergency, where the robots lead the people to the evacuation area. Moreover, obstacle avoidance and dynamic and kinematic constraints are considered in robotics to achieve complete navigation system [100].

Cellular Networks: Location information can be used to address many challenges in cellular networks [101]. The accuracy of location estimation is gradually improved in several generations of cellular networks. For example, the accuracy is improved from hundreds to tens of meters using cell-ID localization technique in second generation cellular networks. In third generation, the accuracy is improved based on timing via synchronization signal and in fourth generation, a reference signal dedicated for localization purpose is used. As well, localization technologies can be used by numerous devices in the future fifth generation cellular system to attain an accuracy of location estimation in the range of centimeter. Basically, in fifth generation cellular networks, it is expected to use precise localization information through all layers of the communication protocol stack [102]. This is due to the prediction of most of the fifth generation cellular user terminals in their mobility patterns knowing that these terminals will be either associated with fixed or controllable units or people [8]. Last but not least, localization is also required for several jobs in cyber-physical systems, like smart transportation systems and robotics in fifth generation cellular system [103,104].

5. Evaluation Criteria for Localization

Evaluating the performance of the localization algorithm is important for researchers, either to validate a new algorithm against the previous state of the art or choosing a localization algorithm that best fit the requirements of the corresponding application scenario. Since different applications will have different needs, it is important for the researcher to decide what performance criteria or evaluation metrics the localization algorithm are to be compared against other algorithms that fits different applications need. A broader set of evaluation criteria are useful both for the developers and the users of the localization algorithms in order to deeply understand the application needs. Examples of the evaluation metrics are localization accuracy, cost, coverage, robustness, scalability, topology etc. These criteria reflect the constraints such as computational complexity and limitations, power consumption, unit cost and network scalability. Some evaluation criteria are binary in nature, such as some algorithms either have some property or they dont have, e.g., anchor based or anchor free; range based or range free; self configuring or not; etc. Binary criteria can be used by researchers to narrow down the comparative evaluation of an algorithm against others. For example, one can narrow down the comparative evaluation by designing self configuring and range free localization algorithm by immediately limiting the number of comparison against range based solutions.

5.1. Accuracy

Accuracy is defined as how well the position estimated by the localization algorithm matches the known, ground truth positions. A good localization algorithm should provide the match as closely as possible. However, positional accuracy is not the only over-riding goal of a good localization algorithm. This is largely application dependent. Different applications will have different requirements on the resolution of the positional accuracy. The granularity of the required positional accuracy depends on the inter-node spacing. If the inter-node spacing is of the order of 100 m, then positional error of 1 m can be tolerable. However, if the inter-node spacing is of the order of 0.5 m, then 1 m error is highly unacceptable. It is also important to measure, how well a localization algorithm achieves good accuracies without a full set of input data. For example, some algorithms such as [105] assume measurements from every node to every other node for the localization algorithm to arrive at a stable estimation. This assumption is totally unrealistic given the realities of deployment environments.

Evaluation should show how the algorithms performance is affected by measurement noise, bias or uncorrelated error in the input data. It should also determine the number of sensor nodes that can actually be localized. Errors in measurement data is important for those algorithms that is designed to work for 2D and assume to work for 3D also. Because in 3D environment, measurement noise can result in flips and reflections of the estimated coordinates of the sensor nodes [106].

The simplest way to calculate accuracy is to determine the residual error between estimated positions and the actual positions for every sensor nodes in the network, sum them and average the result. This is known as mean absolute error [107] and is defined as

$$E_{mae} = \frac{\sum_{i=1}^{n} \sqrt{(x_i - \tilde{x_i})^2 + (y_i - \tilde{y_i})^2 + (z_i - \tilde{z_i})^2}}{n} \tag{6}$$

where, (x_i, y_i, z_i) are actual coordinates and $(\tilde{x}_i, \tilde{y}_i, \tilde{z}_i)$ are estimated coordinates of the sensor node. The total number of sensor nodes in the network is n.

The mean average error has the similarity to the root mean square (rms) error, which is defined as

$$E_{rms} = max_{i=1...n} \sqrt{(x_i - \tilde{x_i})^2 + (y_i - \tilde{y_i})^2 + (z_i - \tilde{z_i})^2} \tag{7}$$

It is also important for the accuracy metric to reflect not only the positional error in terms of the distance, but also in terms of the geometry of the network. If only average node position error is used, then there is a huge difference in the correctness of the relative geometry of the network estimated by the localization algorithm and the relative geometry of the actual network. This problem was identified by [108] and is addressed by defining the following metric known as global energy ratio.

$$GER = \frac{1}{n(n-1)/2}\sqrt{\sum_{i=1}^{n}\sum_{j=i+1}^{n}\left(\frac{\hat{d}_{ij}-d_{ij}}{d_{ij}}\right)^2} \quad (8)$$

The distance error between the estimated distance ($\hat{d}_{ij}$) and the known distance (d_{ij}) is normalized by the known distance (d_{ij}), making the error a percentage of the known distance.

The *GER* metric does not reflect the rms error [109] and is addressed by defining an accuracy metric that better reflects the rms error called global distance error (*GDE*).

$$GDE = \frac{1}{R}\sqrt{\frac{\sum_{i=1}^{n}\sum_{j=i+1}^{n}\left(\frac{\hat{d}_{ij}-d_{ij}}{d_{ij}}\right)^2}{n(n-1)/2}} \quad (9)$$

where, R represents the average radio range of a sensor node. The *GDE* calculates the localization error represented as a percentage of the average distance nodes can communicate over.

5.2. Cost

Cost is defined as how expensive the algorithm is in terms of power consumption, communication overhead, pre-deployment setup (i.e., how many anchor nodes are needed), time taken to localize a sensor node, etc. An algorithm which can minimize several cost constraints is likely to be desirable if maximizing network lifetime is the primary goal. However, cost is an important tradeoff against accuracy and is often motivated by realistic applications requirement. For example, an algorithm may focuses on minimizing communication overhead and complex processing to save power, quick convergence etc., but at the cost of the overall accuracy. Some of the common metrics are described below:

Anchor to Node Ratio: Minimizing the number of anchors is desirable from the equipment cost or deployment point of view. For example, using too many anchor nodes in the network that estimate their positions by global positioning system must be equipped with a GPS device, which is both power hungry and expensive; thus limiting the overall network lifetime. Similarly, predefined anchor positions are difficult to implement if placement of the nodes (including the anchor nodes) are carried out by a vehicle (e.g., from airplane). The anchor to node ratio is defined as the total number of anchor nodes divided by the total number of nodes in the network. This ratio is very important for the design of a localization algorithm. This metric is useful to calculate the trade-off between localization accuracy, the percentage of the nodes that can be localized against the deployment cost. For example, increasing the number of anchor nodes will lead to high accuracy as well as the percentage of the nodes that can be localized. On the other hand, the deployment cost will increase. A good localization algorithm must investigate the minimum number of anchor nodes that is needed for desired accuracy of the application.

Communication Overhead: Since radio communication is considered to be the most power consuming process relative to the overall power consumption of a wireless sensor node, minimizing communication overhead is a paramount in increasing the overall network lifetime. This metric is evaluated with respect to the scaling of the network, i.e., how much do the communication overhead increase as the network increases in size?

Algorithm Complexity: Algorithmic complexity can be described as the standard notions (big O notation) of computational complexity in time and space. That is how long a localization algorithm runs before estimating the positions of all the nodes in the network and how much memory (storage) is needed for such calculations. For example, as a network increase in size, the localization algorithm with O(n^3) complexity is going to take longer time to converge than an algorithm whose complexity is O(n^2). The same is true for space complexity.

Convergence Time: Convergence time is defined as the time taken from gathering localization related data to calculating the position estimates of all the nodes in the network. This metric is evaluated against the network size. That is, how long it takes for a localization algorithm to converge as the network increases in size. This metric is also important for some applications with fixed number of nodes in the network. For example, tracking of a moving target requires fast convergence. So, even if any particular localization algorithm that gives very accurate position estimates but takes long time is useless in this scenario. Similarly, if one or more nodes are mobile in a network, the time taken to update positions may not reflect the current physical state of the network if the algorithm is slow.

5.3. Coverage

Coverage is simply a measure of the percentage of the nodes deployed in the network that can be localized, regardless of the localization accuracy. Some localization algorithms may not be able to localize all the nodes in the network. This depends on the density of the nodes as well as the placement of the anchor nodes in the network. In evaluating coverage performance of localization algorithms, one must try various scenarios/strategies of anchor placements as well as various node densities. One can evaluate how the localization accuracy varies as the number of anchor nodes, placement of anchor nodes or neighbor per nodes varies. There is a saturation point, after which no additional gains in accuracy can be achieved. However, in attempting to minimize the number of anchor nodes or remove them entirely, a localization algorithm may compromise its accuracy and simplicity. Anchor free localization algorithms are frequently centralized and framed as non-linear optimization problem [110]. These approaches may not be feasible to implement in a resource constraint nodes due to computational complexity.

Density: If the density of the node deployment is low, it may be impossible to localize many nodes for a localization algorithm with random topology due to the connectivity problem [111]. Localization algorithm focusing on denser network should also take care of radio traffic, number of packet collisions, and energy consumption of the nodes as these factors will also increase as the number of nodes increase in the network.

Anchor Placement: Position of anchor nodes may have a significant impact on the calculation of the localization accuracy. Localization algorithms assumption of uniform grid or predefined placement of anchor nodes gives them high accuracy but failed to reflect the real world situation. Thus, this assumption is unrealistic for any localization algorithms since they do not take into account the environmental factors such as obstacles (that affect the anchor placement), terrain, signal propagation conditions etc. The geometry of the anchor nodes with respect to the unlocalized sensor nodes can have a varying effect on the calculation of the position estimates [9].

5.4. Topologies

Defining real node deployment topologies in simulations can play an important role when comparing the performance of localization algorithms. Different topologies such as uniform grid, C-shape, S-shape, O-shape topologies have significant effect on localization accuracy. Sensor network topologies can be divided mainly into two categories: even and random. In even topologies, sensor and anchor nodes are placed over the network area in an exact grid. On the other hand, in random topologies, sensor and anchor nodes are placed uniformly and randomly over the network area. Figure 2 shows node deployment in a random topology in an area of 10 m $\times$ 10 m with sensor density 8. Between these two topologies, random topology better reflects the real world deployment scenarios. This is because, in reality, sensor nodes are placed in areas where manual placement is restricted (in forest) or totally impossible (inside volcano). In such cases, sensor nodes are usually scattered in the deployment area from an airplane. So uniform deployment is not guaranteed. For these reasons, random topologies are popular among researchers for evaluating the localization algorithm in simulation and comparison with other state of the arts.

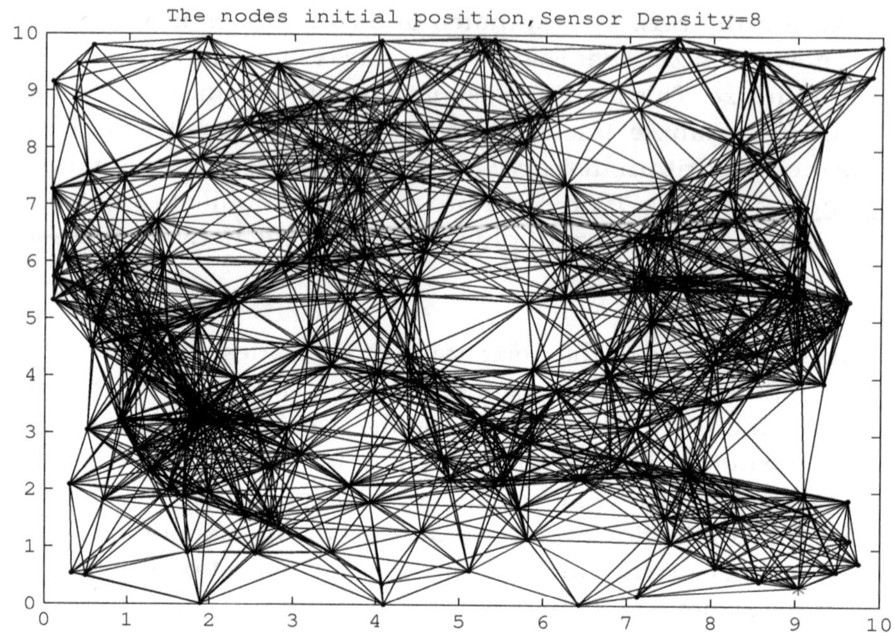

Figure 2. Random uniform topology.

Topologies can be further subdivided into regular and irregular topologies according to the placement strategies of sensor nodes as well as the shape of the obstacles inside the network area.

Regular Topology: In regular topology, nodes are placed uniformly over an area as a grid or randomly. In such deployment strategy, the average node density becomes consistent over each part of the distributed area. Many well known multihop localization algorithms [14] estimate the shortest path distance (number of hops multiplied by the average hop distance) between sensor nodes by utilizing this advantage of deployment strategy and derive the actual Euclidean distance from this to estimate the position of the sensor nodes. This gives very accurate position estimates or at least a bounded value. However, this assumption of regular topologies does not reflect the real world condition due to various factors that restrict the deployment of sensor nodes and thus is not effective at all.

Irregular Topology: In irregular topology, the estimated distance between nodes greatly deviates from the actual Euclidean distance due to the presence of obstacles or other objects inside the network area. Node density in an individual region may greatly deviate from the average node density of the whole region. Depending on obstacle size and shape inside the network area, the shape of the irregular topologies can be C-shaped, S-shaped, L-shaped, O-shaped etc. as can be seen from the Figures 3 and 4 and represent irregular deployment configurations that many applications may find themselves constraint by. Therefore, such topologies are generally useful to compare and stress the various attributes of localization algorithms to prove themselves robust. Note that, in Figures 3 and 4, two nodes can be connected via a detoured path around the obstacles and because of this the difference between the estimated hop distance and the actual Euclidean distance is large. Therefore, individual error in localization algorithms may accumulate, resulting in large localization error in the overall network. Obviously, a localization algorithm that generates accurate results in such topologies are considered to be more robust and useful in many real world applications.

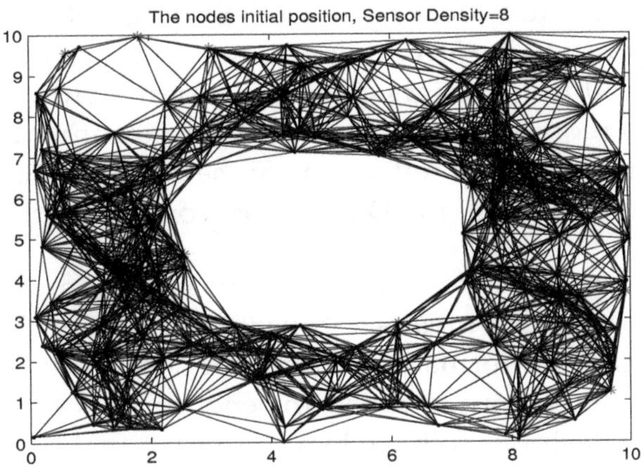

Figure 3. Irregular Topology: O-shape.

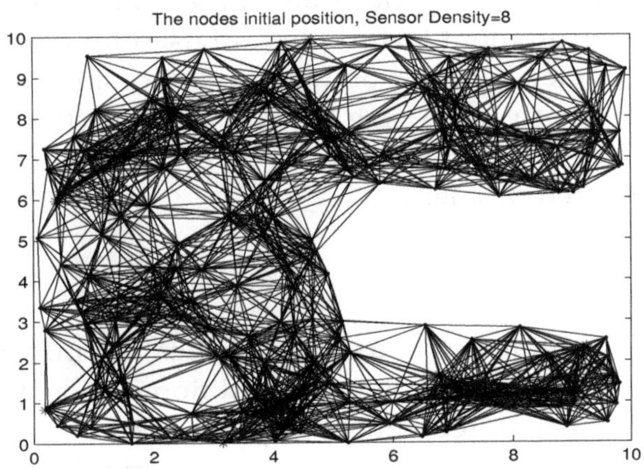

Figure 4. Irregular Topology: C-shape.

6. Open Challenges for Future Study

In this section, we summarize different perspectives and challenges in localization that need to be addressed. The challenges may be quite different in different potential applications. The scale of the network in these applications may be small or large and the environment may be different. Traditional localization methods are not suitable for different applications with different environmental challenges. Following are some challenges that need to be solved:

Combining different non-radio frequency techniques: Use of different non-radio technologies such as visual sensors can compensate for the errors that exist in current localization algorithms. The improved accuracy can be achieved by the additional installation of the costly equipment. Therefore, investigating the cost-effective solution will be a promising future direction for research.

Integration of different solution: Different wireless sensors can be used for the purpose of localization. Different sensor's physical measurement principles are different. Therefore, integrating measurement techniques from different sensors can improve the overall system positioning accuracy.

Scalability: A scalable localization system means, it performs equally well when its scope gets larger. A localization system may usually require scaling on two dimensions: geographical scaling and sensor density scaling. Geographical scaling means increasing the network area size. On the other hand sensor density scaling means increasing the number of sensors in unit area. Increasing the sensor density posses several challenges in localization. One such challenge is the loss of information due to wireless signal collision. Thus, locating sensors in dense environment should consider such collision while computing position information. A third metric in scaling is system dimension. Most of the localization algorithm is designed for 2D system. However, recent recommendations (e.g., FCC recommendations) require localization in 3D environment. Because in 3D environment, measurement noise can result in flips and reflections of the estimated coordinates of the sensor nodes. Thus a localization algorithm works well in 2D may not work perfectly in 3D.

Computational complexity: Localization algorithms have complexity in terms of software and hardware. Computational complexity means software complexity. That is, how fast a localization algorithm can compute the position information of a sensor node. This is a very critical factor when the computation is done in a distributed way. Because, the energy is spent for computation and for a short battery life sensors, it is highly desirable to have less computational complexity localization algorithm. Additionally, representing various localization algorithms computational complexity analytically is a really difficult task for the researcher to be addressed in future.

Accuracy vs. cost effectiveness: Different localization system has different positioning accuracy and is dependent on which measurement techniques are used for distance estimation. In range free localization techniques, the accuracy depends on the number of anchor nodes (preinstalled with GPS device) in the network area. Obviously increasing the number of anchor node will increase the accuracy as well as the cost of the overall system. Thus, how to achieve high accuracy with minimum number of anchor nodes is an open research problem.

7. Conclusions

Localization in WSNs is a fundamental task, where location information can be used for target tracking, location based application, data tagging etc. Traditional range free localization algorithms and protocols in WSNs do not meet the requirement of many applications, where adverse environment and channel conditions call for novel techniques. Recently, a large number of localization techniques have been proposed to meet the requirements to a certain extent. Therefore, in this paper, we have provided a comprehensive survey of various range free localization algorithms, measurement techniques, and evaluation criteria for localization.

We first group the localization algorithms based on the measurement techniques. Then, we further classified the localization techniques into two broad categories: centralized and distributed. Most of the applications in WSNs demand distributed localization method as they are more convenient for online monitoring than centralized system. Centralized and distributed localization system is further subdivided into range based and range free method. Range based methods are more accurate than range free methods. However, accuracy in range based methods are obtained with the cost of additional hardware, which in turn consumes more energy and in many applications is not suitable at all. Thus, range free methods are more desirable in many applications in WSNs. However, obtaining higher accuracy in adverse channel conditions and environments with different obstacles remains a future challenge for range free localization methods. Moreover, to improve the accuracy and robustness of the overall system, fusing the information from different positioning systems with different physical principles lead to the development of hybrid data fusion category.

Furthermore, we have provided a key inside of the challenges for future study. We have highlighted the metric in localization that needs to be addressed to meet the various requirements of various applications in order to get optimal localization accuracy.

Author Contributions: Anup Kumar Paul conducted the survey and wrote the paper. Anup Kumar Paul and Takuro Sato have revised the paper.

References

1. Vossiek, M.; Wiebking, L.; Gulden, P.; Wieghardt, J.; Hoffmann, C.; Heide, P. Wireless local positioning. *IEEE Microw. Mag.* **2003**, *4*, 77–86.
2. Flora, C.D.; Ficco, M.; Russo, S.; Vecchio, V. Indoor and outdoor location based services for portable wireless devices. In Proceedings of the 25th IEEE International Conference on Distributed Computing Systems Workshops, Columbus, OH, USA, 6–10 June 2005; pp. 244–250.
3. Paul, A.K.; Sato, T. Effective Data Gathering and Energy Efficient Communication Protocol in Wireless Sensor Network. In Proceedings of the Wireless Personal Multimedia Communication (WPMC'11), Brest, France, 3–7 October 2011; pp. 1–5.
4. Al-Karaki, J.N.; Kamal, A.E. Routing techniques in wireless sensor networks: A survey. *IEEE Wirel. Commun.* **2004**, *11*, 6–28.
5. Chowdhury, T.; Elkin, C.; Devabhaktuni, V.; Rawat, D.B.; Oluoch, J. Advances on Localization Techniques for Wireless Sensor Networks. *Comput. Netw.* **2016**, *110*, 284–305.
6. Halder, S.; Ghosal, A. A survey on mobile anchor assisted localization techniques in wireless sensor networks. *Wirel. Netw.* **2016**, *22*, 2317–2336.
7. Paul, A.K.; Sato, T. Detour Path Angular Information Based Range Free Localization in Wireless Sensor Network. *J. Sens. Actuator Netw.* **2013**, *2*, 25–45.
8. Yassin, A.; Nasser, Y.; Awad, M.; Al-Dubai, A.; Liu, R.; Yuen, C.; Raulefs, R.; Aboutanios, E. Recent Advances in Indoor Localization: A Survey on Theoretical Approaches and Applications. *IEEE Commun. Surv. Tutor.* **2017**, *19*, 1327–1346.
9. Paul, A.K.; Li, Y.; Sato, T. A Distributed Range Free Sensor Localization with Friendly Anchor Selection Strategy in Anisotropic Wireless Sensor Network. *Trans. Jpn. Soc. Simul. Technol.* **2013**, *4*, 96–106.
10. Wang, C.; Xiao, L. Sensor Localization in Concave Environments. *ACM Trans. Sens. Netw.* **2008**, *4*, doi:10.1145/1325651.1325654.
11. Piccolo, F.L. A New Cooperative Localization Method for UMTS Cellular Networks. In Proceedings of the IEEE Global Telecommunications Conference (IEEE GLOBECOM), New Orleans, LA, USA, 30 November–4 December 2008; pp. 1–5.
12. Mensing, C.; Sand, S.; Dammann, A. Hybrid data fusion and tracking for positioning with GNSS and 3GPP-LTE. *Int. J. Navig. Observ.* **2010**, *2010*, 1–12.
13. Rezazadeh, J.; Moradi, M.; Ismail, A.S.; Dutkiewicz, E. Superior Path Planning Mechanism for Mobile Beacon-Assisted Localization in Wireless Sensor Networks. *IEEE Sens. J.* **2014**, *14*, 3052–3064.
14. Niculescu, D.; Nath, B. Ad hoc positioning system (APS). In Proceedings of the IEEE Global Telecommunications Conference (GLOBECOM '01), San Antonio, TX, USA, 25–29 November 2001; Volume 5, pp. 2926–2931.
15. Zhong, Z.; He, T. RSD: A Metric for Achieving Range-Free Localization beyond Connectivity. *IEEE Trans. Parallel Distrib. Syst.* **2011**, *22*, 1943–1951.
16. He, T.; Huang, C.; Blum, B.M.; Stankovic, J.A.; Abdelzaher, T. Range-free Localization Schemes for Large Scale Sensor Networks. In Proceedings of the 9th Annual International Conference on Mobile Computing and Networking (MobiCom '03), New York, NY, USA, 14–19 September 2003; pp. 81–95.
17. Boukerche, A.; Oliveira, H.A.B.F.; Nakamura, E.F.; Loureiro, A.A.F. DV-Loc: A scalable localization protocol using Voronoi diagrams for wireless sensor networks. *IEEE Wirel. Commun.* **2009**, *16*, 50–55.
18. Gui, L.; Val, T.; Wei, A. Improving Localization Accuracy Using Selective 3-Anchor DV-Hop Algorithm. In Proceedings of the IEEE Vehicular Technology Conference (VTC Fall), San Francisco, CA, USA, 5–8 September 2011; pp. 1–5.
19. Ta, X.; Mao, G.; Anderson, B.D.O. On the Probability of K-hop Connection in Wireless Sensor Networks. *IEEE Commun. Lett.* **2007**, *11*, 662–664.
20. Ta, X.; Mao, G.; Anderson, B.D.O. Evaluation of the Probability of K-Hop Connection in Homogeneous Wireless Sensor Networks. In Proceedings of the IEEE GLOBECOM 2007—IEEE Global Telecommunications Conference, Washington, DC, USA, 26–30 November 2007; pp. 1279–1284.
21. Kleinrock, L.; Silvester, J. Optimum Transmission Raddi for Packet Radio Networks or Why Six Is a Magic Number. In Proceedings of the IEEE National Telecommunications Conference, Birmingham, AL, USA, 4–6 December 1978; pp. 431–435.

22. Chen, Y.S.; Lo, T.T.; Ma, W.C. Efficient localization scheme based on coverage overlapping in wireless sensor networks. In Proceedings of the 5th International ICST Conference on Communications and Networking in China, Beijing, China, 25–27 August 2010; pp. 1–5.
23. Assaf, A.E.; Zaidi, S.; Affes, S.; Kandil, N. Range-free localization algorithm for heterogeneous Wireless Sensor Networks. In Proceedings of the IEEE Wireless Communications and Networking Conference (WCNC), Istanbul, Turkey, 6–9 April 2014; pp. 2805–2810.
24. Vural, S.; Ekici, E. On Multihop Distances in Wireless Sensor Networks with Random Node Locations. *IEEE Trans. Mob. Comput.* **2010**, *9*, 540–552.
25. Kuo, J.C.; Liao, W. Hop Count Distribution of Multihop Paths in Wireless Networks With Arbitrary Node Density: Modeling and Its Applications. *IEEE Trans. Veh. Technol.* **2007**, *56*, 2321–2331.
26. Li, M.; Liu, Y. Rendered Path: Range-Free Localization in Anisotropic Sensor Networks With Holes. *IEEE/ACM Trans. Netw.* **2010**, *18*, 320–332.
27. Zhang, S.; Cao, J.; Li-Jun, C.; Chen, D. Accurate and Energy-Efficient Range-Free Localization for Mobile Sensor Networks. *IEEE Trans. Mob. Comput.* **2010**, *9*, 897–910.
28. Xiao, B.; Chen, L.; Xiao, Q.; Li, M. Reliable Anchor-Based Sensor Localization in Irregular Areas. *IEEE Trans. Mob. Comput.* **2010**, *9*, 60–72.
29. Xiao, Q.; Xiao, B.; Cao, J.; Wang, J. Multihop Range-Free Localization in Anisotropic Wireless Sensor Networks: A Pattern-Driven Scheme. *IEEE Trans. Mob. Comput.* **2010**, *9*, 1592–1607.
30. Singh, M.; Bhoi, S.K.; Khilar, P.M. Geometric Constraint Based Range Free Localization Scheme for Wireless Sensor Networks. *IEEE Sens. J.* **2017**, *17*, 5350–5366.
31. Zaidi Assaf, A.E.; Affes, S.; Kandil, N. Range-free node localization in multi-hop wireless sensor networks. In Proceedings of the IEEE Wireless Communications and Networking Conference, Doha, Qatar, 3–6 April 2016; pp. 1–7.
32. Ahmadi, Y.; Neda, N.; Ghazizadeh, R. Range Free Localization in Wireless Sensor Networks for Homogeneous and Non-Homogeneous Environment. *IEEE Sens. J.* **2016**, *16*, 8018–8026.
33. Zaidi, S.; El Assaf, A.; Affes, S.; Kandil, N. Accurate Range-Free Localization in Multi-Hop Wireless Sensor Networks. *IEEE Trans. Commun.* **2016**, *64*, 3886–3900.
34. Han, S.; Lee, S.; Lee, S.; Park, J.; Park, S. Node distribution-based localization for large-scale wireless sensor networks. *Wirel. Netw.* **2010**, *16*, 1389–1406.
35. Patwari, N.; Hero, A., III; Perkins, M.; Correal, N.; O'Dea, R. Relative Location Estimation in Wireless Sensor Networks. *Trans. Signal Proc.* **2003**, *51*, 2137–2148.
36. Bulusu, N.; Heidemann, J.; Estrin, D. GPS-less low-cost outdoor localization for very small devices. *IEEE Pers. Commun.* **2000**, *7*, 28–34.
37. Singh, M.; Khilar, P.M. An Analytical Geometric Range Free Localization Scheme Based on Mobile Beacon Points in Wireless Sensor Network. *Wirel. Netw.* **2016**, *22*, 2537–2550.
38. Chen, H.; Shi, Q.; Tan, R.; Poor, H.V.; Sezaki, K. Mobile element assisted cooperative localization for wireless sensor networks with obstacles. *IEEE Trans.Wirel. Commun.* **2010**, *9*, 956–963.
39. Galstyan, A.; Krishnamachari, B.; Lerman, K.; Pattem, S. Distributed Online Localization in Sensor Networks Using a Moving Target. In Proceedings of the 3rd International Symposium on Information Processing in Sensor Networks (IPSN '04), New York, NY, USA, 27–27 April 2004; pp. 61–70.
40. Mei, J.; Chen, D.; Gao, J.; Gao, Y.; Yang, L. Range-Free Monte Carlo Localization for Mobile Wireless Sensor Networks. In Proceedings of the International Conference on Computer Science and Service System, Nanjing, China, 11–13 August 2012; pp. 1066–1069.
41. Hu, L.; Evans, D. Localization for Mobile Sensor Networks. In Proceedings of the 10th Annual International Conference on Mobile Computing and Networking (MobiCom '04), New York, NY, USA, 26 September–1 October 2004; pp. 45–57.
42. He, Z.; Ma, Y.; Tafazolli, R. A hybrid data fusion based cooperative localization approach for cellular networks. In Proceedings of the 7th International Wireless Communications and Mobile Computing Conference, Istanbul, Turkey, 4–8 July 2011; pp. 162–166.
43. Ijaz, F.; Yang, H.K.; Ahmad, A.W.; Lee, C. Indoor positioning: A review of indoor ultrasonic positioning systems. In Proceedings of the 15th International Conference on Advanced Communications Technology (ICACT), PyeongChang, Korea, 27–30 January 2013; pp. 1146–1150.

44. Deak, G.; Curran, K.; Condell, J. Review: A Survey of Active and Passive Indoor Localisation Systems. *Comput. Commun.* **2012**, *35*, 1939–1954.
45. Gu, Y.; Lo, A.; Niemegeers, I. A survey of indoor positioning systems for wireless personal networks. *IEEE Commun. Surv. Tutor.* **2009**, *11*, 13–32.
46. Adalja, D.M. A comparative analysis on indoor positioning techniques and systems. *Int. J. Eng. Res. Appl.* **2013**, *3*, 1790–1796.
47. Al-Ammar, M.A.; Alhadhrami, S.; Al-Salman, A.; Alarifi, A.; Al-Khalifa, H.S.; Alnafessah, A.; Alsaleh, M. Comparative Survey of Indoor Positioning Technologies, Techniques, and Algorithms. In Proceedings of the International Conference on Cyberworlds, Santander, Spain, 6–8 October 2014; pp. 245–252.
48. Vo, Q.D.; De, P. A Survey of Fingerprint-Based Outdoor Localization. *IEEE Commun. Surv. Tutor.* **2016,** *18*, 491–506.
49. Khan, H.; Hayat, M.N.; Rehman, Z.U. Wireless sensor networks free-range base localization schemes: A comprehensive survey. In Proceedings of the International Conference on Communication, Computing and Digital Systems (C-CODE), Islamabad, Pakistan, 8–9 March 2017; pp. 144–147.
50. Mao, G.; Fidan, B. *Localization Algorithms and Strategies for Wireless Sensor Networks*; Information Science Reference-Imprint of IGI Publishing: Hershey, PA, USA, 2009.
51. Gezici, S.; Tian, Z.; Giannakis, G.B.; Kobayashi, H.; Molisch, A.F.; Poor, H.V.; Sahinoglu, Z. Localization via ultra-wideband radios: A look at positioning aspects for future sensor networks. *IEEE Signal Process. Mag.* **2005**, *22*, 70–84.
52. Niculescu, D.; Nath, B. DV based positioning in ad hoc networks. *J. Telecommun. Syst.* **2003**, *22*, 267–280.
53. Patwari, N.; Ash, J.N.; Kyperountas, S.; Hero, A.O.; Moses, R.L.; Correal, N.S. Locating the nodes: Cooperative localization in wireless sensor networks. *IEEE Signal Process. Mag.* **2005**, *22*, 54–69.
54. Römer, K. The Lighthouse Location System for Smart Dust. In Proceedings of the 1st International Conference on Mobile Systems, Applications and Services (MobiSys '03), New York, NY, USA, 5–8 May 2003; pp. 15–30.
55. Bahl, P.; Padmanabhan, V.N. RADAR: An in-building RF-based user location and tracking system. In Proceedings of the IEEE INFOCOM 2000 Conference on Computer Communications. Nineteenth Annual Joint Conference of the IEEE Computer and Communications Societies (Cat. No.00CH37064), Tel Aviv, Israel, 26–30 March 2000; Volume 2, pp. 775–784.
56. Krishnan, P.; Krishnakumar, A.S.; Ju, W.H.; Mallows, C.; Gamt, S.N. A system for LEASE: Location estimation assisted by stationary emitters for indoor RF wireless networks. In Proceedings of the Twenty-third AnnualJoint Conference of the IEEE Computer and Communications Societies (IEEE INFOCOM), Hong Kong, China, 7–11 March 2004; Volume 2, pp. 1001–1011.
57. Prasithsangaree, P.; Krishnamurthy, P.; Chrysanthis, P. On indoor position location with wireless LANs. In Proceedings of the 13th IEEE International Symposium on Personal, Indoor and Mobile Radio Communications, Pavilhao Altantico, Lisboa, Portugal, 18 September 2002; Volume 2, pp. 720–724.
58. Ray, S.; Lai, W.; Paschalidis, I.C. Deployment optimization of sensornet-based stochastic location-detection systems. In Proceedings of the IEEE 24th Annual Joint Conference of the IEEE Computer and Communications Societies, Miami, FL, USA, 13–17 March 2005; Volume 4, pp. 2279–2289.
59. Ji, X.; Zha, H. Sensor positioning in wireless ad-hoc sensor networks using multidimensional scaling. In Proceedings of the Twenty-third AnnualJoint Conference of the IEEE Computer and Communications Societies IEEE INFOCOM, Hong Kong, China, 7–11 March 2004; Volume 4, pp. 2652–2661.
60. Doherty, L.; Pister, K.S.; Ghaoui, L.E. Convex position estimation in wireless sensor networks. In Proceedings of the IEEE Conference on Computer Communications, Anchorage, AK, USA, 22–26 April 2001; Volume 3, pp. 1655–1663.
61. Kannan, A.A.; Mao, G.; Vucetic, B. Simulated annealing based localization in wireless sensor network. In Proceedings of the IEEE Conference on Local Computer Networks 30th Anniversary (LCN'05), Sydney, Australia, 17 November 2005.
62. Kannan, A.A.; Mao, G.; Vucetic, B. Simulated Annealing based Wireless Sensor Network Localization with Flip Ambiguity Mitigation. In Proceedings of the IEEE 63rd Vehicular Technology Conference, Melbourne, Victoria, Australia, 7–10 May 2006; Volume 2, pp. 1022–1026.
63. Chen, H.; Sezaki, K.; Deng, P.; So, H.C. An Improved DV-Hop Localization Algorithm for Wireless Sensor Networks. In Proceedings of the 3rd IEEE Conference on Industrial Electronics and Applications, Weihai, China, 20–23 August 2008; pp. 1557–1561.

64. Stoleru, R.; He, T.; Stankovic, J.A. *Range-Free Localization, Secure Localization and Time Synchronization for Wireless Sensor and Ad Hoc Networks*; Springer: Berlin, Germany, 2007; Volume 30, pp. 3–31.
65. Singh, M.; Khilar, P.M. Mobile Beacon Based Range Free Localization Method for Wireless Sensor Networks. *Wirel. Netw.* **2017**, *23*, 1285–1300.
66. Savarese, C.; Rabaey, J.; Langendoen, K. Robust Positioning Algorithm for Distributed Ad Hoc Wireless Sensor Networks. In Proceedings of the USENIX 2002 Annual Technical Conference, Berkeley, CA, USA, 10–15 June 2002 ; pp. 317–327.
67. Gang, P.; Yuanda, C.; Limin, C. Study of Localization Schemes for Wireless Sensor Networks. *J. Comput. Eng. Appl.* **2004**, *40*, 27–29.
68. Chen, H.; Sezaki, K.; Deng, P.; So, H.C. An Improved DV-Hop Localization Algorithm with Reduced Node Location Error for Wireless Sensor Networks. *IEICE Trans. Fundam.* **2008**, *91*, 2232–2236.
69. Liu, M.; Bao, Y.; Liu, H. An Improvement of DV-Hop Algorithm in Wireless Sensor Networks. *J. Microcomput. Inf.* **2009**, *25*, 128–129.
70. Liu, K.; Yan, X.; Hu, F. A modified DV-Hop localization algorithm for wireless sensor networks. In Proceedings of the IEEE International Conference on Intelligent Computing and Intelligent Systems, Shanghai, China, 20–22 November 2009; Vulume 3, pp. 511–514.
71. Langendoen, K.; Reijers, N. Distributed Localization in Wireless Sensor Networks: A Quantitative Comparison. *Comput. Netw.* **2003**, *43*, 499–518.
72. Shang, Y.; Shi, H.; Ahmed, A. Performance Study of Localization Methods for Ad Hoc Sensor Networks. In Proceedings of the IEEE International Conference on Mobile Ad-hoc and Sensor Systems, Fort Lauderdale, FL, USA, 25–27 October 2004; pp. 184–193.
73. Ssu, K.F.; Ou, C.H.; Jiau, H.C. Localization with mobile anchor points in wireless sensor networks. *IEEE Trans. Veh. Technol.* **2005**, *54*, 1187–1197.
74. Lee, S.; Kim, E.; Kim, C.; Kim, K. Localization with a mobile beacon based on geometric constraints in wireless sensor networks. *IEEE Trans. Wirel. Commun.* **2009**, *8*, 5801–5805.
75. Xiao, B.; Chen, H.; Zhou, S. Distributed Localization Using a Moving Beacon in Wireless Sensor Networks. *IEEE Trans. Parallel Distrib. Syst.* **2008**, *19*, 587–600.
76. Dong, L.; Severance, F.L. Position estimation with moving beacons in wireless sensor networks. In Proceedings of the IEEE Wireless Communications and Networking Conference (WCNC), Kowloon, China, 11–15 March 2007; pp. 2319–2323.
77. Pedersen, C.; Pedersen, T.; Fleury, B.H. A variational message passing algorithm for sensor self-localization in wireless networks. In Proceedings of the IEEE International Symposium on Information Theory Proceedings, St. Petersburg, Russia, 31 July–5 August 2011; pp. 2158–2162.
78. Savic, V.; Wymeersch, H.; Penna, F.; Zazo, S. Optimized edge appearance probability for cooperative localization based on tree-reweighted nonparametric belief propagation. In Proceedings of the IEEE International Conference on Acoustics, Speech and Signal Processing (ICASSP), Prague, Czech Republic, 22–27 May 2011; pp. 3028–3031.
79. Noureddine, H.; Gresset, N.; Castelain, D.; Pyndiah, R. A new variant of Nonparametric Belief Propagation for self-localization. In Proceedings of the 17th International Conference on Telecommunications, Doha, Qatar, 4–7 April 2010; pp. 822–827.
80. Hadzic, S.; Bastos, J.; Rodriguez, J. Reference node selection for cooperative positioning using coalition formation games. In Proceedings of the 9th Workshop on Positioning, Navigation and Communication, Dresden, Germany, 15–16 March 2012; pp. 105–108.
81. Zirari, S.; Denis, B. Comparison of links selection criteria for mobile terminal positioning in cooperative heterogeneous networks. In Proceedings of the 20th International Conference on Software, Telecommunications and Computer Networks, Split, Croatia, 11–13 September 2012; pp. 1–6.
82. Kemppi, P.; Rautiainen, T.; Ranki, V.; Belloni, F.; Pajunen, J. Hybrid positioning system combining angle-based localization, pedestrian dead reckoning and map filtering. In Proceedings of the International Conference on Indoor Positioning and Indoor Navigation, Zurich, Switzerland, 15–17 September 2010; pp. 1–7.
83. Tarrío, P.; Besada, J.A.; Casar, J.R. Fusion of RSS and inertial measurements for calibration-free indoor pedestrian tracking. In Proceedings of the 16th International Conference on Information Fusion, Istanbul, Turkey, 9–12 July 2013; pp. 1458–1464.

84. Wang, H.; Lenz, H.; Szabo, A.; Bamberger, J.; Hanebeck, U.D. WLAN-Based Pedestrian Tracking Using Particle Filters and Low-Cost MEMS Sensors. In Proceedings of the 4th Workshop on Positioning, Navigation and Communication, Hannover, Germany, 22 March 2007; pp. 1–7.
85. Woodman, O.; Harle, R. Pedestrian Localisation for Indoor Environments. In Proceedings of the International Conference on Ubiquitous Computing (UbiComp '08), New York, NY, USA, 21–24 September 2008; pp. 114–123.
86. Hattori, K.; Kimura, R.; Nakajima, N.; Fujii, T.; Kado, Y.; Zhang, B.; Hazugawa, T.; Takadama, K. Hybrid Indoor Location Estimation System Using Image Processing and WiFi Strength. In Proceedings of the International Conference on Wireless Networks and Information Systems, Shanghai, China, 28–29 December 2009; pp. 406–411.
87. Germa, T.; Lerasle, F.; Ouadah, N.; Cadenat, V.; Devy, M. Vision and RFID-based person tracking in crowds from a mobile robot. In Proceedings of the IEEE/RSJ International Conference on Intelligent Robots and Systems, St. Louis, MO, USA, 10–15 October 2009; pp. 5591–5596.
88. Schulz, D.; Fox, D.; Hightower, J. People Tracking with Anonymous and ID-sensors Using Rao-Blackwellised Particle Filters. In Proceedings of the 18th International Joint Conference on Artificial Intelligence (IJCAI'03), Acapulco, Mexico, 9–15 August 2003; Morgan Kaufmann Publishers Inc.: San Francisco, CA, USA; pp. 921–926.
89. Pescaru, D.; Curiac, D. Anchor Node Localization for Wireless Sensor Networks Using Video and Compass Information Fusion. *Sensors* **2014**, *14*, 4211–4224.
90. Gao, D.; Zhu, W.; Xu, X.; Chao, H. A hybrid localization and tracking system in camera sensor networks. *Int. J. Commun. Syst.* **2014**, *27*, 606–622.
91. Khan, M.I.; Syrjarinne, J. Investigating effective methods for integration of building's map with low cost inertial sensors and wifi-based positioning. In Proceedings of the International Conference on Indoor Positioning and Indoor Navigation, Montbeliard-Belfort, France, 28–31 October 2013; pp. 1–8.
92. Chen, J.C.; Yao, K.; Hudson, R.E. Source localization and beamforming. *IEEE Signal Process. Mag.* **2002**, *19*, 30–39.
93. Rabbat, M.; Nowak, R. Distributed optimization in sensor networks. In Proceedings of the Third International Symposium on Information Processing in Sensor Networks, Berkeley, CA, USA, 26–27 April 2004; pp. 20–27.
94. Capkun, S.; Hamdi, M.; Hubaux, J.P. GPS-free positioning in mobile ad-hoc networks. In Proceedings of the 34th Annual Hawaii International Conference on System Sciences, Maui, Hawaii, 3–6 January 2001; pp. 3481–3490.
95. Kwon, O.H.; Song, H.J. Localization through Map Stitching in Wireless Sensor Networks. *IEEE Trans. Parallel Distrib. Syst.* **2008**, *19*, 93–105.
96. Qudah, I.; Leijdekkers, P.; Gay, V. Using mobile phones to improve medication compliance and awareness for cardiac patients. In Proceedings of the 13rd International Conference on PErvasive Technologies Related to Assistive Environments, Samos, Greece, 23–25 June 2010; pp. 1–7.
97. Ye, H.; Gu, T.; Tao, X.; Lu, J. F-Loc: Floor localization via crowdsourcing. In Proceedings of the 20th IEEE International Conference on Parallel and Distributed Systems (ICPADS), Hsinchu, Taiwan, 16–19 December 2014; pp. 47–54.
98. Mahfouz, M.R.; Zhang, C.; Merkl, B.C.; Kuhn, M.J.; Fathy, A.E. Investigation of High-Accuracy Indoor 3-D Positioning Using UWB Technology. *IEEE Trans. Microw. Theory Tech.* **2008**, *56*, 1316–1330.
99. URUS: Ubiquitous Networking Robots in Urban Settings. Available online: http://www.iri.upc.edu/files/scidoc/949-URUS:-Ubiquitous-Networking-Robotics-for-Urban-Settings.pdf (accessed on 10 August 2017).
100. Fazenda, P.V.; Lima, P.U. Non-holonomic robot formations with obstacle compliant geometry. *IFAC Proc. Vol.* **2007**, *40*, 439–444.
101. Taranto, R.D.; Muppirisetty, S.; Raulefs, R.; Slock, D.; Svensson, T.; Wymeersch, H. Location-Aware Communications for 5G Networks: How location information can improve scalability, latency, and robustness of 5G. *IEEE Signal Process. Mag.* **2014**, *31*, 102–112.
102. Slock, D. Location aided wireless communications. In Proceedings of the 5th International Symposium on Communications, Control and Signal Processing, Rome, Italy, 2–4 May 2012; pp. 1–6.
103. Dammann, A.; Agapiou, G.; Bastos, J.; Brunelk, L.; Garcia, M.; Guillet, J.; Ma, Y.; Ma, J.; Nielsen, J.J.; Ping, L.; et al. WHERE2 Location Aided Communications. In Proceedings of the 19th European Wireless Conference, Guildford, UK, 16–18 April 2013; pp. 1–8.

104. Daniels, R.C.; Heath, R.W. Link Adaptation with Position/Motion Information in Vehicle-to-Vehicle Networks. *IEEE Trans. Wirel. Commun.* **2012**, *11*, 505–509.
105. Shang, Y.; Ruml, W.; Zhang, Y.; Fromherz, M.P.J. Localization from Mere Connectivity. In Proceedings of the 4th ACM International Symposium on Mobile Ad Hoc Networking & Computing (MobiHoc '03), Annapolis, MD, USA, 1–3 June 2003; ACM: New York, NY, USA; pp. 201–212.
106. Allen, M.; Gaura, E.; Newman, R.; Mount, S. Experimental Localization with MICA2 Motes. In Proceedings of the Technical Proceedings of the 2006 NSTI Nanotechnology Conference and Trade Show, Boston, MA, USA, 1–11 May 2006; pp. 435–440.
107. Broxton, M.; Lifton, J.; Paradiso, J.A. Localization on the Pushpin Computing Sensor Network Using Spectral Graph Drawing and Mesh Relaxation. *SIGMOBILE Mob. Comput. Commun. Rev.* **2006**, *10*, 1–12.
108. Priyantha, N.B.; Balakrishnan, H.; Demaine, E.D.; Teller, S.J. Anchor-Free Distributed Localization in Sensor Networks. In Proceedings of the 1st International Conference on Embedded Networked Sensor Systems (SenSys '03), Los Angeles, CA, USA, 5–7 November 2003; Akyildiz, I.F., Estrin, D., Culler, D.E., Srivastava, M.B., Eds.; ACM: New York, NY, USA; pp. 340–341.
109. Ahmed, A.A.; Shi, H.; Shang, Y. SHARP: A new approach to relative localization in wireless sensor networks. In Proceedings of the IEEE International Conference on Distributed Computing Systems Workshops, Columbus, OH, USA, 6–10 June 2005; pp. 892–898.
110. Girod, L.; Lukac, M.; Trifa, V.; Estrin, D. The Design and Implementation of a Self-calibrating Distributed Acoustic Sensing Platform. In Proceedings of the 4th International Conference on Embedded Networked Sensor Systems (SenSys '06), Boulder, CO, USA, 31 October–3 November 2006; ACM: New York, NY, USA; pp. 71–84.
111. Basaran, C.; Baydere, S.; Kucuk, G. RH+: A Hybrid Localization Algorithm for Wireless Sensor Networks. *IEICE Trans. Commun.* **2008**, *E91.B*, 1852–1861.

11

Compressive Sensing-Based IoT Applications

Hamza Djelouat *, Abbes Amira and Faycal Bensaali

College of Engineering, Qatar University, Doha 2713, Qatar; abbes.amira@qu.edu.qa (A.A.); f.bensaali@qu.edu.qa (F.B.)

* Correspondence: hamza.djelouat@qu.edu.qa

Abstract: The Internet of Things (IoT) holds great promises to provide an edge cutting technology that enables numerous innovative services related to healthcare, manufacturing, smart cities and various human daily activities. In a typical IoT scenario, a large number of self-powered smart devices collect real-world data and communicate with each other and with the cloud through a wireless link in order to exchange information and to provide specific services. However, the high energy consumption associated with the wireless transmission limits the performance of these IoT self-powered devices in terms of computation abilities and battery lifetime. Thus, to optimize data transmission, different approaches have to be explored such as cooperative transmission, multi-hop network architectures and sophisticated compression techniques. For the latter, compressive sensing (CS) is a very attractive paradigm to be incorporated in the design of IoT platforms. CS is a novel signal acquisition and compression theory that exploits the sparsity behavior of most natural signals and IoT architectures to achieve power-efficient, real-time platforms that can grant efficient IoT applications. This paper assesses the extant literature that has aimed to incorporate CS in IoT applications. Moreover, the paper highlights emerging trends and identifies several avenues for future CS-based IoT research.

Keywords: Internet of Things (IoT); compressive sensing (CS); hardware implementation; reconstruction algorithms

---

1. Introduction

Internet of Things (IoT) refers to the large set of smart embedded devices connected to the Internet to provide specific services to meet the users demands [1]. IoT presents the shift from only connecting the end-user devices to the Internet to using the Internet itself to interconnect smart objects (known also as IoT devices) and to communicate with each other and/or with humans to offer a wide range of applications and services [2,3].

IoT platforms usually deploy a large scale of smart objects including wearable sensors, actuators and radio frequency identification (RFID) devices to remotely monitor different physical, environmental and physiological quantities to improve the everyday activities of the end-users [4,5]. In fact, the IoT devices often operate in a long-term mode and communicate wirelessly with each other and with a central fusion node in order to empower diverse remote monitoring platforms. In general, remote sensing devices are battery-driven, hence their performance is prone to the limited battery lifetime leading to both poor integration and user adherence. Thus, to overcome those limitations, the acquired data should be first compressed and then transmitted through optimized paths to the fusion centre to minimize the high energy. However, applying advanced data compression and transmission techniques may also consume considerable onboard energy. Therefore, the adopted compression technique has to sustain a long-term efficient monitoring along with an optimized power consumption.

Compressed sensing (CS) is an emerging signal processing paradigm which aims to acquire directly a compressed form of signals with sparse behavior at the sensing stage and enables a high reconstruction quality at the receiver stage [6,7]. CS presents an alternative paradigm to the traditional acquisition fundamentals that state that the number of measured samples should be at least equal to the number of samples in the original signal to ensure an exact recovery. However, these conditions do not take the structure of the signal into consideration. Thus, if the signal of interest is sparse, i.e., the signal can be represented by a smaller number of non-zero coefficients than its original dimension, CS claims that taking only a few numbers of random linear measurements (projections) of the sparse signal is enough to capture the salient information in the signal to provide an acceptable reconstruction quality. CS aims to shift the complexity from the sensors which are usually resources constrained and self-powered to the receiver side which is usually installed on computing platforms with relaxed constraints.

For most real-world applications, it is always possible to find a sparse or compressible representation for signals of interest using the appropriate transformation. Thus, CS has spread widely in various applications such as radar, image processing, bio-signal compression, wireless communications and many others.

For instance, the healthcare sector has witnessed a tremendous efforts to explore CS in different applications. Experts believe that CS would be beneficial for a wide range application in medical laboratory and pathology testing, particularly where many data are generated. CS may also improve wearable health monitoring platforms, making it smaller, cheaper, and more energy efficient [8]. CS is expected to optimize power and energy used in wireless ambulatory devices, hence extending the sensor lifespan and significantly simplifying hardware design and reducing both the size and cost of the entire healthcare platform. CS-based healthcare applications include medical imaging [9], electrocardiogram (ECG) monitoring [10], EEG compression [11], biometric solutions [12], etc.

Subsequently, IoT platforms have also witnessed the integration of CS into its various application based on two properties that most of the IoT platforms exhibit. First, a wide range of real-world data can be well approximated by a sparse signal using the appropriate transform. For instance, both discrete cosine transform (DCT) and discrete wavelet transform (DWT) provide a good sparse representation for ECG , images, temperature, humidity data, etc. Moreover, it is always possible to form a sparsifying basis by means of dictionary learning methods [13]. Thus, exploring CS has been widely investigated in long-term data acquisition for large-scale wireless sensor networks (WSNs) [14–16].

The second property relies on the sporadic transmission schemes that most of the IoT platforms exhibit. In a sporadic transmission scheme, not all devices transmit their data simultaneously to the fusion node, rather only a small number of devices contribute on the aggregated signal at any given time. i.e., the rate of the active devices at each transmission time slot is very small. Thus, the architecture sparsity can be explored and the knowledge about the nodes activity can be exploited at the cloud level by means of sparsity aware and joint detection protocols to achieve a high data reliability with a small number of transmitting devices.

The objective of this paper is threefold: First, it aims to present a simple and a coherent overview of the fundamentals and the mathematical models that underlie CS concept as well as its well-considered recovery algorithms. Further, the paper highlights the main difference between CS and state-of-the-art compression techniques. Additionally, the paper also reviews the distributed compressive sensing (DCS) which represents the extension of CS to the multi-channel acquisition scenario.

Secondly, the paper surveys the different research efforts in the area of CS-based IoT platforms aiming to emphasize on the importance of deploying joint CS-IoT applications. Since IoT platforms are multidisciplinary paradigms that support the connection between different cross layers from the data sensing to the enabled services and applications. The paper divides the IoT into three main layers, namely, sensing, processing an application. Thereafter, for each layer, the paper delivers a thorough comprehensive discussion for the extant efforts and research directions that have been investigated and quantified in the literature to design and empower a high efficient embedded CS-based IoT platforms. To the best of our knowledge, no such review study has been presented in the literature before.

Third, the paper extends the discussion to illuminate the main research orientations and the emerging trends that have been originally developed out of the IoT scope. However, they hold great potentials for future integration within the context of energy-efficient real-time response CS-based IoT applications. This review paper aims to present a significant contribution to the joint CS-IoT researches by reviewing current work, diffusing insights and highlighting the important research trends that can be devoted to the CS-based IoT applications.

The remaining sections of this paper are organized as follows. Section 2 provides an overall description of general IoT framework and architecture. In Section 3, a brief overview of CS concept is provided by describing the system model, highlighting the general considerations on the design of the sensing matrix and presenting the most considered recovery algorithms associated with CS. Section 4 addresses the main issue of this paper by discussion thoroughly the joint CS-IoT efforts over the different IoT layers. The limitations facing each layer are exposed and the related CS efforts that have been made to tackle them are analyzed. Section 5 presents a discussion about open research issues and avenues for future work in IoT. Section 6 concludes the paper.

2. IoT Framework

The international telecommunication union (ITU) defines the IoT as a global infrastructure for the information society, enabling advanced services by interconnecting (physical and virtual) things based on existing and evolving interoperable information and communication technologies [17].

IoT consists of a network of sensors, actuators, wireless communication protocols and data processing technologies that interact with each other to provide a specific application [18]. IoT is often characterized by a large number of highly dynamic heterogeneous devices each with different and limited communication and computation resources. This heterogeneity in both the software/hardware levels requires a new level of networking/communication protocols as well as adaptation mechanisms [19]. Moreover, several other issues have to be addressed as well in both the integration and the management of these devices, such as scalability, information exchange, power consumption, interoperability and system flexibility for the dynamic change in the network topology [20]. Moreover, IoT has shifted the applications from the scale of a single device to real-time massive deployments of embedded cross-platform and cloud technologies. Therefore, to link everything together, several research groups and standardization bodies have approached and addressed these issues from several perspective [21].

Subsequently, different architectures have been proposed in the literature to establish a universal framework for IoT applications. These architectures take into consideration several parameters related to the IoT devices, the communication protocols, the networking and the targeted applications and services. For instance, a three-layer IoT platform consisting of a perception layer, a communication layer and a service layer has been proposed in [22,23]. Liu et al. considered an IoT architecture of four layers, namely physical, transport, middleware and application layers [24]. A five-layer IoT platform consists of perception layer, processing layer, transport layer, application layer and the business layer has been proposed in [25]. Table 1 summarizes a three-layer IoT platform with a brief description of the functionality of each layer. The IoT platform presented in Figure 1 consists of several sensors that can be deployed to gather information about the different physical phenomenon, these sensors are interconnected through a wireless network and communicate with a local processing unit where the data can be stored and lightweight processing can be performed. Afterwards, data are routed to the application layer hosted on the cloud where different data analytic algorithms can be applied to provide explicit services.

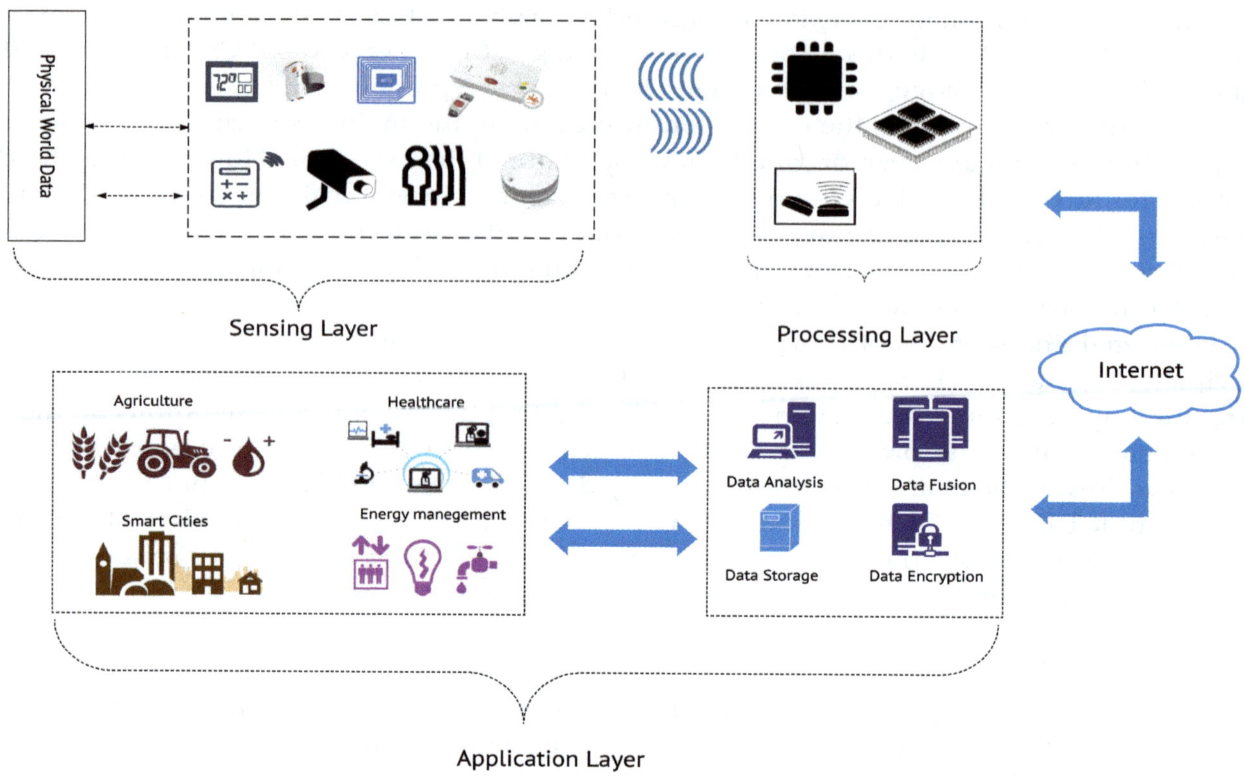

Figure 1. General IoT platform architecture considered in this paper.

Table 1. A three-layer architecture for IoT platform.

IoT Layer	Description
Sensing layer	Collect physical data entities (temperature, ECG blood pressure) using different sensors and convert the collected measurements into digital signal.
Processing layer	The Processing Layer provides mainly transient data storage for the data received from the sensors and performs local data processing, executes real-time actions, and up-links the data to the cloud.
Application layer	The task of the Application Layer is based on the information routed from the processing layer to perform data analysis and develops diverse IoT services.

IoT platforms have evolved rapidly in the last decade by leveraging the interactions among different types of smart devices, such as vehicles, medical sensors, cameras and RFID systems. Subsequently, a wide range of applications has been enabled, for instance, experimental monitoring, automated industry, connected health-care, smart buildings, intelligent transportation systems (ITS), smart grid, etc. [26]. Moreover, growing interest has been dedicated to incorporate IoT platforms in an environmental context. Strong efforts have been made by several national governments and leading information technology (IT) companies to adopt IoT solutions in the development of smart cities [27]. Even though a unified smart city framework has not been clearly defined yet, all the effort to develop this sector of applications aims to fully exploit any public resource to enhance the quality of the services offered to the citizens [28]. Smart cities can provide assistance for public service management. These services include transportation [29–31], city surveillance and emergency management [32,33], preservation of cultural heritage [34] and environment monitoring [4,28,35].

3. Compressive Sensing Overview

CS is a two-stage paradigm that acquires the data in a compressed form at the sensor and reconstructs efficiently the original data from a fewer sample signal at the receiver. Thus, CS simultaneous senses and compresses the data in one single operation.

The well-established traditional sensing and processing theory in WSNs relies mainly on the Nyquist–Shannon sampling theorem which presents the main pillar of the digital processing realm. This theorem states that, given any signal with a bandwidth of W, its information is entirely preserved if it is sampled at secured sampling frequency $f_s \geq 2W$. Further, with the ever-increasing number of digital sensing devices, several challenges have faced this paradigm, from the torrent amount of data generated by the different sensing devices to the high data transmission rate and unbearable constraints on the sampling/processing platforms. Such challenges would largely increase the cost of the wireless transmission and shorten the sensors lifespan. Thus, to address the challenges associated with high-dimensional data, researches depends often on compression techniques by discarding the redundant samples in the signal. The benefits of compression will apply to the whole ubiquitous computing environment, as decreasing data size means shorter communication delays, efficient usage of bandwidth and reduced battery drain. For instance, taking WSN applications, where the energy consumption in the sensing nodes is mainly dominated by wireless communication rather than the onboard processing. State-of-the-art radio transmitters exhibit energy consumption of order of nJ/bit while all the other sensor circuits consume at most only tens of pJ/bit [36]. This cost disparity suggests that some data reduction strategy at the sensor node should be employed to minimize the energy cost of the system. Thus, reducing the number of the transmitted samples is necessary to achieve a long lifespan. However, it should be noted that these compression techniques remain useful as long as they do not consume more energy than the one saved by reducing the transmitted samples. In fact, in WSN, this condition is fulfilled.

Recently, to tackle the issue of acquiring a huge amount of redundant samples, a very interesting theory, namely CS, has been proposed [6,7]. CS suggests that it is possible to surpass the traditional limits of sampling theory for a specific type of structured signal by reducing the sampling rate without any significant loss in the data information. The authors argued that, if the information rate of the signal is less than its bandwidth rate, i.e., the signal is either sparse or compressible, then it is possible to capture all the information without any loss using fewer samples than what the Nyquist–Shannon theorem stats. Hence, CS enables a potentially large reduction in the sampling and computation costs for sensing signals with sparse or compressible behavior.

CS differs significantly from state-of-the-art compression techniques. While the latter exhibit a two stages framework, first, the signal is sampled with respect to its bandwidth rate following the Nyquist sampling frequency high f_s to acquire an N-length signal. Secondly, data compression is performed by means of filtering, coding and extracting the most salient features in the signal to reduce the signal dimension from N to M such that ($M \ll N$). CS, on the other hand, is a one single stage framework that aims to perform both acquisition and compression simultaneously to acquire directly a compressed form of the signal. To this end, CS leverages the fact that most real-world signals exhibit a sparse behavior when represented by the appropriate basis (this basis can be either a fixed transform such DCT and DWT or it can be formed by means of dictionary learning techniques). A sparse behavior indicates that the signal information rate is much smaller than its bandwidth rate. Thus, rather than sampling the signal at a Nyquist rate and then performing compression, CS takes random measures of the signal in a sub-Nyquist rate corresponding to the signal information rate. This process is performed by taking inner products between the signal of interest and a well-designed tall sensing matrix $\boldsymbol{\Phi} \in \mathbb{R}^{M \times N}$. Therefore, to reduce sampling rate and to achieve the objective of joint sensing and compressing, CS is applied prior to sampling and quantization adopting two approaches, Analog CS and Digital CS [37]. Therefore, Applying CS will result in the sensing/transmission of fewer samples. Thus, such paradigm is well suited for WSN applications which require a continuous data.

3.1. Mathematical Overview

Let the orthogonal basis $\{\Psi_i\}_{i=1}^{N}$ span $\mathbb{R}^N$. Then, any signal $\boldsymbol{x} \in \mathbb{R}^N$ can be expressed as a linear combination of the elements of $\boldsymbol{\Psi}$ with the elements of the vector $\boldsymbol{s} = [s_1, s_2, \cdots, s_N]^T \in \mathbb{R}^N$ such that $\boldsymbol{x} = \sum_{i=1}^{N} \Psi_i s_i$. The signal $\boldsymbol{x}$ is said to be k-sparse if the vector $\boldsymbol{s}$ has only $k \ll N$ non-zero entries.

The set of the indices corresponding to the positions of the non-zero entries of s is called the support of s and denoted as Σ_k. In addition, the matrix $\boldsymbol{\Psi}$ is called the sparsifying matrix in the context of CS.

To perform a compressed acquisition on the data, the input signal x is multiplied by a tall random sensing matrix; hence, the acquisition process in CS can be modeled by:

$$y = \boldsymbol{\Phi}\boldsymbol{x} = \boldsymbol{\Phi}\boldsymbol{\Psi}\boldsymbol{s} \tag{1}$$

where $\boldsymbol{\Phi} \in \mathbb{R}^{M \times N}$ represents the sensing matrix used to acquire and compress the data. Moreover, the ratio $\frac{N}{M}$ is defined as the compression factor (CF).

3.2. *Sensing Matrix*

Ideally, the sensing matrix $\boldsymbol{\Phi}$ has to fulfill the following constraints:

- Optimal or near-optimal reconstruction performance: The measured data maintain the salient information of the signal for reconstruction purposes.
- Optimized sensing performance: Only a few measurements are required to obtain an optimal (near-optimal) recovery.
- Universality: The sensing matrix maintains a low coherence with almost all sparsifying matrices.
- Low complexity, fast computation and structure-based processing: These features of the sensing matrix are desired for large-scale, real-time sensing applications.
- Hardware friendly: Easy and efficient implementation on hardware is necessary.

The above-mentioned conditions present the ideal criteria for any sensing matrix. However, it is hard to find matrices that satisfy all these conditions. Thus, two relaxed conditions that can grant a high reconstruction quality have been established for designing the sensing matrix. First, a low coherence between the sensing matrix and the sparsifying matrix is recommended to enable the sensing matrix to capture the salient information of the original signal with the minimum number of projections [38]. Second, the sensing matrix should satisfy the restricted isometry property (RIP) to ensure that the compression preserves the signal information (k-RIP) and to ensure that mapping the signal x from a high dimension space $\mathbb{R}^N$ to a lower dimension one $\mathbb{R}^M$ should be unique, i.e., it should preserve the distance between each pair of two distinct signals x and x' ($2k$-RIP) .

3.3. *Reconstruction Algorithms*

Fast and efficient reconstruction algorithms are the keys to incorporate CS in real-world applications, thus developing such algorithms have been the main concern in the CS community. Several reconstruction algorithm classes have been proposed in the literature. Nevertheless, the well-recognized algorithms fall under the umbrella of two major algorithmic approaches, namely, convex optimization and greedy algorithms.

3.3.1. Convex Optimization

A straightforward approach to recover a sparse vector x is to solve the $\ell 0$ minimization problem:

$$\hat{x} = arg\min ||\boldsymbol{x}||_0 \quad subject \quad to \quad \mathbf{y} = \boldsymbol{\Phi}\boldsymbol{x} \tag{2}$$

However, the $\ell 0$ minimization problem is considered as an NP-hard for large scale matrix [39], i.e., solving Equation (2) for any large matrix $\boldsymbol{\Phi}$ is necessarily computationally intractable. Therefore, convex relaxation to Equation (2) has been considered to overcome the drawback of $\ell 0$, leading to $\ell 1$-minimization.

The $\ell 1$-minimization approach, known as basis pursuit (BP) [40], considers the following solution:

$$\hat{x} = arg\min ||\boldsymbol{x}||_1 \quad subject \quad to \quad \mathbf{y} = \boldsymbol{\Phi}\boldsymbol{x} \tag{3}$$

A noisy acquisition model adapts itself by considering the basis pursuit denoising (BPDN) solution [41]:

$$\hat{x} = arg\min ||\boldsymbol{x}||_1 \quad subject \quad to \quad ||\mathbf{y} - \boldsymbol{\Phi}\boldsymbol{x}||_2 \leq \epsilon \tag{4}$$

where ϵ represents the upper bound on the noise level.

Further, if there is no knowledge about the noise level, least absolute shrinkage and selection operator (LASSO) approach [42] can be explored:

$$\min_{\hat{x}} \frac{1}{2}||\boldsymbol{\Phi}\hat{x} - \boldsymbol{y}||_2 + \gamma||\hat{x}||_1 \tag{5}$$

where $\gamma > 0$ represents a tuning parameter for different levels of sparsity.

3.3.2. Greedy Algorithms

Greedy algorithms have been widely exploited in CS applications due to their relatively simple framework which provides a fast reconstruction with a low implementation cost. These methods enhance iteratively the approximation for the signal by making locally optimal choices. Greedy algorithms consist of two main steps, element(s) selection and coefficients update. These methods are usually initialized with a residual $\boldsymbol{r}^{[0]} = \boldsymbol{y}$, signal estimate $\hat{x}^{[0]} = 0$ and empty support set T = ∅. At each iteration j, single or multiple elements from the sensing matrix are added to the support set, the signal estimate is calculated using a least square approach $\boldsymbol{x}^{[j]} = \boldsymbol{\Phi}_T^{\dagger}\boldsymbol{y}$. Additionally, the residual is minimized $\boldsymbol{r}^{[j]} = \boldsymbol{y} - \boldsymbol{\Phi}_T\boldsymbol{x}^{[j]}$. The algorithms halt when the residual norm is smaller than a predefined threshold.

Currently, the well established greedy algorithms include gradient pursuit [43], matching pursuit (MP) [44,45] and orthogonal matching pursuits (OMP) [46]. OMP offers a fast recovery compared to convex optimization approaches, yet it suffers from bad recovery quality for signals with a low degree of sparsity. Thus, several improved versions of OMP have been proposed, such as compressive sampling matching pursuit (CoSaMP) [47], subspace pursuit (SP) [48], Regularized OMP [49], Stagewise OMP [50] and orthogonal multiple matching pursuit [51].

The performance of these recovery algorithms depends on the targeted application and no unique recovery metric is established to determine the best recovery technique for all scenarios. Thus, both theoretical and experimental performance comparison between the different classes of CS recovery algorithms can be found in [52–55]. Table 2 lists the details regarding the complexity and the minimum number of measurements required for each algorithm to achieve its optimal recovery bound.

Table 2. Complexity and minimum measurements required for CS reconstruction.

Algorithm	Minimum Number of Measurements	Complexity
BP	$k \log N$	$\mathcal{O}(M^2N^{1.5})$
OMP	$k \log N$	$\mathcal{O}(kMN)$
CoSaMP	$k \log N$	$\mathcal{O}(MN)$
SP	$k \log(\frac{N}{k})$	$\mathcal{O}(MN \log k)$
Stagewise OMP	$N \log N$	$\mathcal{O}(N \log N)$

3.4. Distributed Compressive Sensing

Conventional CS theory explores only the intra-signal structures (sparsity and compressibility) at a single sensor. However, if the scenario of interest presents an architecture where the data are not only collected by a single sensor but rather with a distributed network of sensors, then exploiting the collaboration of these sensors would lead to an additional gain in the information collected and improve the design of the application. Such scenario arises in several IoT monitoring applications, where a network of wireless sensors is deployed to measure different scalar physical/physiological quantities, such as temperature, light, humidity, human vital signs, etc. If the sensors are densely deployed, their measurements are correlated and they are likely to share certain structures.

CS-based multi-sensor architecture with J sensors can exhibit two scenarios. In the first one, the acquired signals are independent; hence, the reconstruction problem is equivalent to performing J individual reconstruction operation for each sensor and no collaboration between the sensing nodes can be explored for the reason that each sensor holds no information about the data acquired by the other sensors. However, in the second scenario, which emerges frequently in most IoT applications, the signals acquired from all the sensor are highly correlated, which presents an inter-signal sparsity structure. Therefore, to reconstruct the whole signal ensemble, not only the individual sparsity of each sensor measurements can be exploited, but also the dependency among the samples acquired by different sensors at the same time should be explored to reduce the required number of measurements to assemble the most pertinent information related to the sensed phenomenon. This process ,which we refer to as joint measurements recovery, is the motivation for introducing distributed compressive sensing (DCS) concept [56].

DCS presents a new distributed coding paradigm that exploits both intra- and inter-signal correlation structures. In DCS scenarios, a number of sensors acquire sparse signals which are correlated with each other. Each sensor individually acquires a compressed signal by taking random projections and then transmits it to the fusion node. More importantly, DCS requires no collaboration between the sensors during the acquisition phase. However, the recovery process can exploit the inter-signal correlation between the measurements to reconstruct all the signals simultaneously. The recovery algorithms associated with DCS are derivatives from their CS counter-parts, such as multichannel-BPDN [57], simultaneous OMP (SOMP) [58] and distributed compressive sensing OMP (DCS-SOMP) [56] which is the general form of SOMP.

4. CS-Based IoT Applications

The heart of any IoT platform is the smart devices that build up the sensing layer. Data acquisition is performed using several sophisticated smart IoT devices that are deployed in different locations to collect various types of data over a long period depending on the targeted application. The collected data are usually large and contain some redundant information. Therefore, the data are first transmitted to a local processing unit with adequate storage and computing capacities to perform different pre-processing techniques on the samples to extract different information and features. The processing unit also plays the role of a gateway to route only the useful information to the cloud rather than transmitting the whole collected data, this approach can significantly reduce the network bandwidth.

CS presents a very promising paradigm to be explored in IoT applications. A CS-based IoT framework can be implemented on three levels. First, at the sensing level, where data acquisition and transmission process presents a challenging task in term of power consumption, thus minimizing the latter is the critical issue to tackle. Therefore, CS can be deployed effectively as a compression technique to develop an energy-efficient scheme for data acquisition and transmission. Second, CS can be deployed on the embedded systems implemented on the local processing unit to realize "CS-based Edge computing platform" where the compressed data transmitted from the sensors can be aggregated, stored and reconstructed to extract salient information and important features to be sent to the cloud which hosts the application layer. The latter represents the third level in which CS can be explored along with other data analytic algorithms to execute actions and provide solutions based on the specific tasks that the IoT is supposed to deliver. Figure 2 illustrates the possible scenarios where CS can be deployed over IoT platform.

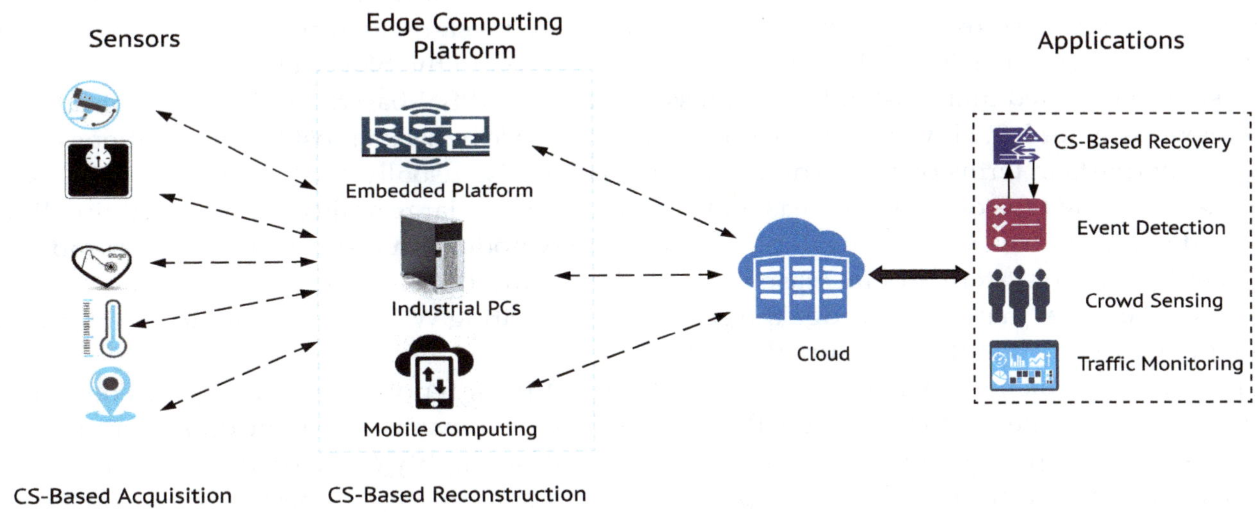

Figure 2. CS-based IOT platform.

4.1. Sensing Layer

The proper integration of sensors, actuators, RFIDs and communication technologies will deliver a strong foundation for the sensing layer of any IoT platforms. An adequate sensing layer that meets the high demands of energy-efficient IoT application should be empowered by exploring efficient hardware/software solutions. The latter consist of incorporating state-of-the-art compression techniques and data routing protocols.

The hardware part of the IoT sensing layer is enabled using two main technologies, RFIDs and WSNs. While RFIDs are mainly used to enable low-cost identification and tracking applications, WSNs can provide the IoT with broad sensing and actuation capabilities. In addition, intensive research and efforts have already been dedicated to expanding the WSNs to cover unlimited set of IoT applications.

RFIDs generally consist of two major elements, an RFID tag embedded on a chip with a unique identification sequence (ID) and a radio scanner device, called RFID reader. RFIDs have gained a lot of attention in IoT applications, they serve to store data, to track objects and to communicate with other devices [59]. Moreover, RFIDs present appealing solutions due to their ability to operate in battery-free mode and their lowest cost and small dimension [23].

For IoT applications, RFIDs have been used in different variations, the tags could be implemented on the human body [60], deployed on the skin [59] or attached to the walls or other objects. Although the reading distance of RFIDs (the distance between the tags and the reader) is not large (5–6 m), they can be deployed efficiently for indoor tracking applications [61].

Several IoT applications have witnessed the integration of RFIDs, for instance, Khan et al. [62] presented the design and the implementation of a GPS-RFID tag that can be effectively deployed for different IoT tracking applications. The designed tag operates in a semi-passive mode (battery assisted) with a reading range up to 5.6 m. In addition, it can be reconfigured to serve as a continuous data transmitting platform in the online mode or as data logging platform in the offline mode. Attran et al. [63] presented new chipless RFID tag using microelectromechanical (MMES) systems technology for IoT applications, the proposed prototype is expected to reduce the cost of the RFID tag as the chip fabrication is eliminated, hence, the cost of the RFID tag will be as low as barcode labels.

On the other hand, WSN-based IoT applications can be divided into two categories: outdoor and indoor. Outdoor monitoring collects environmental data that can be used in the context of smart cities (light, humidity, traffic, GPS tracking, and air monitoring), whereas indoor monitoring is more concerned with smart-home applications that can provide remote healthcare services, in which the sensors can be either distributed over the house to measure humidity, temperature and detect motion or they can be deployed over the human body to acquire vital signs such as electrocardiogram (ECG), electroencephalography (EEG), blood pressure, glucose level in the blood, etc.

It should be noted that most of the progress made in the WSN-based monitoring applications can be adapted easily for IoT. However, there is a crucial difference between most IoT applications and their WSN counterparts in terms of device heterogeneity, i.e., WSNs usually deploy a set of similar sensors that collect a single type of data, whereas, in IoT applications, a large scale of sensors acquire different types of data. Moreover, inadequate deployment of WSN nodes can result in a severe degradation in the reliability of the system and increase the complexity and the cost for the overall platform. Thus, numerous researches that aim to bridge the gap between exciting WSNs and future IoT platforms have been widely discussed in the literature [64–67].

The major problem to tackle in the sensing layer is the high power consumption. Following the several results presented in the literature, the radio frequency transmission of the data over the wireless channel is the most power consuming process [36] and it can highly decrease the performance of the system, mainly if the sensors used are self-powered. Thus, determining the best compression technique in term of minimizing power consumption is the first step to develop an energy-efficient platform.

Besides CS, many research groups have developed low power, low-cost compression techniques exploring state-of-the-art compression algorithms [68–70]. Subsequently, various efforts have been made to establish different frameworks to compare CS with state-of-the-art compression algorithms for different applications to determine the optimum compression technique.

First, Mamaghanian et al. [71] developed a non-adaptive CS-based platform for ECG compression on the Shimmer mote. In addition, they provided a comparison study between the performance of CS and DWT-based ECG compression in terms of reconstruction quality, node lifetime and CPU execution time. Although the results show that DWT outperforms the non-adaptive CS in terms of reconstruction quality (to achieve very good reconstruction quality, DWT requires $M = 0.3 \times N$ samples in the compressed signal, whereas, to achieve the same reconstruction quality, CS requires $M = 0.5 \times N$), CS-based compression has shown to provide a better energy-efficient performance, with a node lifetime extension of 37.1% over DWT and an execution time 20 times faster than the one taken by DWT. Chen et al. [36] implemented and quantified two different implementation approaches for CS, namely, digital CS and analog CS. The analysis showed that for real WSN applications, digital CS provides a better energy efficiency performance. Moreover, they compared CS with the Lempel–Ziv–Welch (LZW) compression technique; the CS compression system offers six times higher compression factor and over ten times lower implementation (storage/power) cost.

Moreover, Razzaque et al. [69] provided a more detailed comparative study including CS, DCS, transform coding (TC), predictive coding (PC) and adaptive sampling (AS) approach [72]. The work has investigated three different datasets including temperature, seismic signals and CO_2 emission data. The overall obtained results reveal that CS improves the energy saving by a factor of 79.4% compared to 62.43% and 34% obtained by TC and AS, respectively. In addition, Abate et al. [73] conducted a comparative study in terms of reconstruction quality and computation time between CS and the segmentation and labelling technique [74] which has been considered for the definition of the new standard of the IEEE 1451 [75]. The obtained results show that, even though CS is more complex, it outperforms remarkably the segmentation and labelling technique in terms of data reconstruction and robustness to noise. Furthermore, a comparison between CS and transform coding (TC) and transmission without compression in terms of energy consumption is presented in [15]. First, the authors established energy dissipation models for both approaches to investigate the energy consumption in a unified framework. The obtained results indicate that CS extends the network lifetime up to four times compared to no transmission without compression processing if the data are highly sparse and the transmission range is between 50 m and 100 m. All these results prove that incorporating CS in IoT applications will provide more improvements in terms of energy-efficiency. Thus, with the relative superiority of conventional CS over classical compression techniques being established, a fundamental question is how to raise the standards and further enhance the exciting results of CS to minimize power consumption. To this end, various approaches have been proposed in the literature to answer this fundamental question.

4.1.1. Adaptive Measurements

The intuitive solution is to minimize the number of the transmitted samples from the sensors to the processing unit by selecting a high CF. However, using a high CF can degrade the performance of data reconstruction. In addition, the optimal value of CF depends heavily on the signal sparsity which may vary in time (at the level of each sensor) and in space (sensors may have different readings for the same signal), hence CF can be high as long as it assures an acceptable recovery performance to meet the desired quality of service (QoS) for a specific application. Therefore, determining the optimal CF has been the quest of different studies by exploring different ideas and approaches. Fragkiadakis et al. [76] proposed an adaptive framework for selecting CF value to overcome the signal sparsity level variations from data-block to another. The proposed framework consists of a network of sensors, where each sensor transmits its data to a central node (CN). First, each sensor transmits a part of its data which can be seen as a training set to the CN. The latter starts a compression/decompression operations with several values of CF to determine the signal sparsity along with its best CF based on a predefined QoS metric (error of reconstruction). In addition, the CN creates a profile that assigns to each range of sparsity the best CF that renders the minimum reconstruction error. Furthermore, Charalampidis et al. [77] proposed a new approach to detect the changes in the signal sparsity using change point methods (CPM) [78] to update the CF value each time the signal sparsity level changes.

4.1.2. Weighted Measurements

Another approach to optimizing the number of transmitted samples is to assess a weight to each sensor depending to its importance in the application (a scenario that often occurs in the heterogeneous WSNs). This idea was first introduced in the context of CS for image reconstruction using wavelet decomposition where different CFs are assigned to different wavelet sub-bands [79]. Subsequently, borrowing the same concept to IoT applications, Lee et al. [80] introduced weighted CoSaMP (wCoSaMP) reconstruction algorithm to enable a new framework for data acquisition in heterogeneous IoT platforms where different type of sensors have different impact on the overall system, for instance, for IoT-based connected health, the ECG data are more crucial than humidity and temperature data. Applying this concept, fewer measurements are needed to achieve the same performance as conventional CS.

4.1.3. CS-Based Data Gathering

Efficient data helps to find suitable solutions for the high energy consumption in large-scale WSNs. Luo et al. [81] proposed a compressive data gathering (CDG) approach to extend the sensors lifetime in multi-hope architecture, where each node performs a CS acquisition and transform its random measurements to the next node. Subsequently, the data at the sink node will be the sum of the different random measurements. CDG approach disperses both communication and computation costs to all sensor nodes, resulting in a natural load balancing. Moreover, based on CDG, Xu et al. [82] presented compressed sparse function (CSF) algorithm for data gathering in WSNs. This approach adopts DCT to sparsify the collected data from each sensor and then transmits the sparse functions to the sink. Then, the reconstruction is performed by means of polynomial interpolation techniques. This approach allows reconstructing the whole ensemble data from an incomplete set of received measurements. Beside CSF and CDG, Xiang et al. [83] proposed minimum energy compressed data aggregation (MECDA) approach. The algorithm determines the best routing path from the sensor nodes to the sink node that consumes the minimum power. To determine this optimum path, the algorithm exploits minimum spanning tree and shortest path forest techniques. Nevertheless, this algorithm maintains its optimum performance as long as the network topology remains fixed. In addition, an opportunistic routing approach called compressive data collection (CDC) has been proposed in [84]. In the proposed approach, each sensor selects randomly one of its neighbours by means of opportunistic routing protocols [85] and forward to its compressed reading.

4.1.4. Sparse Networks

From another perspective, IoT applications are usually deployed using dense WSNs in which the data collected from the sensors are redundant and highly correlated. Thus, exploring the redundancy in the reading each of sensor can be used to minimize its duty cycle, thus, prolonging its lifespan and subsequently the network lifetime. Furthermore, the high correlation between the measurements can be used to enable a scheme where only a few sensors operate in an active mode for each time slot. By taking advantage of these facts, Du et al. [86] presented CS based activation scheme for water distribution in IoT networks, the proposed method selects a few numbers of sensors to send their data to the sink at each time slot, which renders a remarkable reduction in the power consumption. Besides, a recent active node selection framework aims to improve signal acquisition performance, network lifetime, and the use of spectrum resources has been proposed in [87]. The proposed approach exploits the temporal correlation to select the active nodes based on the support of the data reconstructed in the previous time slot. In addition, the correlation between the sensor's measurements were exploited in [88], in which a 1-bit DCS-WSN framework was proposed to achieve more data compression while retaining an acceptable performance using the joint sparsity between the sensors reading. Zhang et al. [89] presented another idea for saving energy in WSN based on regression and CS. The main idea is to divide the sensors into different clusters, where at each cluster only one sensor (reference) node works in a periodically sampling mode, whereas all other nodes in the cluster exhibit a CS-based acquisition. After receiving data from all nodes, the sink node runs prediction algorithm to roughly estimate the signal series of all nodes based on the signal of the reference node.

A comparative study has been conducted [86] to compare the performance of different data gathering approaches in term of the achieved network lifetime. The investigated scenario includes a network of 100 sensors deployed randomly over a specific area, in which only 20 sensors are active at each time slot and all the sensors transmit their data to a single node. The lifetime achieved by the approach proposed in [86] is about 5 and 2.5 times the CDG and CSF lifetime, respectively, whereas the lifetime achieved by the CDC algorithm is about 80% of the achieved lifetime using the proposed algorithm in [86].

4.1.5. CS-Based Routing Protocols

Routing and CS have been jointly addressed in [90] in which the routing path is iteratively built through a greedy choice to minimize the intermediate coherence of the measurement matrix with the sparse matrix. Moreover, based on the results obtained in [81,90], where the authors proved the benefit of their approaches in terms of good reconstruction quality with fewer transmitted samples, Caione et al. [91] introduced and quantified a new data routing approach. The authors proposed a DCS-based data gathering in large-scale WSNs based on CS acquisition, where they have investigated the performance from a node lifetime point of view. The approach considered in [91] provides each node with the ability to take the proper decision about the optimum value of its CF. After compression, the data are routed by exploring a hybrid approach based on DCS and pack and forward (PF) technique. The aim of this hybrid is to reduce the overall number of transmitted packets. The obtained results reveal an extension in the network lifetime compared to the use of conventional DCS even with a large number of sensors in the network. Moreover, Masoum et al. [92] built up a probabilistic model for the ensemble of the signal using a Bayesian approach and adopted belief propagation (BP) algorithm to recover the ensemble measurements. Their results show an improved reconstruction quality as well as a reduction in the energy consumption up to 20% compared to Multichannel BP associated with DCS.

4.2. Processing Layer

With the deployment of huge, dense WSNs to collect different types of information. Big data is generated from the sensing objects and routed to the cloud. Even though the big data provides a comprehensive knowledge about the sensed phenomena, they require very powerful storage, processing and information retrieving mechanisms. Moreover, big data is usually so massive that it exceeds the capabilities of commonly used computing environments. Thus, it restrains the performance of the IoT platforms by increasing the high-level service requirements (massive storage, expensive processing power, high latency, etc). Subsequently, a major challenge in IoT platforms design is to decide the layer in which the data can be processed. In fact, this decision depends on several factors such as real-time constraints, energy efficiency, communication bandwidth, delay, etc.

Therefore, for large-scale IoT platforms in which real-time response and power efficiency are strict requirements, opting for a device-centric processing does not seem to be effective. In a device-centric approach, the micro-controllers embedded within the sensors are exploited to process the data. Nevertheless, the computation/communication resources of these sensors are limited which would lead to poor system performances. On the other hand, opting for a cloud-centric processing can satisfy the computing requirements, however, the transmission of a substantial amount of data over the wireless channel incurs a heavy communication overhead. Consequently, cloud processing usually faces issues of scalability, high energy cost, latency and bandwidth availability [93]. Thus, a trade-off solution can be endorsed by adopting a gateway/fog centric architecture. The IoT gateway devices which are used to provide an interoperability between the different heterogeneous cross platforms are usually installed on powerful computing embedded platforms such as ARM Cortex-A [94]. In addition, smartphones can be used as processing units for gateway centric IoT [95].

Moreover, gateway/fog computing does not only bring the computing platform to the edge of the network but it also provides a low latency, location awareness, smart geographical distribution and on-line analysis [96]. Therefore, moving the computation from the cloud to the edge near the IoT devices would help to meet the computation requirements, to provide real-time response and to improve the scalability of the network. This approach is known as edge computing.

Edge computing presents the new trend in IoT infrastructure development; it aims to analyze the time-sensitive data at a platform in a proximity to where the data were initially collected rather than sending many data to the cloud. Edge computing platform does not eliminate the role of cloud computing, however, it will only send selected information and historical analysis for long-term storage and deep data analytic operations. Edge computing can be well suited for CS-based IoT applications, where the compressed data can be reconstructed at the gateway unit, then only the extracted features, the reconstruction reports and other useful service-specific information can be sent to the cloud for further deep learning, classification and objects/events detection.

Edge computing leverages the powerful local processing units to shift the data computation and storage from the cloud clusters to the embedded cross platforms. Edge computing can be accomplished using PCs, mobile computing devices, field programmable gate array (FPGAs), etc. Moreover, Edge computing offers real-time analysis and response, low-cost management as well as fewer data transferred to the cloud.

From the several available computing platforms, FPGA technology establishes itself as an appealing environment to implement such edge computing platforms. FPGAs would provide energy efficiency, interactivity, powerful computation abilities and reconfigurable embedded system to IoT. Moreover, compared to other computing platforms such as Raspberry Pi, FPGAs provide a better performance [97]. Although the implementation of CS-based edge computing on the FPGA for IoT applications have never been addressed in the literature, we present herein some of the efforts which can be explored to provide powerful computing platforms.

For any CS-based edge computing platform, efficient implementation of data reconstruction algorithms is the main task to realize. While a software implementation of these algorithms can be time-consuming due to the massive matrix multiplications requirements, an FPGA-based solution presents an attractive alternative to accelerate the reconstruction process by leveraging the high level of parallelism of FPGA platforms.

Therefore, the hardware implementation for the different reconstruction algorithms has been investigated by the CS research community. In fact, greedy algorithms gained most of the attention compared to convex optimization approaches due to the high complexity of the latter. Nevertheless, from all the greedy algorithms, OMP has been the focus of most researchers due to its computing efficiency, stability and relative simplicity. Besides, in contrast to advanced greedy algorithms such as SP and CoSaMP, OMP does not require any inputs except for the compressed signal and the sensing matrix. OMP recovery algorithm includes two computationally expensive steps, namely, coefficients selection and signal estimation. Coefficients selection step requires intensive matrix-vector multiplication operations in order to find the closely correlated atoms to form the signal proxy. For high dimensional signals, matrix-vector multiplication is a very time-consuming process. The signal estimation step consists of solving a least squares problem which is both time and resources consuming as it includes both matrix inversion and matrix-vector multiplication. Therefore, all efforts have been made to provide an FPGA-based implementation for OMP rely on the optimization of these two fundamental steps.

Septimus et al. presented the first implementation of OMP on Xilinx Virtex-5 FPGA [98] where the implementation was validated on a signal of length $N = 128$ and a maximum sparsity of 5. Although the size of the signal and its sparsity are small, the obtained results showed that adopting FPGA-based OMP implementation has great promises for real-world applications. Furthermore, the same authors further optimized their implementation by adopting Q-R decomposition (QRD) to solve the LS problem in [99]. The obtained results revealed three times faster execution time than their first implementation. Thereafter, various efforts have been made to provide more efficient FPGA implementation for OMP. Bai et al. [100] proposed a solution to the LS problem based on QR decomposition using a vector multiplication unit (VMU) which consists of a large number of parallel computing channels. The proposed solution has been implemented on a Xilinx Virtex-6 FPGA. The authors evaluated their implementation on an image block of size 32×32 pixels achieving reconstructed quality with signal-to-noise ratio (SNR) of 23.5 dB with an execution time of only 0.63 ms. In [101] single-precision floating-point CS reconstruction engine implemented on a Kintex-7 FPGA is presented. The author fully explored the maximum hardware resources available to achieve high performance by deploying highly parallel architecture that shares computing resources among different tasks of OMP by using configurable processing elements (PEs). Rabah et al. [102] presented a descriptive analysis to OMP algorithm where they conducted a detailed study on the complexity of each step in the OMP algorithm. In addition, they adopted a Cholesky factorization method to solve the LS problem. A four-block architecture has been implemented on MATLAB-SIMULINK and Xilinx system generator (XSG) has been used to map the system-level design to a high-level description language (HDL). Subsequently, more efforts have been made to implement OMP for real-world signals. Kulkarni et al. quantified the performance of OMP implementation on FPGA Vertex-7 for ECG signal [103] and image data [104]. Recently, Quan et al. presented an efficient solution for the FPGA implementation of OMP [105]. The proposed approach uses fast Fourier transform (FFT) to implement the coefficient selection step which is represented by a correlation operation. Additional efforts for FPGA-based implementation of OMP can be found in [106–109].

Beside FPGAs, the hardware implementation of CS reconstruction algorithms on other computing platforms such as graphics processing units (GPUs) and application specific integrated circuits (ASICs) have been presented in the literature. A hardware implementation of three CS recovery algorithms on ASIC have been proposed in [110] for channel estimation in wireless network. The authors provided a comparative study of the architecture, the complexity and the cost of MP, OMP and gradient pursuit. Fang et al. [111] presented a GPU implementation of OMP where they proposed a matrix inversion update method to minimize the computation complexity of the matrix inversion problem which leads to speed up the reconstruction.

In [112], the authors presented a comparative study between the implementation of the OMP for image reconstruction using a high parallel computation with an LU decomposition for solving the LS problem. The proposed architecture has been implemented on different platforms including general purpose CPUs, GPUs, a Virtex-7 FPGA and a domain-specific many-core. The implementation results showed that reconstruction time on FPGA is improved by two times compared to the results in [99]. In addition, the GPU based implementation has shown to be three times faster than results obtained in [111].

Table 3 expatriates on the comparative results between several works presented in the literature on the implementation of OMP on different platforms.

Table 3. Comparison between FPGA implementation for OMP presented in the literature.

References	Platform	Data Size	Reconstruction Time	Max Frequency (MHz)
[100]	FPGA (Virtex6)	512	360 µs	100
[99]	FPGA (Virtex5)	128×128	13.7 µs	165
[105]	FPGA (Virtex7)	128	391.8 µs	165
[101]	FPGA (Kintex-7)	1024	39.9 µs	53.7
[104]	FPGA (Virtex5)	256×256	9.32 µs	N/A
[102]	FPGA (Virtex6)	1024	340 µs	119
[109]	ASIC	256	10.17 µs	196
[113]	GPU (NVIDIA GTX480)	81922	15 ms	N/A

4.3. *Application Layer*

Although CS was developed as a sensing/compression paradigm, it can be extended to other domains such as signal detection and sparse problem optimization. Thus, for IoT platform scenario, the application layer relies on the data routed from the local processing layer (edge computing platform) to provide both object-based and event-based data analytics, such as data aggregation, classification, active detection, signal identification, smart cities application, cloud status monitoring and even CS-based recovery.

4.3.1. Multi-User Detection and Identification

Regarding users/devices activity, CS-based event detection approaches exploit the fact that the rate of the active sensing devices over the overall number of the sensing nodes is usually small, which can be seen as a sparse network architecture. Therefore, CS techniques have been exploited not only to detect the signal of interest but also to detect and identify the number of the active devices/users that contribute on the data aggregated at the cloud level. Zhu et al. [114] introduced a CS multi-user detection (MUD) detectors for communication systems with code division multiple access (CDMA) by adopting BP algorithm. Furthermore, CS-based MUD using greedy CS algorithms has been proposed in [115]. The authors studied the efficiency and reliability of applying CS to detect both the users activity and to identify data in a CDMA transmission. The obtained results in [115] indicate that the orthogonal least squares (OLS) algorithm is reliable for sparse MUD even in overloaded CDMA systems. However, it is shown in [115] that the symbol errors of the OLS are mainly caused by incorrect activity detection. These results imply that improving user activity detection would enhance the overall system performance. Thus, Schepker et al. [116] introduced block-wise orthogonal least squares (BOLS) detection as sparse MUD for sporadic communication using CDMA.

In [117,118] a CS-based two-stages framework for joint data identification and detection is proposed for IoT applications using multi-carrier CDMA (MC-CDMA) with data transmitting scheme based on sparse index multiple access (SIMA) [119]. First, CS is exploited to estimate the channel state information (CSI) to determine the number of the devices/users that are actually transmitting data using OMP. Next, using the estimated CSI, the transmitted data are detected using expectation-maximization (EM) algorithm. In addition, a proposed solution to detect the active users in large scale IoT by combining linear detection and CS recovery algorithm is presented in [120]. This approach aims to improve the receiver performance relying on linear detection and reducing the computational complexity by exploring efficient CS recovery algorithms.

Furthermore, Wang et al. [121] addressed the problem of MUD in massive IoT networks. The authors adopted structure matching pursuit (SMP) algorithm to jointly detect the common active users at all times, and then to detect the dynamic active users at each time slot individually. However, if the number of devices scales up remarkably, this method will suffer from a high system complexity and a slow reconstruction time. Therefore, Liu et al. [122] proposed a scalable CS-based MUD scheme to tackle this issue based on single measurement vector CS (SMV-CS) [123,124] which uses group OMP (GOMP) [125] to divide the data into different sub-blocks and reconstruct each sub-block individually.

4.3.2. CS-Based Cloud Storage

On the other hand, receiving big data has shifted the storing paradigm from single server to multiple mega-server and it is continuously growing. The massive number of data that are collected on a daily basis requires large-scale data processing engines to analyze. However, applications are rarely interested in raw data records. Instead, only the salient features are required for deep data analytics. CS can be explored efficiently to store the cloud data by reducing the data size and discarding the non-relevant features to facilitate different types of applications. Many IoT data analytic applications are executed by sending queries to the cloud server to retrieve data. For instance, Top-k, range, and keyword queries present the most popular types of queries for IoT and WSN applications. Top-k query means that the k highest (or lowest) data values are retrieved from a dataset [126]. Thus, by leveraging the fact that the CS compresses and preserves an approximate form of the data, Zhang et al. [127] proposed a new impression store cloud architecture based on CS that does not store all the raw data, but rather some features that can yield to provide a successful application in term of QoS. This approach saves storage capacity and bandwidth and allows for efficient and scalable parallel updates and queries. This approach consists of compressing the data to the minimum factor that allows the recovery of only the application-defined principal components. In addition, Chen et al. [128] a two-stage compression scheme. First, the data are compressed at the sensor level with an M samples. Afterwards, they wait for a query from the cloud to send only the $k \ll M$ samples that are required for a specific defined application.

4.3.3. Mobile Crowd Sensing

CS can be deployed efficiently for applications related to smart cities in order to monitor environmental and urban conditions using the so-called mobile crowd sensing (MCS) approaches. MCS explores the user's mobility; the sensors embedded within the users smart-phones and the existing wireless infrastructures are used to sense and collect environmental data. The collected data can relate to a number of phenomena, including air quality, noise level, street surface and pavement conditions. MCS consists of dividing a specific area to N different cells, where, at each cell, single or multiple users are allocated with different sensing tasks. The main drawback of conventional MCS approaches is that they incur a high sensing cost in term of energy consumption to achieve a high sensing quality. In addition, many participants are required to collect and transmit data which puts high constraints in term of bandwidth occupancy. Thus, sparse MCS presents an alternative solution to the conventional methods. Sparse MCS consists of allocating only a small number of cells ($M \ll N$) with participants to collect data. The aim of this approach is to estimate the data of the all N cells from the collected data from M cells. This can be seen as a CS basic problem by estimating an N-length data from an incomplete set of measurements. Sparse MCS approaches face two key challenges, first, how to select the cells that provide the optimum coverage, second, how to assess the estimated data quality without apriori knowledge of the data of the non-sensed cells. To address these challenges, wang et al. [129] presented a framework for sparse MCS application with an iterative task allocation process. In this framework, cells allocation operation is determined using uncertainty-based approach [130]. After transmitting data from the selected cells, the whole area conditions are estimated using spatiotemporal CS recovery algorithms [131]. The obtained results show a significant reduction in the sensing cost while still guaranteeing the overall data quality for urban sensing. Thereafter, several efforts have been made to further enhance the performance of sparse MCS in terms of privacy [132], cell allocations [133,134] and reducing the number of participants required for data collection [135].

4.3.4. Traffic Monitoring

Traffic monitoring is the crucial task that can provide great insights for the design and the management of infrastructures in highly urban cities. The state-of-the-art traffic monitoring approaches exploit roving vehicles and the mobility of smart-phones to provide periodic reports regarding traffic status, driving speeds and the flow direction to estimate the whole traffic conditions [136]. These approaches often employ an intensive number of users (probes) to cover the roads of interest. Nevertheless, their performances are usually limited by the number of users, energy expenditure of the sensing devices, privacy as well as the high cost associated with hiring many participants to provide the complete traffic map. However, the high dimensional datasets for traffic conditions can be well approximated with low-rank matrices by mining the hidden structure of this datasets. Zhu et al. [137] used principal component analysis to mine a large set of traffic data collected in China. Their analysis showed that the energy is concentrated in just a few principal components which underpin the applicability of CS-based traffic sensing. On the other hand, individual traffic reports are usually error-prone; for instance, if the car goes through obstructions (urban canyons and underpasses), the reliability of their GPS data would fade. In addition, with the continuous mobility of the probes, multi-path propagation phenomena will degrade the quality of wireless communication which will lead to huge amount of missing data. Thus, CS can be used to estimate the entire data (including the missing one) from the partially received traffic data points. The approach proposed in [137] has proven to provide an estimation error less than 20% even though 80% of the data were not received. Furthermore, by leveraging the fact that the traffic conditions among interconnected roads are highly dependent, the correlation model between these traffic conditions can be explored to reduce the number of users, while maintaining the same quality for the entire city traffic map. Liu et al. [138] presented a multiple linear regression (MLR) method based on CS to estimate the entire road traffic from a small set of measurements. The idea is to divide the map to N blocks relying on the intersecting points. Then, only a small number of blocks $M \ll N$ are assigned with j vehicles (probes) to measure the average velocity v_j and the distance to travel each road $L_j = [l_j^1, l_j^2, \cdots, l_j^M]$. Hence, to estimate the entire road velocities from the set of observed velocities $\{v_j\}_{j=0}^{j=M}$, CS recovery algorithms can be adopted. The authors compared their approach with state-of-the-art solutions based on singular value decomposition (SVD), the results showed that the proposed method achieved an estimation accuracy up to 80% while the SVD approach achieved only a 60% estimation accuracy.

Finally, more CS-based data analytic approaches can be addressed in that future. For instance, by leveraging the reconstruction reports, data mining and incremental learning algorithms [139] can be deployed at the cloud to determine the optimal number of measurements every node has to acquire to provide the best reconstruction performance.

Table 4 provides a summary for the extant efforts to integrate CS into IoT applications.

Table 4. Summary of the Joint CS-Based approaches for IoT applications.

IoT Layer	Approach	Features	Main Attribute
Sensing Layer	Adaptive measurements [76,77]	Optimize CF value	• Sparsity change detection based on QoS [76] • Sparsity change detection using CPM [77]
	CS-based sparse Networking [86–89]	Energy efficiency High reconstruction quality	• Alternate the selection of the active sensors at each time slot [86] • Nodes activation based on data reconstruct in the previous time slot [87] • Divide network into clusters, each with only one single reference node [89].
	Weighted measurements [79,80]	Optimize CF value	• Applicable on heterogeneous IoT platform • Assign for each sensor a CF value depending on its importance in the application
	CS-based data gathering [81,83,90,92]	Data gathering Low communication cost Low error rate	• CDG for multi-hope: Compress and route to the next node [81] • CSF: each node sparsifies the data and route the sparse functions to the sink [82] • (MECDA) approach: select the best route that consumes the minimum power [83]. • CDC: each sensor selects its route based on opportunistic routing protocols [84].
	CS-based Routing [90–92,140]	Extend nodes lifetime Better signal quality	• Build routing path by choosing sensing matrix that minimize the coherence with sparsyfing basis [90] • Utilize BP algorithm to recover the ensemble measurements [92] • Build the Sensing matrix by minimizing the coherence • Hybrid routing approach using PF and DCS [91] • Random walk (RN)with DCS [140]
Processing Layer	FPGA-based OMP implementation [99,100,102–109]	Fast execution High performance	• Adopt Q-R decomposition (QRD) to solve the LS problem [99] • Implement a parallel vector multiplication unit (VMU) to solve LS problem [100] • Cholesky factorization method to solve the LS problem to [102] • FFT approach to perform the correlation operation (coefficients selection) [105]
	CPU/GPU-based OMP implementation [99,110–112]	Easy implementation	• Matrix inversion update method [111] • LU decomposition for solving the LS problem [112]
Application layer	Multi-user detection [114–118,120,121]	Active users identification Enhance data detection	• User detection in networks with CDMA based transmission [114–116] • Joint identification and detection in massive IoT with MC-CDMA [117,118,120,121]
	Cloud data Storing [126–128]	Efficient storage	• CS-based Top-k query data retrieval approach [127] • A 2-stage compression architecture [128]
	Traffic Monitoring [136–138]	Reduce number of required probes low cost	• Multiple linear regression (MLR) mode based on CS to estimate the entire road traffic from a small set of measurements • Explore the correlation between the road status to reduce required probe of entire road map estimation
	Sparse MCS [89,132–135]	Low cost Reduce energy expenditure User privacy	• Allocate small of cells with specific monitoring tasks • Estimate the whole cell ensemble information from the small set of collected data • User privacy enforcement [132] • Reduce number of required participants for data collection [135] • Minimize the number of cells to collect information [133,134]

5. Challenges and Research Trends

Exploring the techniques and the approaches presented in Section 4 would provide an efficient CS-based IoT platform that can be used to enable numerous applications. However, many studies related to CS can be further incorporated. This section provides some of the research orientations and challenges that the CS research community is trying to investigate and to address. Although these efforts are introduced out of the IoT scope, they can fit well in the design and the development of an energy-efficient, real-time and secure IoT platforms. This section highlights the advantages of incorporating such techniques in IoT applications, where four of most prominent research orientations that can be well explored to enhance the IoT performance over the three different layer are discussed. For the sensing layer, using both energy-efficient CS encoders and structured sensing matrices would decrease the system complexity, hence provide energy efficiency. For the processing layer, using gateway embedded on many-core platforms instead of single-core can boost up the speed and the quality which will help meet the real-time requirements. Finally, incorporating CS-based encryption methods helps to achieve a high level of security and privacy for specific IoT application.

5.1. CS Encoder Design

The different techniques that explore CS at the sensing layer can contribute significantly to the design of energy-efficient CS-based IoT system by using the CS as a compression technique. However, to fully exploit CS theory, the sensing nodes have to collect directly a compressed form of the signal. Therefore, it is of high importance to design CS-based sensing nodes.

In fact, one of the main attributes of CS is to shift the high complexity burden from the IoT sensing devices to the gateways that are usually embedded on platforms with much-relaxed computation, communication and energy efficiency abilities. CS is expected to reduce the power consumption of the sensing and the transmission processes. Unlike conventional compression techniques which require some processing to extract the salient information, CS encoding can be implemented in two different modes, analog or digital domains. In analog CS, the linear projection is applied in the analog domain prior to digitization, whereas the digital encoder first quantizes the signal samples and then performs the equivalent modulations using digital logic. CS-based IoT applications require sensing nodes with low-power consumption design, thus they could benefit from research that aims to develop a low rate-energy efficient CS acquisition sensors [36,37,141,142]:

- Analog CS [143,144] represents the hardware implementation of the CS acquisition model (Equation (1)) in which the signals are acquired at the sub-Nyquist rate. The CS-based acquired signal is the inner product between the input signal and M random vectors. In fact, the analog CS encoder is designed using random modulator (RM) [145]. The RM is implemented using a mixer, an integrator, and an analog to digital converter (ADC). The mixer performs the inner product between the signal and the measurement matrix in a sub-Nyquist rate. The integrator accumulates the output voltage of the mixer, and it has to be reset after each sample is taken. Finally, the signal is sampled at rate $1/M$ using the ADC. For more illustrations and implementations, the reader can refer to [146,147].
- Digital CS is performed by sampling the input signal following the Shannon–Nyquist theorem, and then performing M random modulation. In addition, a non-uniform sampler (NUS) [148] technique can be used, where the CS encoder picks an M samples randomly from the whole N dimension vector after the digital conversion. The NUS can be seen as an RM modulator with binary sensing matrix with elements $\{0,1\}$.
- Bellasi et al. [37] examined the implementation of both analog and digital encoders and showed that an inexpensive and energy-efficient digital logic is most suitable to implement CS-based data reduction. Moreover, they investigated two scenarios that can occur often in WSN. In the first scenario, the energy consumption of storage/transmission is dominating the total power balance, thus the superior compression performance of analog CS leads to a significant advantage

over digital CS. Instead, in scenarios where signal acquisition and processing are dominant, digital encoders are indeed more energy-efficient. The results hold great promises to extend CS application of CS encoder from small WSNs to massive IoT platforms.

5.2. Structured Sensing Matrix

A crucial task in CS is to select the appropriate sensing matrix, which has a crucial impact on the reconstruction performance. Moreover, selecting inappropriate sensing matrix would lead to additional system complexity and may provide unacceptable reconstruction quality.

In CS, the sensing matrix is usually designed using a dense random matrix with entries drawn from an independent identical sub-Gaussian distribution [149,150]. Gaussian processes are widely used as sensing matrices in CS; they offer a low coherence with most of the natural basis used in signal processing [148]. Moreover, they satisfy the RIP condition with high probability [151]. However, they suffer from several limitations in term of implementation on hardware due to their dense nature. Consequently, they can be implemented using a non-chip random seed to generate all the entries or they can be generated offline and stored in a large memory [152]. In addition, they slow down the measurement process of CS because it has to deal with all non zero entries. Thus, the random matrix measurement operator must be replaced by more structured sensing architectures that correspond to the characteristics of feasible acquisition hardware.

Therefore, using structured sensing matrices would improve both the sensing and the reconstruction processes. At the sensing level, deploying sparse Bernoulli matrices with entries $\{0, 1\}$ can reduce the number of multiplication operations by replacing the inner product by just an accumulation process [153], which can be seen as an NUS approach.

Furthermore, structured sensing matrices would accelerate the reconstruction time and minimize the onboard computation on the receiver side. Therefore, incorporating such techniques aligns perfectly with the vision of providing efficient CS-based IoT platform.

Using Bernoulli matrices with entries $\{1, -1\}$ [71] can be seen as good fit for FPGA reconstruction, where the multiplication process can be replaced by data accumulation only. Several works have also focused on structured sensing matrices such as Toeplitz, circular, block diagonal and permuted block diagonal (BPBD) [154] and deterministic binary bloc-diagonal (DBBD) [152]. These structured matrices have been shown to be hardware friendly by discarding the need to implement on-chip random seed for random matrices [152].

5.3. Implementation on Multi-Core Platforms

With the fast increase in the number of IoT devices, the big data transmitted to the computing platforms will put a lot of pressure on the real-time requirements and the processing will exceed the capabilities of single-core platforms. This scenario occurs in several IoT applications that require real-time video streaming. Subsequently, single-core platforms will soon fade and be replaced with multi-core platforms that have more relaxed computation and communication features [155]. Moreover, most IoT applications would utilize the end-user personal computer and mobile as the edge computing platform; thus, the implementation of reconstruction algorithms on multi-core platforms that are similar to the ones held by the end-user has been addressed in the literature for different applications. For instance, taking advantage of parallel computing, Borghi et al. presented in [106] the design and investigated the performance of BP algorithm on four different multi-core platforms, namely, Intel Core 2 Quad, Intel Core I7, PS3 cell and NVIDIA 8800 GTS. In addition, medical imaging applications have witnessed several implementation of CS reconstruction algorithms on multi-core platforms [156–159]. More recently, Bortolotti et al. presented a hardware implementation of OMP on a heterogeneous mobile SoC based on the ARM BIG.LITTLE TM architecture for CS-based ECG monitoring [160]. The proposed architecture shows to be able to provide a real-time reconstruction for ECG signals which can fit well into IoT applications for long-term healthcare monitoring.

5.4. Data Security

Enabling secure communications is an imperative task in IoT applications. In the design of IoT platforms, the sensitive data (image, video, Medical records, etc.) transmitted over the wireless channel have to be secured against unauthorized access. Subsequently, data privacy is a paramount challenge that has to be addressed in IoT applications. A reliable solution can be realized by means of sophisticated state-of-the-art data encryption techniques. Nevertheless, CS can also be well suited to act as an encryption mechanism by exploiting the fact that randomly mapping the original data to another one would encrypt the information. Exploring CS as an encryption technique by using the random sensing matrix as the key has been investigated in [161], where the authors showed that even-though a perfect secrecy is not guaranteed, computational secrecy can be achieved as long as the sensing matrix satisfies the RIP and the number of random projections is at least twice the sparsity of the signal. In addition, Kailkhura et al. [162] showed that the design of an optimal sensing matrix in the context of full secrecy is not universal but it depends mainly on the structure of the signal of interest.

In addition, WSNs have witnessed some applications where CS has been used as both acquisition and encryption scheme. Wang et al. proposed a two-level CS-based security platform to transmit the data from the sensors to the cloud [163]. The first level of encryption is performed by acquiring data following the CS concept. Afterwards, the sensors further encrypt the compressed data without changing its dimension so no additional bandwidth cost is required. Recently, Peng et al. [164] proposed a chaotic CS (CCS) scheme to tackle both energy saving and data security problems in large-scale wireless body area networks (WBANs). The aforementioned schemes are designed mainly to prevent external access to the data. However, to face attacks that can occur within the system (inference attacks) such as controllable event triggering attacks (CETA) and random event triggering attacks (RETA), Hu et al. [165] presented a method called secure compressive data gathering (SCDG). The presented approach is based on changing the sensing matrix coefficients at each time slot.

In addition, cloud security present as well a hot topic to address [166] and CS techniques have shown to disclose great opportunities. For instance, a CS-based encryption scheme for crowd sensing to enable efficient remote sensing system (RSS) maps are presented in [167]. In addition, Wang et al. [168] presented a CS-based framework for preserving the privacy of reconstructed image when operating in outsourcing mode. Cloud outsourcing refers to shifting the data management from the embedded servers to the cloud servers. This framework would be very beneficial for application of connected health-care where the medical images with diagnostic results for different patients are privacy-sensitive. Reader should refer to [169–173] for variety of CS-based frameworks to enable secure cloud operations.

6. Conclusions

IoT applications require a huge number of heterogeneous smart devices to sense, communicate and collaborate with each other in order to gather the maximum information to enable numerous types of services. A large number of sensors will generate big redundant data that would cause unnecessary network traffic, thus degrading the overall system performance. Therefore, exploring CS can reduce the number of data collected and can release the pressure on the wireless communication.

This paper provides a brief overview on CS theory, starting from the basic model of the signal acquisition to the criteria that should be satisfied while designing the sensing matrix, as well as listing some of the most considered reconstruction algorithms used in the literature.

Moreover, this paper reviews the extant work presented in the literature that addresses the integration of CS techniques in the IoT platforms. To this end, and aiming to make the discussion clear and easy to follow, the IoT platform is divided into three layers and the related CS researches for each layer is discussed thoroughly.

The acquisition layer consists of the huge number of sensors that communicate with each other and with a central fusion node. High power consumption presents the most important issue to tackle. Thus, CS techniques can be deployed as a compression technique to reduce the power consumption.

In addition, compressive gathering, sparse networking and hybrid joint DCS-routing protocols can also be explored to achieve energy-efficient sensing layer.

At the processing unit, implementing edge computing platform on FPGAs presents an appealing solution to achieve real-time performance. Several design and architectures for CS reconstruction algorithms have already been proposed and investigated in the literature.

For the application layer, the utilization of CS can be extended from being just a compression paradigm to a broad range of applications by exploring the recovery algorithms associated with CS to enable several applications related to smart cities applications and next generation communication systems.

While it is hard to cover all of the CS software and hardware research that can be integrated with the realm of IoT, this paper highlights the open issues and future trends that can be addressed exploring CS techniques along with some powerful computing platforms to stimulate new research orientation in the field of IoT. The realization of CS encoder that provides directly a compressed signal acquisition is a hot topic to investigate as it enables the next era of digital sampling. In addition, using structured sensing matrices would simplify the acquisition operation and allow real-time data reconstruction. Moreover, providing a CS implementation on commercial multi-core platforms can be well fitted for different IoT applications related to smart home and health care. At the top of all that, CS can be used to enable secure and encrypted transmission scheme by exploring the randomness accommodated with acquisition process without introducing any additional processing.

Acknowledgments: We would like to thank Hamza Baali from Hamad Bin Khalifa University (HBKU), Qatar, for his valuable discussion related to this paper.

References

1. Alaba, F.A.; Othman, M.; Hashem, I.A.T.; Alotaibi, F. Internet of things security: A survey. *J. Netw. Comput. Appl.* **2017**, *88*, 10–28. [CrossRef]
2. Kortuem, G.; Kawsar, F.; Sundramoorthy, V.; Fitton, D. Smart objects as building blocks for the Internet of things. *IEEE Internet Comput.* **2010**, *14*, 44–51. [CrossRef]
3. Holler, J.; Tsiatsis, V.; Mulligan, C.; Avesand, S.; Karnouskos, S.; Boyle, D. *From Machine-to-Machine to the Internet of Things: Introduction to a New Age of Intelligence*; Academic Press: Cambridge, MA, USA, 2014.
4. Lazarescu, M.T. Design of a WSN platform for long-term environmental monitoring for IoT applications. *IEEE J. Emerg. Sel. Top. Circuits Syst.* **2013**, *3*, 45–54. [CrossRef]
5. Yang, G.; Xie, L.; Mäntysalo, M.; Zhou, X.; Pang, Z.; Da Xu, L.; Kao-Walter, S.; Chen, Q.; Zheng, L.R. A health-iot platform based on the integration of intelligent packaging, unobtrusive bio-sensor, and intelligent medicine box. *IEEE Trans. Ind. Inform.* **2014**, *10*, 2180–2191. [CrossRef]
6. Donoho, D.L. Compressed sensing. *IEEE Trans. Inf. Theory* **2006**, *52*, 1289–1306. [CrossRef]
7. Candès, E.J.; Romberg, J.; Tao, T. Robust uncertainty principles: Exact signal reconstruction from highly incomplete frequency information. *IEEE Trans. Inf. Theory* **2006**, *52*, 489–509. [CrossRef]
8. Zhang, Z.; Jung, T.P.; Makeig, S.; Rao, B.D. Compressed sensing for energy-efficient wireless telemonitoring of noninvasive fetal ECG via block sparse Bayesian learning. *IEEE Trans. Biomed. Eng.* **2013**, *60*, 300–309. [CrossRef] [PubMed]
9. Jaspan, O.N.; Fleysher, R.; Lipton, M.L. Compressed sensing MRI: A review of the clinical literature. *Br. J. Radiol.* **2015**, *88*, 20150487. [CrossRef] [PubMed]
10. Djelouat, H.; Baali, H.; Amira, A.; Bensaali, F. IoT Based Compressive Sensing for ECG Monitoring. In Proceedings of the 2017 IEEE International Conference on Internet of Things (iThings) and IEEE Green Computing and Communications (GreenCom) and IEEE Cyber, Physical and Social Computing (CPSCom) and IEEE Smart Data (SmartData), Exeter, UK, 21–23 June 2017; IEEE: Piscataway, NJ, USA, 2017; pp. 183–189.

11. Djelouat, H.; Baali, H.; Amira, A.; Bensaali, F. An Adaptive Joint Sparsity Recovery for Compressive Sensing Based EEG System. *Wirel. Commun. Mob. Comput.* **2017**, *2017*, 9823684. [CrossRef]
12. Djelouat, H.; Zhai, X.; Al Disi, M.; Amira, A.; Bensaali, F. System-on-Chip Solution for Patients Biometric: A Compressive Sensing-Based Approach. *IEEE Sens. J.* **2018**. [CrossRef]
13. Duarte-Carvajalino, J.M.; Sapiro, G. Learning to sense sparse signals: Simultaneous sensing matrix and sparsifying dictionary optimization. *IEEE Trans. Image Process.* **2009**, *18*, 1395–1408. [CrossRef] [PubMed]
14. Li, S.; Da Xu, L.; Wang, X. Compressed sensing signal and data acquisition in wireless sensor networks and Internet of things. *IEEE Trans. Ind. Inform.* **2013**, *9*, 2177–2186. [CrossRef]
15. Karakus, C.; Gurbuz, A.C.; Tavli, B. Analysis of energy efficiency of compressive sensing in wireless sensor networks. *IEEE Sens. J.* **2013**, *13*, 1999–2008. [CrossRef]
16. Fazel, F.; Fazel, M.; Stojanovic, M. Random access compressed sensing for energy-efficient underwater sensor networks. *IEEE J. Sel. Areas Commun.* **2011**, *29*, 1660–1670. [CrossRef]
17. *ITU-T Y.2060: Overview of the Internet of Things*; ITU: Geneva, Switzerland, 2012.
18. Da Xu, L.; He, W.; Li, S. Internet of things in industries: A survey. *IEEE Trans. Ind. Inform.* **2014**, *10*, 2233–2243.
19. Autili, M.; Inverardi, P.; Tivoli, M. Choreography realizability enforcement through the automatic synthesis of distributed coordination delegates. *Sci. Comput. Program.* **2018**, *160*, 3–29. [CrossRef]
20. Zahariadis, T.; Papadimitriou, D.; Tschofenig, H.; Haller, S.; Daras, P.; Stamoulis, G.D.; Hauswirth, M. Towards a future Internet architecture. In *The Future Internet Assembly*; Springer: Beilin, Germany, 2011; pp. 7–18.
21. Miorandi, D.; Sicari, S.; De Pellegrini, F.; Chlamtac, I. Internet of things: Vision, applications and research challenges. *Ad Hoc Netw.* **2012**, *10*, 1497–1516. [CrossRef]
22. Domingo, M.C. An overview of the Internet of Things for people with disabilities. *J. Netw. Comput. Appl.* **2012**, *35*, 584–596. [CrossRef]
23. Jia, X.; Feng, Q.; Fan, T.; Lei, Q. RFID technology and its applications in Internet of Things (IoT). In Proceedings of the 2012 2nd International Conference on Consumer Electronics, Communications and Networks (CECNet), Yichang, China, 21–23 April 2012; IEEE: Piscataway, NJ, USA, 2012; pp. 1282–1285.
24. Liu, C.H.; Yang, B.; Liu, T. Efficient naming, addressing and profile services in Internet-of-Things sensory environments. *Ad Hoc Netw.* **2014**, *18*, 85–101. [CrossRef]
25. Wu, M.; Lu, T.J.; Ling, F.Y.; Sun, J.; Du, H.Y. Research on the architecture of Internet of things. In Proceedings of the 2010 3rd International Conference on Advanced Computer Theory and Engineering (ICACTE), Chengdu, China, 20–22 August 2010; IEEE: Piscataway, NJ, USA, 2010; Volume 5, pp. V5–V484.
26. Al-Fuqaha, A.; Guizani, M.; Mohammadi, M.; Aledhari, M.; Ayyash, M. Internet of things: A survey on enabling technologies, protocols, and applications. *IEEE Commun. Surv. Tutor.* **2015**, *17*, 2347–2376. [CrossRef]
27. Schaffers, H.; Komninos, N.; Pallot, M.; Trousse, B.; Nilsson, M.; Oliveira, A. Smart cities and the future Internet: Towards cooperation frameworks for open innovation. In *The Future Internet Assembly*; Springer: Beilin, Germany, 2011; pp. 431–446.
28. Zanella, A.; Bui, N.; Castellani, A.; Vangelista, L.; Zorzi, M. Internet of things for smart cities. *IEEE Internet Things J.* **2014**, *1*, 22–32. [CrossRef]
29. He, W.; Yan, G.; Da Xu, L. Developing vehicular data cloud services in the IoT environment. *IEEE Trans. Ind. Inform.* **2014**, *10*, 1587–1595. [CrossRef]
30. Chunli, L. Intelligent transportation based on the Internet of things. In Proceedings of the 2012 2nd International Conference on Consumer Electronics, Communications and Networks (CECNet), Yichang, China, 21–23 April 2012; IEEE: Piscataway, NJ, USA, 2012; pp. 360–362.
31. Huang, L.; Liu, C. The application mode in urban transportation management based on Internet of things. In Proceedings of the 2nd International Conference on Computer Science and Electronics Engineering, ICCSEE, Hangzhou, China, 22–23 March 2013; pp. 1226–1229.
32. Ji, Z.; Anwen, Q. The application of Internet of things(IOT) in emergency management system in China. In Proceedings of the 2010 IEEE International Conference on Technologies for Homeland Security (HST), Waltham, MA, USA, 8–10 November 2010; pp. 139–142.

33. Liu, Z.; Yan, T. Study on multi-view video based on IOT and its application in intelligent security system. In Proceedings of the 2013 International Conference on Mechatronic Sciences, Electric Engineering and Computer (MEC), Shenyang, China, 20–22 December 2013; IEEE: Piscataway, NJ, USA, 2013; pp. 1437–1440.
34. Chianese, A.; Piccialli, F. Designing a smart museum: When cultural heritage joins IoT. In Proceedings of the 2014 Eighth International Conference on Next Generation Mobile Apps, Services and Technologies (NGMAST), Oxford, UK, 10–12 September 2014; IEEE: Piscataway, NJ, USA, 2014; pp. 300–306.
35. Kos, A.; Sedlar, U.; Sterle, J.; Volk, M.; Bešter, J.; Bajec, M. Network monitoring applications based on IoT system. Network and Optical Communications (NOC). In Proceedings of the 18th European Conference on Network and Optical Communications (NOC 2013), 8th Conference on Optical Cabling and Infrastructure (OC&I 2013), Graz, Austria, 10–12 July 2013; IEEE: Piscataway, NJ, USA, 2013; pp. 69–74.
36. Chen, F.; Chandrakasan, A.P.; Stojanovic, V.M. Design and analysis of a hardware-efficient compressed sensing architecture for data compression in wireless sensors. *IEEE J. Solid-State Circuits* **2012**, *47*, 744–756. [CrossRef]
37. Bellasi, D.E.; Benini, L. Energy-efficiency analysis of analog and digital compressive sensing in wireless sensors. *IEEE Trans. Circuits Syst. I Regul. Pap.* **2015**, *62*, 2718–2729. [CrossRef]
38. Donoho, D.L.; Huo, X. Uncertainty principles and ideal atomic decomposition. *IEEE Trans. Inf. Theory* **2001**, *47*, 2845–2862. [CrossRef]
39. Muthukrishnan, S. *Data Streams: Algorithms and Applications*; Now Publishers Inc.: Delft, The Netherlands, 2005.
40. Chen, S.S.; Donoho, D.L.; Saunders, M.A. Atomic decomposition by basis pursuit. *SIAM J. Sci. Comput.* **1998**, *20*, 33–61. [CrossRef]
41. Candes, E.J. The restricted isometry property and its implications for compressed sensing. *Comptes Rendus Math.* **2008**, *346*, 589–592. [CrossRef]
42. Candes, E.; Tao, T. The Dantzig selector: Statistical estimation when p is much larger than n. *Ann. Stat.* **2007**, *35*, 2313–2351. [CrossRef]
43. Figueiredo, M.A.; Nowak, R.D.; Wright, S.J. Gradient projection for sparse reconstruction: Application to compressed sensing and other inverse problems. *IEEE J. Sel. Top. Signal Process.* **2007**, *1*, 586–597. [CrossRef]
44. Mallat, S.G.; Zhang, Z. Matching pursuits with time-frequency dictionaries. *IEEE Trans. Signal Process.* **1993**, *41*, 3397–3415. [CrossRef]
45. Krstulovic, S.; Gribonval, R. MPTK: Matching pursuit made tractable. In Proceedings of the 2006 IEEE International Conference on Acoustics, Speech and Signal Processing (ICASSP), Toulouse, France, 14–19 May 2006; IEEE: Piscataway, NJ, USA, 2006; Volume 3, p. III.
46. Tropp, J.A.; Gilbert, A.C. Signal recovery from random measurements via orthogonal matching pursuit. *IEEE Trans. Inf. Theory* **2007**, *53*, 4655–4666. [CrossRef]
47. Needell, D.; Tropp, J.A. CoSaMP: Iterative signal recovery from incomplete and inaccurate samples. *Appl. Comput. Harmon. Anal.* **2009**, *26*, 301–321. [CrossRef]
48. Dai, W.; Milenkovic, O. Subspace pursuit for compressive sensing signal reconstruction. *IEEE Trans. Inf. Theory* **2009**, *55*, 2230–2249. [CrossRef]
49. Needell, D.; Vershynin, R. Uniform uncertainty principle and signal recovery via regularized orthogonal matching pursuit. *Found. Comput. Math.* **2009**, *9*, 317–334. [CrossRef]
50. Donoho, D.L.; Tsaig, Y.; Drori, I.; Starck, J.L. Sparse solution of underdetermined systems of linear equations by stagewise orthogonal matching pursuit. *IEEE Trans. Inf. Theory* **2012**, *58*, 1094–1121. [CrossRef]
51. Liu, E.; Temlyakov, V.N. The orthogonal super greedy algorithm and applications in compressed sensing. *IEEE Trans. Inf. Theory* **2012**, *58*, 2040–2047. [CrossRef]
52. Blanchard, J.D.; Tanner, J. Performance comparisons of greedy algorithms in compressed sensing. *Numer. Linear Algebra Appl.* **2015**, *22*, 254–282. [CrossRef]
53. Cartis, C.; Thompson, A. A new and improved quantitative recovery analysis for iterative hard thresholding algorithms in compressed sensing. *IEEE Trans. Inf. Theory* **2015**, *61*, 2019–2042. [CrossRef]
54. Foucart, S.; Rauhut, H. *A Mathematical Introduction to Compressive Sensing*; Birkhäuser: Basel, Switzerland, 2013; Volume 1.

55. Avonds, Y.; Liu, Y.; Van Huffel, S. Simultaneous greedy analysis pursuit for compressive sensing of multi-channel ECG signals. In Proceedings of the 2014 36th Annual International Conference of the IEEE Engineering in Medicine and Biology Society (EMBC), Chicago, IL, USA, 6–30 August 2014; IEEE: Piscataway, NJ, USA, pp. 6385–6388.
56. Baron, D.; Duarte, M.F.; Wakin, M.B.; Sarvotham, S.; Baraniuk, R.G. Distributed compressive sensing. *arXiv* **2009**, arXiv:0901.3403.
57. Tropp, J.A. Algorithms for simultaneous sparse approximation. Part II: Convex relaxation. *Signal Process.* **2006**, *86*, 589–602. [CrossRef]
58. Tropp, J.A.; Gilbert, A.C.; Strauss, M.J. Algorithms for simultaneous sparse approximation. Part I: Greedy pursuit. *Signal Process.* **2006**, *86*, 572–588. [CrossRef]
59. He, D.; Zeadally, S. An analysis of RFID authentication schemes for Internet of things in healthcare environment using elliptic curve cryptography. *IEEE Internet Things J.* **2015**, *2*, 72–83. [CrossRef]
60. Rogers, A.; Jones, E.; Oleynikov, D. Radio frequency identification (RFID) applied to surgical sponges. *Surg. Endosc.* **2007**, *21*, 1235–1237. [CrossRef] [PubMed]
61. Amendola, S.; Lodato, R.; Manzari, S.; Occhiuzzi, C.; Marrocco, G. RFID technology for IoT-based personal healthcare in smart spaces. *IEEE Internet Things J.* **2014**, *1*, 144–152. [CrossRef]
62. Khan, M.S.; Islam, M.S.; Deng, H. Design of a reconfigurable RFID sensing tag as a generic sensing platform toward the future Internet of things. *IEEE Internet Things J.* **2014**, *1*, 300–310. [CrossRef]
63. Attaran, A.; Rashidzadeh, R. Chipless Radio Frequency Identification Tag for IoT Applications. *IEEE Internet Things J.* **2016**, *3*, 1310–1318. [CrossRef]
64. Islam, S.M.R.; Kwak, D.; Kabir, M.H.; Hossain, M.; Kwak, K.S. The Internet of Things for Health Care: A Comprehensive Survey. *IEEE Access* **2015**, *3*, 678–708. [CrossRef]
65. Mainetti, L.; Patrono, L.; Vilei, A. Evolution of wireless sensor networks towards the Internet of things: A survey. In Proceedings of the 2011 19th International Conference on Software, Telecommunications and Computer Networks (SoftCOM), Split, Croatia, 15–17 September 2011; IEEE: Piscataway, NJ, USA, 2011; pp. 1–6.
66. Christin, D.; Reinhardt, A.; Mogre, P.S.; Steinmetz, R. Wireless sensor networks and the Internet of things: Selected challenges. In Proceedings of the 2009 8th GI/ITG Kuvs FachgesprÄch Drahtlose Sensornetze, Hamburg, Germany, 13–14 August 2009; pp. 31–34.
67. Colitti, W.; Steenhaut, K.; De Caro, N. Integrating wireless sensor networks with the web. In Proceedings of the Extending Internet Low Power Lossy Netw (IP+ SN 2011), Chicago, IL, USA, 11 April 2011.
68. Harrison, R.R.; Watkins, P.T.; Kier, R.J.; Lovejoy, R.O.; Black, D.J.; Greger, B.; Solzbacher, F. A Low-Power Integrated Circuit for a Wireless 100-Electrode Neural Recording System. *IEEE J. Solid-State Circuits* **2007**, *42*, 123–133. [CrossRef]
69. Razzaque, M.A.; Bleakley, C.; Dobson, S. Compression in wireless sensor networks: A survey and comparative evaluation. *ACM Trans. Sens. Netw. (TOSN)* **2013**, *10*, 5. [CrossRef]
70. Verma, N.; Shoeb, A.; Bohorquez, J.; Dawson, J.; Guttag, J.; Chandrakasan, A.P. A micro-power EEG acquisition SoC with integrated feature extraction processor for a chronic seizure detection system. *IEEE J. Solid-State Circuits* **2010**, *45*, 804–816. [CrossRef]
71. Mamaghanian, H.; Khaled, N.; Atienza, D.; Vandergheynst, P. Compressed sensing for real-time energy-efficient ECG compression on wireless body sensor nodes. *IEEE Trans. Biomed. Eng.* **2011**, *58*, 2456–2466. [CrossRef] [PubMed]
72. Gedik, B.; Liu, L.; Yu, P.S. ASAP: An Adaptive Sampling Approach to Data Collection in Sensor Networks. *IEEE Trans. Parallel Distrib. Syst.* **2007**, *18*, 1766–1783. [CrossRef]
73. Abate, F.; Huang, V.K.L.; Monte, G.; Paciello, V.; Pietrosanto, A. A Comparison Between Sensor Signal Preprocessing Techniques. *IEEE Sens. J.* **2015**, *15*, 2479–2487. [CrossRef]
74. Monte, G. Sensor signal preprocessing techniques for analysis and prediction. In Proceedings of the 2008 34th Annual Conference of IEEE Industrial Electronics, IECON 2008, Orlando, FL, USA, 11–13 November 2008; IEEE: Piscataway, NJ, USA, 2008; pp. 1788–1793.
75. Monte, G.; Huang, V.; Liscovsky, P.; Marasco, D.; Agnello, A. Standard of things, first step: Understanding and normalizing sensor signals. In Proceedings of the 2013 39th Annual Conference of the IEEE Industrial Electronics Society, IECON, Vienna, Austria, 10–13 November 2013; IEEE: Piscataway, NJ, USA, 2013; pp. 118–123.

76. Fragkiadakis, A.; Charalampidis, P.; Tragos, E. Adaptive compressive sensing for energy efficient smart objects in IoT applications. In Proceedings of the 2014 4th International Conference on Wireless Communications, Vehicular Technology, Information Theory and Aerospace & Electronic Systems (VITAE), Aalborg, Denmark, 11–14 May 2014; IEEE: Piscataway, NJ, USA, 2014; pp. 1–5.
77. Charalampidis, P.; Fragkiadakis, A.G.; Tragos, E.Z. Rate-adaptive compressive sensing for iot applications. In Proceedings of the 2015 IEEE 81st Vehicular Technology Conference (VTC Spring), Glasgow, Scotland, 11–14 May 2015; IEEE: Piscataway, NJ, USA, 2015; pp. 1–5.
78. Peel, L.; Clauset, A. Detecting change points in the large-scale structure of evolving networks. *arXiv* **2014**, arXiv:1403.0989.
79. Lee, H.; Oh, H.; Lee, S.; Bovik, A.C. Visually weighted compressive sensing: Measurement and reconstruction. *IEEE Trans. Image Process.* **2013**, *22*, 1444–1455. [PubMed]
80. Lee, H.; Jo, S.K.; Lee, N.; Lee, H.W. A method for co-existing heterogeneous IoT environments based on compressive sensing. In Proceedings of the 2016 18th International Conference on Advanced Communication Technology (ICACT), Pyeongchang, Korea, 31 January–31 February 2016; IEEE: Piscataway, NJ, USA, 2016; pp. 206–209.
81. Luo, C.; Wu, F.; Sun, J.; Chen, C.W. Efficient Measurement Generation and Pervasive Sparsity for Compressive Data Gathering. *IEEE Trans. Wirel. Commun.* **2010**, *9*, 3728–3738. [CrossRef]
82. Xu, L.; Qi, X.; Wang, Y.; Moscibroda, T. Efficient data gathering using Compressed Sparse Functions. In Proceedings of the 2013 Proceedings IEEE INFOCOM, Turin, Italy, 14–19 April 2013; pp. 310–314.
83. Xiang, L.; Luo, J.; Vasilakos, A. Compressed data aggregation for energy efficient wireless sensor networks. In Proceedings of 2011 8th Annual IEEE Communications Society Conference on Sensor, Mesh and Ad Hoc Communications and Networks (SECON), Salt Lake City, UT, USA, 27–30 June 2011; IEEE: Piscataway, NJ, USA, 2011; pp. 46–54.
84. Liu, X.Y.; Zhu, Y.; Kong, L.; Liu, C.; Gu, Y.; Vasilakos, A.V.; Wu, M.Y. CDC: Compressive data collection for wireless sensor networks. *IEEE Trans. Parallel Distrib. Syst.* **2015**, *26*, 2188–2197. [CrossRef]
85. Biswas, S.; Morris, R. Opportunistic routing in multi-hop wireless networks. *ACM Sigcomm Comput. Commun. Rev.* **2004**, *34*, 69–74. [CrossRef]
86. Du, R.; Gkatzikis, L.; Fischione, C.; Xiao, M. Energy efficient sensor activation for water distribution networks based on compressive sensing. *IEEE J. Sel. Areas Commun.* **2015**, *33*, 2997–3010. [CrossRef]
87. Chen, W.; Wassell, I.J. Optimized node selection for compressive sleeping wireless sensor networks. *IEEE Trans. Veh. Technol.* **2016**, *65*, 827–836. [CrossRef]
88. Cao, D.Y.; Yu, K.; Zhuo, S.G.; Hu, Y.H.; Wang, Z. On the implementation of compressive sensing on wireless sensor network. In Proceedings of the 2016 IEEE First International Conference on Internet-of-Things Design and Implementation (IoTDI), Berlin, Germany, 4–8 April 2016; IEEE: Piscataway, NJ, USA, 2016; pp. 229–234.
89. Zhang, B.; Liu, Y.; He, J.; Zou, Z. An energy efficient sampling method through joint linear regression and compressive sensing. In Proceedings of the 2013 Fourth International Conference on Intelligent Control and Information Processing (ICICIP), Beijing, China, 9–11 June 2013; pp. 447–450.
90. Lee, S.; Pattem, S.; Sathiamoorthy, M.; Krishnamachari, B.; Ortega, A. *Compressed Sensing and Routing in Multi-Hop Networks*; University of Southern California CENG Technical Report; University of Southern California: Los Angeles, CA, USA, 2009.
91. Caione, C.; Brunelli, D.; Benini, L. Distributed compressive sampling for lifetime optimization in dense wireless sensor networks. *IEEE Trans. Ind. Inform.* **2012**, *8*, 30–40. [CrossRef]
92. Masoum, A.; Meratnia, N.; Havinga, P.J. A distributed compressive sensing technique for data gathering in wireless sensor networks. *Procedia Comput. Sci.* **2013**, *21*, 207–216. [CrossRef]
93. Zhang, B.; Mor, N.; Kolb, J.; Chan, D.S.; Lutz, K.; Allman, E.; Wawrzynek, J.; Lee, E.A.; Kubiatowicz, J. The Cloud is Not Enough: Saving IoT from the Cloud. In Proceedings of the HotCloud 2015, Santa Clara, CA, USA, 6–7 July 2015.
94. Datta, S.K.; Bonnet, C.; Nikaein, N. An IoT gateway centric architecture to provide novel M2M services. In Proceedings of the 2014 IEEE World Forum on Internet of Things (WF-IoT), Seoul, Korea, 6–8 March 2014; IEEE: Piscataway, NJ, USA, 2014; pp. 514–519.
95. Piyare, R. Internet of things: Ubiquitous home control and monitoring system using android based smart phone. *Int. J. Internet Things* **2013**, *2*, 5–11.

96. Samie, F.; Bauer, L.; Henkel, J. IoT technologies for embedded computing: A survey. In Proceedings of the 2016 International Conference on Hardware/Software Codesign and System Synthesis (CODES + ISSS), Pittsburgh, PA, USA, 2–7 October 2016; IEEE: Piscataway, NJ, USA, 2016; pp. 1–10.
97. Brzoza-Woch, R.; Nawrocki, P. FPGA-Based Web Services–Infinite Potential or a Road to Nowhere? *IEEE Internet Comput.* **2016**, *20*, 44–51. [CrossRef]
98. Septimus, A.; Steinberg, R. Compressive sampling hardware reconstruction. In Proceedings of the 2010 IEEE International Symposium on Circuits and Systems (ISCAS), Paris, France, 30 May–2 June 2010; IEEE: Piscataway, NJ, USA, 2010; pp. 3316–3319.
99. Stanislaus, J.L.; Mohsenin, T. High performance compressive sensing reconstruction hardware with QRD process. In Proceedings of the 2012 IEEE International Symposium on Circuits and Systems (ISCAS), Seoul, Korea, 20–23 May 2012; IEEE: Piscataway, NJ, USA, 2012; pp. 29–32.
100. Bai, L.; Maechler, P.; Muehlberghuber, M.; Kaeslin, H. High-speed compressed sensing reconstruction on FPGA using OMP and AMP. In Proceedings of the 2012 19th IEEE International Conference on Electronics, Circuits and Systems (ICECS), Seville, Spain, 9–12 December 2012; IEEE: Piscataway, NJ, USA, 2012; pp. 53–56.
101. Ren, F.; Dorrace, R.; Xu, W.; Marković, D. A single-precision compressive sensing signal reconstruction engine on FPGAs. In Proceedings of the 2013 23rd International Conference on Field Programmable Logic and Applications (FPL), Porto, Portugal, 2–4 September 2013; IEEE: Piscataway, NJ, USA, 2013; pp. 1–4.
102. Rabah, H.; Amira, A.; Mohanty, B.K.; Almaadeed, S.; Meher, P.K. FPGA implementation of orthogonal matching pursuit for compressive sensing reconstruction. *IEEE Trans. Very Large Scale Integr. (VLSI) Syst.* **2015**, *23*, 2209–2220. [CrossRef]
103. Kulkarni, A.; Jafari, A.; Shea, C.; Mohsenin, T. CS-based secured big data processing on FPGA. In Proceedings of the 2016 IEEE 24th Annual International Symposium on Field-Programmable Custom Computing Machines (FCCM), Washington, DC, USA, 1–3 May 2016; IEEE: Piscataway, NJ, USA, 2016; p. 201.
104. Kulkarni, A.; Stanislaus, J.L.; Mohsenin, T. Parallel heterogeneous architectures for efficient omp compressive sensing reconstruction. In Proceedings of the 2014 SPIE Sensing Technology + Applications, International Society for Optics and Photonics, Baltimore, MD, USA, 5–9 May 2014; p. 91090G.
105. Quan, Y.; Li, Y.; Gao, X.; Xing, M. FPGA Implementation of real-time Compressive Sensing with partial Fourier Dictionary. *Int. J. Antennas Propag.* **2016**, *2016*. [CrossRef]
106. Borghi, A.; Darbon, J.; Peyronnet, S.; Chan, T.F.; Osher, S. A simple compressive sensing algorithm for parallel many-core architectures. *J. Signal Process. Syst.* **2013**, *71*, 1–20. [CrossRef]
107. El-Sayed, M.; Koch, P.; Moullec, Y.L. Architectural design space exploration of an FPGA-based compressed sampling engine: Application to wireless heart-rate monitoring. In Proceedings of the 2015 Nordic Circuits and Systems Conference (NORCAS): NORCHIP International Symposium on System-on-Chip (SoC), Oslo, Norway, 26–28 October 2015; pp. 1–5.
108. Yu, Z.; Su, J.; Yang, F.; Su, Y.; Zeng, X.; Zhou, D.; Shi, W. Fast compressive sensing reconstruction algorithm on FPGA using Orthogonal Matching Pursuit. In Proceedings of the 2016 IEEE International Symposium on Circuits and Systems (ISCAS), Montreal, QC, Canada, 22–25 May 2016; pp. 249–252.
109. Kulkarni, A.; Mohsenin, T. Low Overhead Architectures for OMP Compressive Sensing Reconstruction Algorithm. *IEEE Trans. Circuits Syst. I Regul. Pap.* **2017**, *64*, 1468–1480. [CrossRef]
110. Maechler, P.; Greisen, P.; Sporrer, B.; Steiner, S.; Felber, N.; Burg, A. Implementation of greedy algorithms for LTE sparse channel estimation. In Proceedings of the 2010 Conference Record of the Forty Fourth Asilomar Conference on Signals, Systems and Computers (ASILOMAR), Pacific Grove, CA, USA, 7–10 November 2010; IEEE: Piscataway, NJ, USA, 2010; pp. 400–405.
111. Fang, Y.; Chen, L.; Wu, J.; Huang, B. GPU implementation of orthogonal matching pursuit for compressive sensing. In Proceedings of the 2011 IEEE 17th International Conference on Parallel and Distributed Systems (ICPADS), Tainan, Taiwan, 7–9 December 2011; IEEE: Piscataway, NJ, USA, 2011; pp. 1044–1047.
112. Kulkarni, A.; Mohsenin, T. Accelerating compressive sensing reconstruction OMP algorithm with CPU, GPU, FPGA and domain specific many-core. In Proceedings of the 2015 IEEE International Symposium on Circuits and Systems (ISCAS), Lisbon, Portugal, 24–27 May 2015; pp. 970–973.

113. Blache, P.; Rabah, H.; Amira, A. High level prototyping and FPGA implementation of the orthogonal matching pursuit algorithm. In Proceedings of the 2012 11th International Conference on Information Science, Signal Processing and their Applications (ISSPA), Montreal, QC, Canada, 2–5 July 2012; pp. 1336–1340.
114. Zhu, H.; Giannakis, G.B. Exploiting sparse user activity in multiuser detection. *IEEE Trans. Commun.* **2011**, *59*, 454–465. [CrossRef]
115. Schepker, H.F.; Dekorsy, A. Sparse multi-user detection for CDMA transmission using greedy algorithms. In Proceedings of the 2011 8th International Symposium on Wireless Communication Systems (ISWCS), Aachen, Germany, 6–9 November 2011; IEEE: Piscataway, NJ, USA, 2011; pp. 291–295.
116. Schepker, H.F.; Dekorsy, A. Compressive sensing multi-user detection with block-wise orthogonal least squares. In Proceedings of the 2012 IEEE 75th Vehicular Technology Conference (VTC Spring), Yokohama, Japan, 6–9 May 2012; IEEE: Piscataway, NJ, USA, 2012; pp. 1–5.
117. Choi, J. Joint device identification and data detection in physical layer for many devices in IoT. In Proceedings of the 2016 International Conference on Information and Communication Technology Convergence (ICTC), Jeju Island, Korea, 19–21 October 2016; IEEE: Piscataway, NJ, USA, 2016; pp. 85–89.
118. Choi, J. Two-Stage Multiple Access for Many Devices of Unique Identifications over Frequency-Selective Fading Channels. *IEEE Internet Things J.* **2016**, *4*, 16–171. [CrossRef]
119. Choi, J. Sparse index multiple access for multi-carrier systems with precoding. *J. Commun. Netw.* **2016**, *18*, 439–445.
120. Choi, J.W.; Shim, B. Sparse Detection of Non-Sparse Signals for Large-Scale Wireless Systems. *arXiv* **2015**, arXiv:1512.01683.
121. Wang, B.; Mir, T.; Jiao, R.; Dai, L. Dynamic multi-user detection based on structured compressive sensing for IoT-oriented 5G systems. In Proceedings of the 2016 URSI Asia-Pacific Radio Science Conference (URSI AP-RASC), Seoul, Korea, 21–25 August 2016; IEEE: Piscataway, NJ, USA, 2016; pp. 431–434.
122. Liu, J.; Cheng, H.Y.; Liao, C.C.; Wu, A.Y.A. Scalable compressive sensing-based multi-user detection scheme for Internet-of-Things applications. In Proceedings of the 2015 IEEE Workshop on Signal Processing Systems (SiPS), Hangzhou, China, 14–16 October 2015; IEEE: Piscataway, NJ, USA, 2015; pp. 1–6.
123. Monsees, F.; Bockelmann, C.; Wübben, D.; Dekorsy, A. Sparsity aware multiuser detection for machine to machine communication. In Proceedings of the 2012 IEEE Globecom Workshops (GC Wkshps), Anaheim, CA, USA, 3–7 December 2012; IEEE: Piscataway, NJ, USA, 2012; pp. 1706–1711.
124. Bockelmann, C.; Schepker, H.F.; Dekorsy, A. Compressive sensing based multi-user detection for machine-to-machine communication. *Trans. Emerg. Telecommun. Technol.* **2013**, *24*, 389–400. [CrossRef]
125. Swirszcz, G.; Abe, N.; Lozano, A.C. Grouped orthogonal matching pursuit for variable selection and prediction. In Proceedings of the 2009 Advances in Neural Information Processing Systems, Vancouver, BC, Canada, 7–10 December 2009, pp. 1150–1158.
126. Anagnostopoulos, T.; Zaslavsy, A.; Medvedev, A.; Khoruzhnicov, S. Top—k Query Based Dynamic Scheduling for IoT-enabled Smart City Waste Collection. In Proceedings of the 2015 16th IEEE International Conference on Mobile Data Management, Pittsburgh, PA, USA, 15–18 June 2015; Volume 2, pp. 50–55.
127. Zhang, J.; Yan, Y.; Chen, L.J.; Wang, M.; Moscibroda, T.; Zhang, Z. Impression Store: Compressive Sensing-based Storage for Big Data Analytics. In Proceedings of the HotCloud 2014, Philadelphia, PA, USA, 17–20 June 2014.
128. Chen, Y.S.; Tsou, Y.T. Compressive Sensing-Based Adaptive Top-k Query over Compression Domain in Wireless Sensor Networks. In Proceedings of the 2017 IEEE Wireless Communications and Networking Conference (WCNC), San Francisco, CA, USA, 19–22 March 2017; pp. 1–6.
129. Wang, L.; Zhang, D.; Wang, Y.; Chen, C.; Han, X.; M'hamed, A. Sparse mobile crowdsensing: challenges and opportunities. *IEEE Commun. Mag.* **2016**, *54*, 161–167. [CrossRef]
130. Hsieh, H.P.; Lin, S.D.; Zheng, Y. Inferring air quality for station location recommendation based on urban big data. In Proceedings of the 21th ACM SIGKDD International Conference on Knowledge Discovery and Data Mining, Sydney, Australia, 10–13 August 2015; ACM: New York, NY, USA, 2015; pp. 437–446.
131. Kong, L.; Xia, M.; Liu, X.Y.; Wu, M.Y.; Liu, X. Data loss and reconstruction in sensor networks. In Proceedings of the 2013 Proceedings IEEE INFOCOM, Turin, Italy, 14–19 April 2013; pp. 1654–1662.

132. Wang, L.; Zhang, D.; Yang, D.; Lim, B.Y.; Ma, X. Differential location privacy for sparse mobile crowdsensing. In Proceedings of the 2016 IEEE 16th International Conference on Data Mining (ICDM), Barcelona, Spain, 12–15 December 2016; IEEE: Piscataway, NJ, USA, 2016; pp. 1257–1262.
133. Wang, L.; Zhang, D.; Pathak, A.; Chen, C.; Xiong, H.; Yang, D.; Wang, Y. CCS-TA: Quality-guaranteed online task allocation in compressive crowdsensing. In Proceedings of the 2015 ACM International Joint Conference on Pervasive and Ubiquitous Computing, Osaka, Japan, 7–11 September 2015; ACM: New York, NY, USA, 2015; pp. 683–694.
134. Chen, H.; Guo, B.; Yu, Z.; Chen, L.; Ma, X. A generic framework for constraint-driven data selection in mobile crowd photographing. *IEEE Internet Things J.* **2017**, *4*, 284–296. [CrossRef]
135. Xu, L.; Hao, X.; Lane, N.D.; Liu, X.; Moscibroda, T. More with less: Lowering user burden in mobile crowdsourcing through compressive sensing. In Proceedings of the 2015 ACM International Joint Conference on Pervasive and Ubiquitous Computing, Osaka, Japan, 7–11 September 2015; ACM: New York, NY, USA, 2015; pp. 659–670.
136. Yang, B.; Kaul, M.; Jensen, C.S. Using Incomplete Information for Complete Weight Annotation of Road Networks. *IEEE Trans. Knowl. Data Eng.* **2014**, *26*, 1267–1279. [CrossRef]
137. Zhu, Y.; Li, Z.; Zhu, H.; Li, M.; Zhang, Q. A compressive sensing approach to urban traffic estimation with probe vehicles. *IEEE Trans. Mob. Comput.* **2013**, *12*, 2289–2302. [CrossRef]
138. Liu, Z.; Li, Z.; Li, M.; Xing, W.; Lu, D. Mining Road Network Correlation for Traffic Estimation via Compressive Sensing. *IEEE Trans. Intell. Transp. Syst.* **2016**, *17*, 1880–1893. [CrossRef]
139. Najafabadi, M.M.; Villanustre, F.; Khoshgoftaar, T.M.; Seliya, N.; Wald, R.; Muharemagic, E. Deep learning applications and challenges in big data analytics. *J. Big Data* **2015**, *2*, 1. [CrossRef]
140. Nguyen, M.T.; Teague, K.A. Compressive sensing based random walk routing in wireless sensor networks. *Ad Hoc Netw.* **2017**, *54*, 99–110. [CrossRef]
141. Caione, C.; Brunelli, D.; Benini, L. Compressive sensing optimization for signal ensembles in WSNs. *IEEE Trans. Ind. Inform.* **2014**, *10*, 382–392. [CrossRef]
142. Mamaghanian, H.; Khaled, N.; Atienza, D.; Vandergheynst, P. Design and exploration of low-power analog to information conversion based on compressed sensing. *IEEE J. Emerg. Sel. Top. Circuits Syst.* **2012**, *2*, 493–501. [CrossRef]
143. Pareschi, F.; Albertini, P.; Frattini, G.; Mangia, M.; Rovatti, R.; Setti, G. Hardware-algorithms co-design and implementation of an analog-to-information converter for biosignals based on compressed sensing. *IEEE Trans. Biomed. Circuits Syst.* **2016**, *10*, 149–162. [CrossRef] [PubMed]
144. Gangopadhyay, D.; Allstot, E.G.; Dixon, A.M.; Natarajan, K.; Gupta, S.; Allstot, D.J. Compressed sensing analog front-end for bio-sensor applications. *IEEE J. Solid-State Circuits* **2014**, *49*, 426–438. [CrossRef]
145. Kirolos, S.; Laska, J.; Wakin, M.; Duarte, M.; Baron, D.; Ragheb, T.; Massoud, Y.; Baraniuk, R. Analog-to-information conversion via random demodulation. In Proceedings of the 2006 IEEE Dallas/CAS Workshop on Design, Applications, Integration and Software, Madrid, Spain, 29–30 October 2006; IEEE: Piscataway, NJ, USA, 2006; pp. 71–74.
146. Chen, X.; Yu, Z.; Hoyos, S.; Sadler, B.M.; Silva-Martinez, J. A sub-Nyquist rate sampling receiver exploiting compressive sensing. *IEEE Trans. Circuits Syst. I Regul. Pap.* **2011**, *58*, 507–520. [CrossRef]
147. Baheti, P.K.; Garudadri, H. An ultra low power pulse oximeter sensor based on compressed sensing. In Proceedings of the 2009 Sixth International Workshop on Wearable and Implantable Body Sensor Networks, BSN 2009, Berkeley, CA, USA, 3–5 June 2009; IEEE: Piscataway, NJ, USA, 2009; pp. 144–148.
148. Candès, E.J.; Wakin, M.B. An introduction to compressive sampling. *IEEE Signal Process. Mag.* **2008**, *25*, 21–30. [CrossRef]
149. Mendelson, S.; Pajor, A.; Tomczak-Jaegermann, N. Uniform uncertainty principle for Bernoulli and subgaussian ensembles. *Constr. Approx.* **2008**, *28*, 277–289. [CrossRef]
150. Candes, E.J.; Tao, T. Near-optimal signal recovery from random projections: Universal encoding strategies? *IEEE Trans. Inf. Theory* **2006**, *52*, 5406–5425. [CrossRef]
151. Davenport, M.A. Random Observations on Random Observations: Sparse Signal Acquisition and Processing. Ph.D. Thesis, Rice University, Houston, TX, USA, 2010.
152. Ravelomanantsoa, A.; Rabah, H.; Rouane, A. Simple deterministic measurement matrix: Application to EMG signals. In Proceedings of the 2014 26th International Conference on Microelectronics (ICM), Doha, Qatar, 14–17 December 2014; pp. 76–79.

153. Bah, B.; Tanner, J. Vanishingly sparse matrices and expander graphs, with application to compressed sensing. *IEEE Trans. Inf. Theory* **2013**, *59*, 7491–7508. [CrossRef]
154. He, Z.; Ogawa, T.; Haseyama, M. The simplest measurement matrix for compressed sensing of natural images. In Proceedings of the 2010 IEEE International Conference on Image Processing, Hong Kong, China, 26–29 September 2010; pp. 4301–4304.
155. Munir, A.; Gordon-Ross, A.; Ranka, S. Multi-core embedded wireless sensor networks: Architecture and applications. *IEEE Trans. Parallel Distrib. Syst.* **2014**, *25*, 1553–1562. [CrossRef]
156. Murphy, M.; Alley, M.; Demmel, J.; Keutzer, K.; Vasanawala, S.; Lustig, M. Fast l1 SPIRiT Compressed Sensing Parallel Imaging MRI: Scalable Parallel Implementation and Clinically Feasible Runtime. *IEEE Trans. Med Imaging* **2012**, *31*, 1250–1262. [CrossRef] [PubMed]
157. Chang, C.H.; Ji, J. Compressed sensing MRI with multichannel data using multicore processors. *Magn. Reson. Med.* **2010**, *64*, 1135–1139. [CrossRef] [PubMed]
158. Smith, D.S.; Gore, J.C.; Yankeelov, T.E.; Welch, E.B. Real-time compressive sensing MRI reconstruction using GPU computing and split Bregman methods. *Int. J. Biomed. Imaging* **2012**, *2012*. [CrossRef] [PubMed]
159. Kim, D.; Trzasko, J.; Smelyanskiy, M.; Haider, C.; Dubey, P.; Manduca, A. High-performance 3D compressive sensing MRI reconstruction using many-core architectures. *J. Biomed. Imaging* **2011**, *2011*, 2. [CrossRef] [PubMed]
160. Bortolotti, D.; Mangia, M.; Bartolini, A.; Rovatti, R.; Setti, G.; Benini, L. Energy-Aware Bio-signal Compressed Sensing Reconstruction on the WBSN-gateway. *IEEE Trans. Emerg. Top. Comput.* **2016**. [CrossRef]
161. Rachlin, Y.; Baron, D. The secrecy of compressed sensing measurements. In Proceedings of the 2008 46th Annual Allerton Conference on Communication, Control, and Computing, Monticello, IL, USA, 23–26 September 2008; pp. 813–817.
162. Kailkhura, B.; Liu, S.; Wimalajeewa, T.; Varshney, P.K. Measurement Matrix Design for Compressed Detection With Secrecy Guarantees. *IEEE Wirel. Commun. Lett.* **2016**, *5*, 420–423. [CrossRef]
163. Wang, C.; Zhang, B.; Ren, K.; Roveda, J.M.; Chen, C.W.; Xu, Z. A privacy-aware cloud-assisted healthcare monitoring system via compressive sensing. In Proceedings of the 2014 IEEE INFOCOM, Toronto, ON, Canada, 27 April–2 May 2014; IEEE: Piscataway, NJ, USA, 2014; pp. 2130–2138.
164. Peng, H.; Tian, Y.; Kurths, J.; Li, L.; Yang, Y.; Wang, D. Secure and Energy-Efficient Data Transmission System Based on Chaotic Compressive Sensing in Body-to-Body Networks. *IEEE Trans. Biomed. Circuits Syst.* **2017**, *11*, 558–573. [CrossRef] [PubMed]
165. Hu, P.; Xing, K.; Cheng, X.; Wei, H.; Zhu, H. Information leaks out: Attacks and countermeasures on compressive data gathering in wireless sensor networks. In Proceedings of the 2014 IEEE INFOCOM, Toronto, ON, Canada, 27 April–2 May 2014; IEEE: Piscataway, NJ, USA, 2014; pp. 1258–1266.
166. Singh, S.; Jeong, Y.S.; Park, J.H. A survey on cloud computing security: Issues, threats, and solutions. *J. Netw. Comput. Appl.* **2016**, *75*, 200–222. [CrossRef]
167. Wu, X.; Yang, P.; Tang, S.; Zheng, X.; Xiong, Y. Privacy preserving RSS map generation for a crowdsensing network. *IEEE Wirel. Commun.* **2015**, *22*, 42–48. [CrossRef]
168. Wang, C.; Zhang, B.; Ren, K.; Roveda, J.M. Privacy-Assured Outsourcing of Image Reconstruction Service in Cloud. *IEEE Trans. Emerg. Top. Comput.* **2013**, *1*, 166–177. [CrossRef]
169. Kang, L.W.; Muchtar, K.; Wei, J.D.; Lin, C.Y.; Chen, D.Y.; Yeh, C.H. Privacy-preserving multimedia cloud computing via compressive sensing and sparse representation. In Proceedings of the 2012 International Conference on Information Security and Intelligence Control (ISIC), Yunlin, Taiwan, 14–16 August 2012; IEEE: Piscataway, NJ, USA, 2012; pp. 246–249.
170. Wang, Q.; Zeng, W.; Tian, J. A compressive sensing based secure watermark detection and privacy preserving storage framework. *IEEE Trans. Image Process.* **2014**, *23*, 1317–1328. [CrossRef] [PubMed]
171. Zhang, Y.; Zhou, J.; Zhang, L.Y.; Chen, F.; Lei, X. Support-set-assured parallel outsourcing of sparse reconstruction service for compressive sensing in multi-clouds. In Proceedings of the 2015 International Symposium on Security and Privacy in Social Networks and Big Data (SocialSec), Hangzhou, China, 16–18 November 2015; IEEE: Piscataway, NJ, USA, 2015; pp. 1–6.

172. Jyothish, L.G.; Veena, V.; Soman, K. A cryptographic approach to video watermarking based on compressive sensing, arnold transform, sum of absolute deviation and SVD. In Proceedings of the 2013 Annual International Conference on Emerging Research Areas and 2013 International Conference on Microelectronics, Communications and Renewable Energy (AICERA/ICMiCR), Kerala, India, 4–6 June 2013; IEEE: Piscataway, NJ, USA, 2013; pp. 1–5.
173. Bianchi, T.; Bioglio, V.; Magli, E. Analysis of one-time random projections for privacy preserving compressed sensing. *IEEE Trans. Inf. Forensics Secur.* **2016**, *11*, 313–327. [CrossRef]

Permissions

All chapters in this book were first published in MDPI; hereby published with permission under the Creative Commons Attribution License or equivalent. Every chapter published in this book has been scrutinized by our experts. Their significance has been extensively debated. The topics covered herein carry significant findings which will fuel the growth of the discipline. They may even be implemented as practical applications or may be referred to as a beginning point for another development.

The contributors of this book come from diverse backgrounds, making this book a truly international effort. This book will bring forth new frontiers with its revolutionizing research information and detailed analysis of the nascent developments around the world.

We would like to thank all the contributing authors for lending their expertise to make the book truly unique. They have played a crucial role in the development of this book. Without their invaluable contributions this book wouldn't have been possible. They have made vital efforts to compile up to date information on the varied aspects of this subject to make this book a valuable addition to the collection of many professionals and students.

This book was conceptualized with the vision of imparting up-to-date information and advanced data in this field. To ensure the same, a matchless editorial board was set up. Every individual on the board went through rigorous rounds of assessment to prove their worth. After which they invested a large part of their time researching and compiling the most relevant data for our readers.

The editorial board has been involved in producing this book since its inception. They have spent rigorous hours researching and exploring the diverse topics which have resulted in the successful publishing of this book. They have passed on their knowledge of decades through this book. To expedite this challenging task, the publisher supported the team at every step. A small team of assistant editors was also appointed to further simplify the editing procedure and attain best results for the readers.

Apart from the editorial board, the designing team has also invested a significant amount of their time in understanding the subject and creating the most relevant covers. They scrutinized every image to scout for the most suitable representation of the subject and create an appropriate cover for the book.

The publishing team has been an ardent support to the editorial, designing and production team. Their endless efforts to recruit the best for this project, has resulted in the accomplishment of this book. They are a veteran in the field of academics and their pool of knowledge is as vast as their experience in printing. Their expertise and guidance has proved useful at every step. Their uncompromising quality standards have made this book an exceptional effort. Their encouragement from time to time has been an inspiration for everyone.

The publisher and the editorial board hope that this book will prove to be a valuable piece of knowledge for researchers, students, practitioners and scholars across the globe.

List of Contributors

Christian Sauer and Eike Lyczkowski
SEW-EURODRIVE GmbH & Co. KG, 76646 Bruchsal, Germany

Marco Schmidt
Bochum University of Applied Sciences, Campus Velbert/Heiligenhaus, 42579 Heiligenhaus, Germany

Yamin Han and Liangliang Zhang
Department of Computer Science, The University of Suwon, Hwaseong 18323, Korea

Heejung Byun
Department of Information and Technology, The University of Suwon, Hwaseong 18323, Korea

Simone Grimaldi, Aamir Mahmood and Mikael Gidlund
Department of Information Systems and Technology, Mid Sweden University, 851 70 Sundsvall, Sweden

Chiara Bedon, Enrico Bergamo, Matteo Izzi and Salvatore Noè
Department of Engineering and Architecture, University of Trieste, Piazzale Europa 1, 34127 Trieste, Italy

Khandakar Ahmed
Discipline of IT, College of Engineering & Science, Victoria University, Footscray Park Campus, Melbourne, VIC 3011, Australia

Jan O. Blech and Heinz W. Schmidt
School of Science (Computer Science), RMIT University, Melbourne, VIC 3001, Australia

Mark A. Gregory
School of Engineering (Electrical and Computer Systems), RMIT University, Melbourne, VIC 3001, Australia

Alex Adim Obinikpo and Burak Kantarci
School of Electrical Engineering and Computer Science, University Ottawa, Ottawa, ON K1N 6N5, Canada

Marco Cattani, Carlo Alberto Boano and Kay Römer
Institute for Technical Informatics, Graz University of Technology, Graz 8010, Austria

Beiyu Lin and Diane J. Cook
School of Electrical Engineering & Computer Science, Washington State University, Spokane Street, Pullman, WA 99163, USA

Yibo Huangfu, Bertram Jobson, Patrick O'Keeffe, Shelley N. Pressley, Von Walden and Brian Lamb
Department of Civil & Environmental Engineering, Washington State University, 2001 Grimes Way, Pullman, WA 99163, USA

Nathan Lima and Max Kirk
School of Design & Construction, Washington State University, Spokane Street, Pullman, WA 99163, USA

Antonio José Calderón Godoy and Isaías González Pérez
Department of Electrical Engineering, Electronics and Automation, University of Extremadura, Avenida de Elvas, s/n, 06006 Badajoz, Spain

Anup Kumar Paul
Department of Electronics and Communications Engineering, East West University, Dhaka 1212, Bangladesh
Waseda University, Tokyo 169-8050, Japan

Takuro Sato
Waseda University, Tokyo 169-8050, Japan

Hamza Djelouat, Abbes Amira and Faycal Bensaali
College of Engineering, Qatar University, Doha 2713, Qatar

Index

Printed in the USA
CPSIA information can be obtained
at www.ICGtesting.com
LVHW080054211023
761596LV00010BA/2059

9 798888 360705